T0328928

A D A P T I V E O P T I C S I N A S T R O N O M Y

Adaptive optics is a powerful new technique used to sharpen telescope images blurred by the Earth's atmosphere. This authoritative book is the first dedicated to the use of adaptive optics in astronomy.

Mainly developed for defense applications, the technique of adaptive optics has only recently been introduced to astronomy. Already it has allowed ground-based telescopes to produce images with sharpness rivalling those from the Hubble Space Telescope. The technique is expected to revolutionize the future of ground-based optical astronomy.

Written by an international team of experts who have pioneered the development of the field, this timely volume provides both a rigorous introduction to the technique and a comprehensive review of current and future systems. It is set to become the standard reference for graduate students, researchers and optical engineers in astronomy and other areas of science where adaptive optics is finding exciting new applications.

ADAPTIVE OPTICS IN ASTRONOMY

Edited by

FRANÇOIS RODDIER

Institute for Astronomy, University of Hawaii, USA

CAMBRIDGE
UNIVERSITY PRESS

PUBLISHED BY THE PRESS SYNDICATE OF THE UNIVERSITY OF CAMBRIDGE
The Pitt Building, Trumpington Street, Cambridge, United Kingdom

CAMBRIDGE UNIVERSITY PRESS
The Edinburgh Building, Cambridge CB2 2RU, UK
40 West 20th Street, New York NY 10011–4211, USA
477 Williamstown Road, Port Melbourne, VIC 3207, Australia
Ruiz de Alarcón 13, 28014 Madrid, Spain
Dock House, The Waterfront, Cape Town 8001, South Africa

http://www.cambridge.org

First published 1999
First paperback edition 2004

Typeset in Times 11/14, in 3B2 [KT]

A catalogue record for this book is available from the British Library

ISBN 0 521 55375 X hardback
ISBN 0 521 61214 4 paperback

Contents

Contributors

Jacques M. Beckers
National Optical Astronomy Observatories, NSO Division, Tucson, Arizona, USA

Pierre Léna and Olivier Lai
Observatoire de Paris, DESPA, Meudon, France

Pierre-Yves Madec, Gérard Rousset and Marc Séchaud
ONERA, Paris, France

Malcolm J. Northcott and François Roddier
Institute for Astronomy, University of Hawaii, USA

Jean-Luc Beuzit and François Rigaut
Canada-France-Hawaii Telescope Corporation, Hawaii, USA

David G. Sandler
Thermotrex Corporation, San Diego, California, USA

Part one

Introductory background

1

Historical context

FRANÇOIS RODDIER

Institute for Astronomy, University of Hawaii, USA

Turbulence in the Earth's atmosphere produces inhomogeneities in the air refractive index, which affect the image quality of ground-based telescopes. Adaptive optics (AO) is a means for real time compensation of the image degradation. The technique was first proposed by Babcock (1953) and later independently by Linnick (1957) to improve astronomical images. It consists of using an active optical element such as a deformable mirror to correct the instantaneous wave-front distortions. These are measured by a device called a wave-front sensor which delivers the signals necessary to drive the correcting element. Although both Babcock and Linnick described methods that could be employed to achieve this goal, development cost was too prohibitive at that time to allow the construction of an AO system for astronomy.

The invention of the laser soon triggered both experimental and theoretical work on optical propagation through the turbulent atmosphere. Studies on laser speckle led Labeyrie (1970) to propose speckle interferometry as a means to reconstruct turbulence degraded images. Following Labeyrie, astronomers focused their efforts on developing 'post-detection' image processing techniques to improve the resolution of astronomical images. Meanwhile, defense-oriented research started to use segmented mirrors to compensate the effect of the atmosphere in attempts to concentrate laser beams on remote targets. This was done by trial-and-error (multidither technique). As artificial satellites were sent on orbit, the need came to make images of these objects for surveillance, and attempts were made to use similar techniques for imaging (Buffington *et al.* 1977). The first adaptive optics system able to sharpen two-dimensional images was built at Itek by Hardy and his coworkers (Hardy *et al.* 1977). A larger version was installed in 1982 at the Air-Force Maui Optical Site (AMOS) on Haleakala. By the end of the 1970s AO systems were widely developed by industry for defense applications (Pearson *et al.* 1979). AO systems with more than a thousand degrees-of-freedom have now been built (Cuellar *et al.* 1991).

Owing to this success, astronomers became interested in applying AO to astronomy. Unfortunately, most interesting astronomical sources are much dimmer than artificial satellites. Astronomers are used to expressing the brightness of a star in terms of magnitudes, a logarithmic scale in which an increase of five magnitudes describes a decrease in brightness by a factor 100. On a clear dark night, stars up to magnitude 6 are visible to the naked eye. The AMOS adaptive-optics system goes barely beyond. Astronomers often observe objects as faint as mag 20 or fainter. Therefore, they need more sensitive systems. They soon realized that compensation requirements are less severe for infrared imaging, in which case one can use a fainter 'guide' source to sense the wave front. In the 1980s, great progress was being made in developing InSb and HgCdTe detector arrays for imaging in the near infrared, and infrared astronomy was blooming. Therefore two important astronomical institutions decided to sponsor a development program on AO for infrared astronomy: the European Southern Observatory (ESO) and the US National Optical Astronomy Observatories (NOAO).

The ESO effort, involved astronomers, defense research experts, and industry, and led to the construction of a prototype instrument called 'COME-ON' (Rousset *et al.* 1990; Rigaut *et al.* 1991). An upgraded version of COME-ON called 'COME-ON-PLUS' has since been regularly used for astronomical observations at the ESO 3.6-m telescope in Chile (Rousset *et al.* 1993). A new version called ADONIS is now a user instrument facility. These developments are described in Chapter 8. The technology is basically the same as that developed for defense applications (Shack–Hartmann wave front sensor and thin-plate piezo-stacks deformable mirror), but the number of different aberrations that can be corrected (number of degrees-of-freedom) is smaller than that of defense systems. It allows these systems to sense the wave front with fainter 'guide' sources (up to mag 14 on a 3.6-m telescope), but limits their application to the near infrared.

The NOAO effort was discontinued after 4 years, but work soon started at the Institute for Astronomy of the University of Hawaii (UH) on a novel technique that had been conceived at NOAO. It involved the development of a new type of wave front sensor called a 'curvature sensor', and a new type of deformable mirror called 'bimorph' (Roddier 1988; Roddier *et al.* 1991). The technique was believed to be better adapted to astronomical observations. An experimental instrument was built and successfully tested at the coudé focus of the Canada–France–Hawaii telescope (CFHT) on Mauna Kea. The first astronomical observations were made in December 1993 (Roddier *et al.* 1995). A Cassegrain-focus visitor instrument was then built and first used in December 1994 at the CFHT $f/35$ focus. It has since been widely used for astronomical observations. Compared with the European AO systems, the UH system uses an array of high performance detectors called avalanche photo-diodes (APDs), which allows the sensing of the wave front to be performed on fainter guide sources (up to magnitude 17 on a 3.6-m telescope). A consequence is that a larger number of objects is accessible to wave front compensation. A user AO system based on this technique has now been built for the 3.6-m CFHT. Another one is under construction for the Japanese 8-m Subaru Telescope (Takami *et al.* 1996). Several other institutions are also considering the use of this type of system. These developments are described in Chapter 9.

Because the brightness of the source severely limits the sensing of the wave front, Foy and Labeyrie (1985) proposed the use of laser beacons to create artificial sources with light back-scattered by the atmosphere. We now know that the same idea had been independently proposed earlier by US defense researchers, and was already being developed as classified research (McCall and Passner 1978, Benedict *et al.* 1994). In 1991, after the political changes in Russia, the US National Science Foundation (NSF) convinced government authorities of the importance of the technique to astronomy, and obtained its declassification. The current state of US defense technology was presented in an open meeting held in Albuquerque in March 1992 (Fugate 1992). As a result, many US groups joined the effort and pursued the development of artificial laser guide sources for astronomical applications (see the January and February 1994 issues of *J. Opt. Soc. Am.*). Although the technique has not yet matched astronomers' expectations, encouraging results have been obtained. These are presented in Chapters 11 to 13 together with a description of the difficulties encountered.

At the time of writing, a growing number of observatories are becoming equipped with AO systems to be used with or without laser beacons. The purpose of this book is to describe the current state of the art with its potential and limitations. We hope it will be useful to both engineers in charge of designing and building astronomical instrumentation, and to astronomers whose observations will benefit from it. Chapter 2 describes the statistical properties of the turbulence-induced wave front distortions to be compensated, and their deleterious effects on images. Part II (Chapter 3 to 7) gives performance goals that a theoretically ideal AO system would achieve, describes practical means which have been developed to approach these goals, and shows how to estimate their real performance. Concepts discussed in part II are illustrated in part III by a description of the COME-ON/ADONIS systems (Chapter 8), and of the UH/CFHT systems (Chapter 9). It has not been possible to do justice here to all the work done with natural guide sources at other astronomical institutions such as Durham University or Mt Wilson Observatory. However, we have included a section on solar astronomy (Chapter 10). Part IV (Chapters 11–13) introduces laser beacons and their application to astronomy. Part V (Chapters 14–16) discusses the practical aspects of astronomical observations with AO, shows examples taken from among the wealth of successful results that have now been obtained, and discusses the impact of AO on future observations and instrumentation.

References

Babcock, H. W. (1953) The possibility of compensating astronomical seeing. *Pub. Astr. Soc. Pac.* **65**, 229–36.

Benedict, R. Jr, Breckinridge, J. B. and Fried, D. (1994) Atmospheric-Compensation Technology. Introduction to the special issue of *J. Opt. Soc. Am.* **11**, 257–60.

Buffington, A., Crawford, F. S., Muller, R. A. and Orth, C. D. (1977) First observatory results with an image-sharpening telescope. *J. Opt. Soc. Am.* **67**, 304–5.

Cuellar, L., Johnson, P. and Sandler, D. G. (1991) Performance test of a 1500 degree-of-freedom adaptive optics system for atmospheric compensation. In: *Active and Adaptive Optical Systems*, ed. M. Ealey, Proc. SPIE Conf. 1542, pp. 468–71.

Foy, R. and Labeyrie, A. (1985) Feasibility of adaptive telescope with laser probe. *Astron. Astrophys.* **152**, L29–L31.

Fugate, R. Q., ed. (1992) Laser Guide Star Adaptive Optics. Proc. Workshop, March 10–12 1992, Starfire Optical Range, Phillips Lab./LITE, Kirtland AFB, NM 87117-6008.

Hardy, J. W., Lefebvre, J. E. and Koliopoulos, C. L. (1977) Real-time atmospheric compensation. *J. Opt. Soc. Am.* **67**, 360–69.

Labeyrie, A. (1970) Attainment of diffraction-limited resolution in large telescopes by Fourier analysing speckle patterns in star images. *Astron. Astrophys.* **6**, 85–7.

Linnick, V. P. (1957) On the possibility of reducing the influence of atmospheric seeing on the image quality of stars (in Russian). *Optics and Spectroscopy* **3**, 401–2.

McCall, S. L. and Passner, A. (1978) Adaptive optics in astronomy. *Physics of Quantum Electronics* **6**, 149–74.

Pearson, J. E., Freeman, R. H. and Reynolds H. C. Jr, (1979) Adaptive optical techniques for wave-front correction. In: *Applied Optics and Optical Engineering*, Vol. 7, Chapter 8, pp. 245–340. Academic Press, New York.

Rigaut, F., Rousset, G., Kern, P., Fontanella, J. C., Gaffard, J. P., Merkle, F., *et al.* (1991) Adaptive optics on a 3.6-m telescope: results and performance. *Astron. Astrophys.* **250**, 280–90.

Roddier, F. (1988) Curvature sensing and compensation: a new concept in adaptive optics. *Appl. Opt.* **27**, 1223–5.

Roddier, F., Northcott, M. and Graves, J. E. (1991) A simple low-order adaptive optics system for near-infrared applications. *Pub. Astr. Soc. Pac.* **103**, 131–49.

Roddier, F., Roddier, C., Graves, J. E. and Northcott, M. J. (1995) Adaptive optics imaging of proto-planetary nebulae: Frosty Leo and the Red Rectangle. *Astrophys. J.* **443**, 249–60.

Rousset, G., Fontanella, J. C., Kern, P., Gigan, P., Rigaut, F., Lena, P., *et al.* (1990) First diffraction-limited astronomical images with adaptive optics. *Astron. Astrophys.* **230**, L29–L32.

Rousset, G., Beuzit, J. L., Hubin, N., Gendron, E., Boyer, C., Madec, P. Y., *et al.*

(1993) The COME-ON-PLUS adaptive optics system: results and performance. In: *Active and Adaptive Optics*, ed. F. Merkle, Proc. ICO-16 Satellite Conf., ESO, pp. 65–70.

Takami, H., Iye, M., Takato, N., Otsubo, M. and Nakashima, K. (1996) Subaru adaptive optics program. In: OSA topical meeting on Adaptive Optics, Maui (Hawaii) July 8–12, 1996. Tech. Digest Series 13, pp. 25–7.

2

Imaging through the atmosphere

FRANÇOIS RODDIER

Institute for Astronomy, University of Hawaii, USA

The designing of an AO system requires a good appreciation of the characteristics of the wave-front aberrations that need to be compensated, and of their
effect on image quality. Since these aberrations are random, they can only be
described statistically, using statistical estimates such as variances, or covar-
iances. These estimates define the so-called seeing conditions. We are dealing
here with a non-stationary random process. Seeing conditions evolve with time.
Therefore, one also needs to know the statistics of their evolution, mean value
and standard deviation for a given telescope. A good knowledge of the seeing
conditions during the observations is also important for the observing strategy.
This chapter summarizes our knowledge on the statistics of the air refractive
index fluctuations. From these, are derived the statistics of the wave-front
distortions one seeks to compensate, and their effect on the intensity distribution in the image plane. A more detailed description of this material can be
found in several review papers (Roddier 1981; Roddier 1989; Fried 1994).

2.1 Air refractive index fluctuations

Fluctuations in the air refractive index are essentially proportional to fluctuations in the air temperature. These are found at the interface between different
air layers. Wind shears produce turbulence which mixes layers at different
temperature, and therefore produces temperature inhomogeneities. The statistics of refractive index inhomogeneities follows that of temperature inhomogeneities, which are governed by the Kolmogorov–Obukhov law of turbulence.
We are not interested in the absolute value of the refractive index, but mainly
in the difference between its value $n(\mathbf{r})$ at a point $\mathbf{r}$, and its value $n(\mathbf{r} + \rho)$ at a
nearby point some distance $\rho = |\rho|$ apart. Vectors $\mathbf{r}$ and ρ represent three-

dimensional positions and separations. The variance of the difference between the two values of the refractive index is given by

$$D_N(\rho) = \langle |n(\mathbf{r}) - n(\mathbf{r} + \rho)|^2 \rangle = C_N^2 \rho^{2/3}, \qquad (2.1)$$

where the brackets $\langle \rangle$ represent an ensemble average. $D_N(\rho)$ is called the index structure function. To a first approximation, it depends only upon the separation ρ but not the position $\mathbf{r}$, that is the random process is considered as homogeneous (at least locally). Moreover, it depends only on the modulus of the vector ρ independently of its direction, that is the process is isotropic. The quantity C_N^2 is called the index structure coefficient. It is a measure of the local amount of inhomogeneities. It does vary over distances much larger than the scale of the inhomogeneities. Its integral along the light propagation path gives a measure of the total amount of wave-front degradation or 'seeing'.

Equation (2.1) is an approximation only valid as long as ρ is smaller than some value called the turbulence outerscale. Indeed at long distances the fluctuations $n(\mathbf{r})$ and $n(\mathbf{r} + \rho)$ are expected to become eventually uncorrelated, in which case the structure function becomes independent of ρ and equal to twice the variance of n. According to Eq. (2.1), the variance of n would be infinite which is unphysical. The value of the outerscale has been highly debated. Values in the literature span from a few tens of centimeters to kilometers. Experimentally, the power law in Eq. (2.1) has been found to be quite accurate over distances less than 1 meter. Beyond that, evidence of deviations have often been found. It means that Eq. (2.1) is certainly valid for small telescopes. For large telescopes it is likely to be inaccurate.

Equation (2.1) statistically describes the spatial distribution of the inhomogeneities at a given time t. We also need to know how fast the index fluctuates with time, at a fixed point $\mathbf{r}$ along the line of sight. The temporal evolution can be similarly described by a temporal structure function defined as the variance of the difference between the index at time t and the index at a later time $t + \tau$:

$$D_N(\tau) = \langle |n(\mathbf{r}, t) - n(\mathbf{r}, t + \tau)|^2 \rangle. \qquad (2.2)$$

Experience shows that the lifetimes of air temperature inhomogeneities are much longer than the time it takes for a wind-driven inhomogeneity to cross the line of sight. This is true for open air turbulence under most wind conditions. If $\mathbf{v}$ is the wind velocity, then

$$n(\mathbf{r}, t + \tau) = n(\mathbf{r} - \mathbf{v}\tau, t). \qquad (2.3)$$

This approximation is called the Taylor approximation. Putting Eq. (2.3) into Eq. (2.2) gives the following expression for the temporal structure function

$$D_N(\tau) = \langle |n(\mathbf{r}, t) - n(\mathbf{r} - \mathbf{v}\tau, t)|^2 \rangle = C_N^2 \cdot |\mathbf{v}\tau|^{2/3}. \qquad (2.4)$$

Hence, the temporal structure function is simply obtained by substituting $|\mathbf{v}\tau|$ for ρ in the spatial structure function (Eq. (2.1)).

Within our wavelength range of interest, which extends from the red to the thermal infrared, the air refractive index is fairly wavelength independent. Therefore, in the following we will assume that the structure functions defined by Eq. (2.1) and Eq. (2.4) are wavelength independent quantities.

2.2 Wave-front phase distortions

Waves are best described by means of a complex number Ψ, called the wave complex amplitude. It is defined as

$$\Psi = A \exp(i\varphi), \tag{2.5}$$

where A and φ are real numbers representing respectively the amplitude and the phase of the field fluctuation. A surface over which φ takes the same value is called a wave-front surface. Before entering the atmosphere, light from very far away sources such as stars forms plane waves (flat wave-front surfaces). However, inside the atmosphere the speed of light will vary as the inverse of the refractive index. Light propagating through regions of high index will be delayed compared to light propagating through other regions. The resulting wave-front surface is no longer flat but corrugated. The deformation of the wave-front surface is given by the optical path fluctuation

$$\delta = \int n(z)\,\mathrm{d}z, \tag{2.6}$$

where $n(z)$ is the refractive index fluctuation along the beam. To a first approximation called the near-field approximation, the integral in Eq. (2.6) can be simply taken along the line of sight. It is important to note that, since $n(z)$ is fairly independent of wavelength, the deformation of the wave-front surface (generally expressed in microns, or nanometers) is also a wavelength independent quantity. Hence, it can be compensated at all wavelengths, by means of a deformable mirror having the same deformation as the incoming wave-front surface but with only half the amplitude. Beam areas which have been most delayed by the atmosphere have less distance to travel before being reflected. Other beam areas have further to travel. They are delayed by an amount equal to the time necessary for the light to travel twice (back and forth) across the depth of the mirror deformation.

The wave-front phase fluctuation is related to the wave-front surface deformation by the following relation:

$$\varphi = k \int n(z)\,\mathrm{d}z, \tag{2.7}$$

where k is the wave number. It varies as the inverse of the wavelength λ

$$k = 2\pi/\lambda. \tag{2.8}$$

Hence, the wave-front phase fluctuation is not achromatic. At long wavelengths, fluctuations are smaller and, as we shall see, less detrimental to image quality. Again, we are not interested in any absolute wave-front phase, but rather in the difference between the phase $\varphi(\mathbf{x})$ at a point $\mathbf{x}$ on the telescope entrance aperture and the phase $\varphi(\mathbf{x} + \boldsymbol{\xi})$ at a nearby point a distance $\xi = |\boldsymbol{\xi}|$ apart. The variance of the difference is the structure function of the phase

$$D_\varphi(\boldsymbol{\xi}) = \langle |\varphi(\mathbf{x}) - \varphi(\mathbf{x} + \boldsymbol{\xi})|^2 \rangle. \tag{2.9}$$

It should be noted that $\mathbf{x}$ and $\boldsymbol{\xi}$ are now two-dimensional vectors. Putting Eq. (2.7) into Eq. (2.9) makes it possible after some manipulation to express the phase structure function in terms of index structure functions integrated along the line of sight. Using Eq. (2.1) for the index structure function and performing the integration yields

$$D_\varphi(\boldsymbol{\xi}) = 2.91 k^2 \int C_N^2(z)\,\mathrm{d}z\, \xi^{5/3}. \tag{2.10}$$

The remaining integral in Eq. (2.10) is along the line of sight. Because of the integration, the 2/3 power in Eq. (2.1) has now become a 5/3 power. The atmosphere is generally considered to be stratified in plane parallel layers, that is C_N^2 depends only on the height h above the ground. In this case, Eq. (2.10) can be rewritten

$$D_\varphi(\boldsymbol{\xi}) = 2.91 k^2 (\cos\gamma)^{-1} \int C_N^2(h)\,\mathrm{d}h\, \xi^{5/3}, \tag{2.11}$$

where γ is the angular distance of the source from zenith. The quantity $(\cos\gamma)^{-1}$ is called the air mass.

Eq. (2.11) is often found in the literature under a slightly different form

$$D_\varphi(\boldsymbol{\xi}) = 6.88(\xi/r_0)^{5/3}, \tag{2.12}$$

where

$$r_0 = \left[0.423 k^2 (\cos\gamma)^{-1} \int C_N^2(h)\,\mathrm{d}h \right]^{-3/5} \tag{2.13}$$

is a length called the Fried parameter (Fried 1965). It characterizes the effect of seeing at a particular wavelength. According to Eq. (2.13), r_0 increases as the 6/5 power of wavelength, and decreases as the $-3/5$ power of the air mass. Additional properties of r_0 will be given below.

Equation (2.10) or (2.12) allows us to calculate quantities such as the mean square wave-front phase distortion over a circular area of diameter d, defined as

$$\sigma_1^2 = \left\langle \frac{4}{\pi d^2} \iint_{\text{area}} |\varphi(\mathbf{x}) - \varphi_0(\mathbf{x})|^2 \, d\mathbf{x} \right\rangle, \tag{2.14}$$

where

$$\varphi_0 = \frac{4}{\pi d^2} \iint_{\text{area}} \varphi(\mathbf{x}) \, d\mathbf{x} \tag{2.15}$$

is the wave-front phase averaged over the area. According to Fried (1965) and Noll (1976)

$$\sigma_1^2 = 1.03(d/r_0)^{5/3}. \tag{2.16}$$

Hence, an interesting property of r_0 is that the root mean square (rms) phase distortion over a circular area of diameter r_0 is about 1 radian.

Let us again consider a wave front over a circular area of diameter d. If a plane wave is fitted to the wave front over this area, and its phase is subtracted from the wave-front phase (wave-front tip and tilt removal), then the mean square phase distortion reduces to

$$\sigma_3^2 = 0.134(d/r_0)^{5/3}. \tag{2.17}$$

Subscript 3 is used here for consistency with the notations used in Section 3.2.1 (Eq. (3.20)). Let us assume that a segmented mirror is used to compensate the wave front, and let us approximate each segment as a circular flat mirror of diameter d. Each segment will do its best to compensate the wave front. A 'piston' motion of the segment will compensate the mean value of the phase distortion averaged over the segment area. Tilting the segment will also compensate the mean wave-front slope. Such a compensation is said to be 'zonal'. The residual mean square phase error, also called wave-front fitting error, is given by Eq. (2.17). Such a compensation requires the control of three parameters (piston, tip and tilt), that is at least three actuators per mirror segment. The mean actuator spacing is therefore $r_s = d/\sqrt{3}$, and the fitting error can be written

$$\sigma_{\text{fit}}^2 = 0.335(r_s/r_0)^{5/3}. \tag{2.18}$$

Equation (2.18) shows that the variance of the wave-front fitting error decreases as the 5/3 power of the actuator spacing. This is a general characteristic of zonal compensation. Equation (2.18) also applies to continuous facesheet mirrors (see Chapter 4) with a coefficient in the 0.3–0.4 range, depending on the exact shape of the actuator influence function (Hudgin 1977). The total number N of independently controlled parameters is about $N = (D/r_s)^2$. Expressed as a function of N, the fitting error becomes

$$\sigma_{\text{fit}}^2 \simeq 0.335(D/r_0)^{5/3} N^{-5/6}. \tag{2.19}$$

Equation (2.19) shows that for a zonal compensation, the mean square residual error decreases as the $-5/6$ power of the number of controlled parameters inside a given aperture. In Section 3.1, it will be shown that a slightly steeper decrease can be obtained by controlling global wave-front modes rather than local zones (modal compensation).

Equation (2.12) describes the spatial distribution of wave-front distortions. It allows us to determine the number of parameters we need to control the wave-front surface and the amplitude of the correction to be applied, but it does not tell us how fast this must be done. To do this, we need an expression for the temporal structure function of the wave-front phase

$$D_\varphi(\tau) = \langle |\varphi(\mathbf{x}, \, t) - \varphi(\mathbf{x}, \, t + \tau)|^2 \rangle. \tag{2.20}$$

In Section 2.1, we saw that air refractive index inhomogeneities are essentially wind driven. If the whole turbulent atmosphere were propagating at the same velocity, then the wave-front phase distortion would also propagate at that velocity without noticable deformation while crossing the telescope aperture. However, most of the time wave fronts are affected by more than one turbulent layer, and these propagate at different speeds in different directions. As a result, the wave-front phase still propagates with a velocity $\bar{v}$ which is a weighted average of the layers' velocities, but also deforms very rapidly. This rapid deformation of the wave front is sometimes referred to as wave-front 'boiling'. Wave-front boiling time dictates the time evolution of the intensity distribution in the image plane, but propagation still dictates the response time for adaptive optics (Roddier *et al.* 1982b). Hence, when dealing with adaptive optics, one often assumes propagation with a single velocity $\bar{v}$. This rather rough approximation is often also called a 'Taylor' approximation. It should be clearly distinguished from the more accurate approximation described by Eq. (2.3). Assuming a mean propagation velocity with modulus $\bar{v}$, the temporal structure function of the wave-front phase is simply obtained by substituting $\bar{v}\tau$ for ξ in Eq. (2.12), that is

$$D_\varphi(\tau) = 6.88(\bar{v}\tau/r_0)^{5/3}. \tag{2.21}$$

This expression can be used to calculate the effect of the finite response time of an AO system. As seen from Eq. (2.20), it gives the mean square phase error σ_{time}^2 associated with a pure delay τ in which the phase is measured at time t but the correction is applied at time $t + \tau$,

$$\sigma_{\text{time}}^2(\tau) = 6.88(\bar{v}\tau/r_0)^{5/3}. \tag{2.22}$$

It should be noted that whereas the fitting error σ_{fit}^2 depends on the single atmospheric parameter r_0 (Eq. (2.18)), the time delay error σ_{time}^2 depends on two parameters $\bar{v}$ and r_0. Both vary with time independently of each other ($\bar{v}$ is an instantaneous spatial average not a time average). The knowledge of r_0 alone is insufficient to determine the characteristics of an AO system.

2.3 Image formation

When designing an AO system, it is important to understand the relationship between the wave-front phase in the telescope aperture plane, and the distribution of intensity in the telescope focal plane. As we shall see, the relation is non-linear. Phase perturbations with an amplitude below a threshold of about 1 radian have little effect on image quality and therefore seldom need to be compensated. However, a linear increase in the amplitude of the perturbation produces exponential effects on image quality. We have already seen that rms phase distortions are equal to about 1 radian over a circular area of diameter r_0 (Eq. (2.16)). Hence, there is little need to correct wave-front perturbations at a smaller scale. On a telescope aperture of diameter D, the number of such areas is $(D/r_0)^2$. This is an order of magnitude for the number of parameters one needs to control. For instance, if a segmented mirror is used to compensate the wave front, the residual mean square phase error is given by Eq. (2.19). To bring this value below 1 radian requires the control of a number N of actuators at least equal to

$$N_0 = 0.27(D/r_0)^2. \tag{2.23}$$

Since r_0 increases as $\lambda^{6/5}$ (see Eq. (2.13)), the number of actuators one needs to control decreases as $\lambda^{-12/5}$. This makes AO much easier at longer wavelengths. Using the above definition of N_0, one can rewrite Eq. (2.19) in the form

$$\sigma_{\text{fit}}^2 = (N_0/N)^{5/6}. \tag{2.24}$$

At the end of Section 2.2, we saw that a pure time delay τ in the control loop produces a mean square wave-front error given by Eq. (2.22). In a similar way, one can use this expression to determine an acceptable time delay τ_0 for the control loop. For the mean square phase error to be less than 1 radian, the delay must be less than

$$\tau_0 = (6.88)^{-3/5}\frac{r_0}{\overline{v}} = 0.314\frac{r_0}{\overline{v}}. \tag{2.25}$$

This delay has been called the Greenwood time delay (Fried 1990). The required frequency bandwidth of the control system is called the Greenwood frequency (Greenwood 1977). Since r_0 increases as $\lambda^{6/5}$, the control bandwidth decreases as $\lambda^{-6/5}$. This again makes AO easier at longer wavelengths. Using the above definition of τ_0, Eq. (2.22) can be rewritten

$$\sigma_{\text{time}}^2(\tau) = (\tau/\tau_0)^{5/3}. \tag{2.26}$$

In practice, the required frequency bandwidth also depends on the number of controlled parameters. A more thorough analysis of the effect of the finite bandwidth in an AO system will be presented in Chapter 3.

To get more physical insight into the relationship between the wave front and the image, we will consider first a particular set-up, called a Fizeau interferometer, in which the telescope aperture is masked except for two small circular subapertures a distance ξ apart. The diameter of each subaperture is taken to be small compared to r_0, so that turbulence does not substantially affect the intensity distribution in the diffracted beam. Assuming a monochromatic point source, the superposition of the two diffracted beams produces interference fringes. Let Ψ_1 and Ψ_2 be the complex amplitudes produced by each subaperture in the image plane. The resulting complex amplitude is

$$\Psi = \Psi_1 + \Psi_2, \tag{2.27}$$

and the intensity is

$$I = |\Psi|^2 = |\Psi_1 + \Psi_2|^2 = |\Psi_1|^2 + |\Psi_2|^2 + 2\,\mathrm{Re}\,(\Psi_1\Psi_2^*), \tag{2.28}$$

where * denotes a complex conjugate. The first two terms on the right hand side of Eq. (2.28) are the intensities that each subaperture would produce alone. The last term describes the interference pattern. The fringe amplitude is given by the modulus of $\Psi_1\Psi_2^*$, the fringe phase by its argument. Because in astronomy the image plane is conjugate to the sky, and sky coordinates are angles, in all that follows we will use angles as coordinates in the image plane. Expressed as an angle, the fringe spacing is λ/ξ in radians. The fringe spatial frequency is ξ/λ in radian^{-1}.

A wave-front phase perturbation simply adds a random phase term φ_1 to the argument of Ψ_1, and a random phase term φ_2 to the argument of Ψ_2. That is the complex quantity $\Psi_1\Psi_2^*$ is multiplied by $\exp(\varphi_1 - \varphi_2)$. The fringes are randomly shifted with a phase shift $\varphi_1 - \varphi_2$, but their instantaneous amplitude is not affected. According to Eq. (2.12), the variance of the fringe phase shift is

$$\langle|\varphi_1 - \varphi_2|^2\rangle = 6.88(\xi/r_0)^{5/3}. \tag{2.29}$$

However, if we record the fringes with a long exposure time compared with the characteristic time of the fringe motion, the recorded fringes will be blurred. In a very long exposure, the fringe pattern is described by the ensemble average $\Psi_1\Psi_2^*\langle \exp(\varphi_1 - \varphi_2)\rangle$. Assuming phase perturbations with Gaussian statistics,

$$\langle\exp(\varphi_1 - \varphi_2)\rangle = \exp(-\tfrac{1}{2}\langle|\varphi_1 - \varphi_2|^2\rangle). \tag{2.30}$$

The long exposure fringes are no longer shifted, but their amplitude is attenuated. Putting Eq. (2.29) into Eq. (2.30) gives the following attenuation coefficient for the fringe amplitude

$$\langle \exp(\varphi_1 - \varphi_2) \rangle = \exp[-3.44(\xi/r_0)^{5/3}]. \tag{2.31}$$

If the two subapertures are separated by a distance equal to r_0, the long exposure fringe amplitude is only 3.2×10^{-2} times that of the instantaneous fringes. In other words r_0 is a typical subaperture separation beyond which long exposure fringes disappear. Since, in optics, fringe visibility is a measure of coherence, Eq. (2.31) is often called the coherence function. It describes the loss of fringe visibility in long exposures taken through atmospheric turbulence.

One can think of a full aperture as an array of N small subapertures each producing a complex amplitude Ψ_k in the telescope focal plane. The resulting amplitude is

$$\Psi = \sum_{k=1}^{N} \Psi_k, \tag{2.32}$$

and the resulting intensity is

$$I = |\Psi|^2 = \sum_{k=1}^{N} |\Psi_k|^2 + \sum_{k \neq j} \sum_{j} \Psi_k \Psi_j^*. \tag{2.33}$$

Hence the intensity in the image plane is the sum of the intensities produced by each subaperture plus the sum of the interference terms produced by each pair of subapertures (each pair being counted twice). It shows that the intensity in the image plane can be synthesized by measuring the interference terms with a pair of movable subapertures. This is the basis of aperture synthesis in radio-astronomy. Each interference term describes a fringe pattern, that is a sinusoidal function. Hence Eq. (2.33) describes the intensity as a sum of sinusoidal terms, also called Fourier components, with spatial frequency $\mathbf{f} = \xi/\lambda$ in radian^{-1}. Assuming no turbulence, each Fourier component is weighted by a coefficient equal to the number of identical subaperture vector separations ξ inside the telescope aperture. This number is easily shown to be proportional to the overlap area between two shifted apertures. In other words the weights are given by the autocorrelation of the aperture transmission, where distances are measured in wavelength units. The weighting function of the Fourier components is called the optical transfer function $T(\mathbf{f})$. Its spatial frequency cut-off is $f_c = D/\lambda$. Its Fourier transform is the intensity distribution in a point source image, also called the point-spread function. The Fourier transform of the aperture autocorrelation function describes the image of a point source produced by a perfect diffraction-limited telescope, or Airy disk. Its angular width is about λ/D.

In the presence of turbulence, the Fourier components of a long exposure image suffer from an additional attenuation described by the fringe attenuation coefficient (Eq. (2.31)). The optical transfer function, we now call $G(\mathbf{f})$, is the product of two factors, the autocorrelation of the aperture transmission, or telescope transfer function $T(\mathbf{f})$, and an atmospheric transfer function $A(\mathbf{f})$

$$G(\mathbf{f}) = T(\mathbf{f})A(\mathbf{f}). \tag{2.34}$$

The atmospheric transfer function is obtained by replacing ξ by $\lambda\mathbf{f}$ in Eq. (2.31) (with $f = |\mathbf{f}|$),

$$A(\mathbf{f}) = \exp[-3.44(\lambda f/r_0)^{5/3}]. \tag{2.35}$$

For large astronomical telescopes at visible wavelengths, $A(\mathbf{f})$ decreases much faster than $T(\mathbf{f})$. The angular resolution of the telescope is essentially turbulence limited. To a good approximation, the point-spread function of a turbulence degraded image is simply the Fourier transform of $A(\mathbf{f})$. Since $A(\mathbf{f})$ is almost a Gaussian (Eq. (2.35)), its Fourier transform is also almost a Gaussian, only the wings fall off less steeply. Its angular width is about λ/r_0. Since r_0 varies as $\lambda^{6/5}$, this width decreases as $\lambda^{-1/5}$, that is the telescope angular resolution increases at longer wavelengths, until r_0 becomes of the order of D. At this point, the maximum resolution is achieved. Beyond that the telescope angular resolution is essentially diffraction-limited. The image width increases again as λ/D and the angular resolution decreases.

When the atmospheric phase distortion is partially compensated by an AO system, Eq. (2.34) remains valid with $A(\mathbf{f})$ given by Eq. (2.30), that is

$$A(\mathbf{f}) = \exp[-\tfrac{1}{2}D_\varphi(\lambda\mathbf{f})], \tag{2.36}$$

but the phase structure function D_φ is no longer given by Eq. (2.29). AO systems mostly compensate the large scale wave-front distortions, which have the largest amplitude. This has the effect of levelling off the structure function at some level $2\sigma^2$, where σ^2 is the variance of the remaining (uncorrelated) small scale wave-front distortions. The better the compensation, the smaller is σ^2. At low frequencies $A(\mathbf{f})$ decreases as without compensation but instead of rapidly converging toward zero, it converges toward a constant

$$A(\infty) = \exp(-\sigma^2). \tag{2.37}$$

Hence, $A(\mathbf{f})$ can be described as the sum of the above constant term plus a low frequency term. As a consequence, the point-spread function, which is the Fourier transform of Eq. (2.34), is also the sum of two terms, an Airy disk plus a halo due to light diffracted by the remaining small scale wave-front phase errors. Equation (2.37) represents the fraction of light in the Airy disk. It is a good measure of the quality of the compensation. In optics, a traditional criterion of image quality is the Strehl ratio R, which is the ratio of the maximum intensity in the point spread to that in a theoretically perfect point source image (Airy disk). For a large telescope, with a diameter D much larger than r_0, and under good compensation conditions ($R > 0.2$) the contribution of the halo to the central intensity is small. Hence, to a good approximation,

$$R \approx \exp(-\sigma^2). \tag{2.38}$$

As stated at the begining of this section, image quality degrades exponentially with the variance of the wave-front distortion. The threshold of 1 radian we considered corresponds to $R = 1/e = 0.37$. Experience shows that images with a higher Strehl ratio look fairly good, whereas images with a lower Strehl ratio look poor.

Putting Eq. (2.24) into Eq. (2.38) shows how the Strehl ratio increases with the number N of controlled actuators

$$R \approx \exp[-(N_0/\mathrm{N})^{5/6}]. \tag{2.39}$$

Hence N_0 is the number of actuators needed to obtain a Strehl ratio of the order of 0.37 (assuming no other source of errors). Similarly, putting Eq. (2.26) into Eq. (2.38) gives the decrease of the Strehl ratio as a function of the time delay in the servo loop

$$R \propto \exp[-(\tau/\tau_0)^{5/3}]. \tag{2.40}$$

It shows that τ_0 is the delay over which the Strehl ratio of a compensated image is divided by a factor e.

2.4 Non-isoplanicity and other effects

From the previous sections, it should be clear that the optical quality of an AO-compensated image is necessarily limited. The two main limitations are the finite number of parameters one can control, and the finite speed at which one can control them. In this section, we discuss a few additional limitations. These are related to the fact that current AO systems compensate wave-front phase distortions in the telescope aperture plane, whereas some turbulence layers occur well above in the atmosphere.

In Section 2.2, we have neglected propagation effects by integrating the refractive index fluctuations along the line of sight (near-field approximation). However, index inhomogeneities both refract and diffract light. This has two types of effects. A first effect is that after propagation over some distance, illumination is no longer uniform. It is this variation of illumination which causes the eye to see a star twinkling. The phenomenon is called scintillation. In this case, the atmospheric transfer function is still given by Eq. (2.31), but it now takes the form

$$A(\mathbf{f}) = \exp[-\tfrac{1}{2}D(\lambda\mathbf{f})], \tag{2.41}$$

where $D(\lambda\mathbf{f})$, called the wave structure function, is the sum of two terms

$$D(\lambda\mathbf{f}) = D_\varphi(\lambda\mathbf{f}) + D_\chi(\lambda\mathbf{f}). \qquad (2.42)$$

$D_\varphi(\lambda\mathbf{f})$ is the structure function of the phase defined by Eq. (2.9), and $D_\chi(\lambda f)$ is the structure function of the log amplitude χ in the pupil plane

$$D_\chi(\lambda\mathbf{f}) = \langle |\chi(\mathbf{r}) - \chi(\mathbf{r} + \lambda\mathbf{f})|^2 \rangle. \qquad (2.43)$$

When f increases, $D_\chi(\lambda\mathbf{f})$ quickly saturates at values of the order of unity, whereas $D_\varphi(\lambda\mathbf{f})$ keeps growing and becomes the most important term. For this reason, fluctuations of the wave-front amplitude contribute much less to image degradation than the wave-front phase. However, if the wave-front phase is compensated by an AO system, there will be a residual image degradation due to the fluctuations of the wave-front amplitude. The loss in Strehl ratio is only a few percent in the infrared but may reach 10–15% at visible wavelengths (Roddier and Roddier 1986).

A second effect is due to the fact that refraction and diffraction are both wavelength dependent. It occurs when one tries to compensate images at one wavelength while sensing the wave front at another wavelength. Although we assume that refractive index fluctuations are wavelength independent, the diffraction of light by high altitude layers produces wavelength-dependent effects on the wave front. Again the effect is small. If we assume that sensing is done in the visible, the effect is curiously maximum in the very near infrared (I and J band), where the loss of Strehl ratio is a few percent, and decreases at longer wavelengths. Refraction effects occur only away from the zenith. Because of refraction, light rays follow different paths at different wavelengths. Assuming again that the sensing is done in the visible, the effect is also maximum in the near infrared (I and J band), but should become noticeable only 60° or more from zenith (Roddier and Roddier 1986).

The most important limitation related to the height of turbulence layers, is a limitation in the compensated field-of-view. It occurs because of differences between wave fronts coming from different directions. This effect is called anisoplanicity. If a particular source called a 'guide' source is used to sense the wave front, the compensation will be good only for objects close enough to the guide source. As the angular distance θ between the object and the guide source increases, image quality decreases. For a single turbulent layer at a distance $h/\cos\gamma$, the mean square error σ_{aniso}^2 on the wave front is obtained by simply replacing ξ with $\theta h/\cos\gamma$ in Eq. (2.12). In practice, several turbulence layers contribute to image degradation. In Eq. (2.22) we used a weighted average $\bar{v}$ of the layer velocities to calculate the time delay error. We can similarly use a weighted average $\bar{h}$ of the layer altitudes to calculate the

Table 2.1. *Wavelength and zenith angle dependence of turbulence-related AO parameters*

Parameter	Wavelength	Zenith angle
Fried's parameter (r_0)	$\lambda^{6/5}$	$(\cos \gamma)^{3/5}$
Seeing angle (λ/r_0)	$\lambda^{-1/5}$	$(\cos \gamma)^{-3/5}$
Number of control parameters (N_0)	$\lambda^{-12/5}$	$(\cos \gamma)^{-6/5}$
Actuator distance (r_s)	$\lambda^{6/5}$	$(\cos \gamma)^{3/5}$
Greenwood time delay (τ_0)	$\lambda^{6/5}$	$(\cos \gamma)^{3/5}$
Isoplanatic angle (θ_0)	$\lambda^{6/5}$	$(\cos \gamma)^{8/5}$

anisoplanicity error σ^2_{aniso} (Fried 1982; Roddier *et al.* 1982a; Roddier and Roddier 1986). This gives

$$\sigma^2_{\text{aniso}}(\theta) = 6.88 \left(\frac{\theta \overline{h}}{r_0 \cos \gamma} \right)^{5/3}. \tag{2.44}$$

Like the time delay error σ^2_{time} (Eq. (2.22)), the anisoplanicity error σ^2_{aniso} depends on two atmospheric independent parameters $\overline{h}$ and r_0. Again, the knowledge of r_0 alone is insufficient to estimate the isoplanicity error. One can use Eq. (2.44) to calculate an acceptable angular distance θ_0. For the rms error to be less than 1 radian, the angular distance must be less than

$$\theta_0 = (6.88)^{-3/5} \frac{r_0 \cos \gamma}{\overline{h}} = 0.314 \frac{r_0 \cos \gamma}{\overline{h}}. \tag{2.45}$$

This angular distance is called the isoplanatic angle. Like r_0, it increases as the 6/5 power of the wavelength, which again makes AO easier at longer wavelengths. It also decreases as the $-8/5$ power of the air mass. In practice, the acceptable angular distance also depends on the number of controlled parameters. A more thorough analysis of the effect of anisoplanicity in an AO system will be presented in Chapter 3.

Using the above definition of θ_0, one can rewrite Eq. (2.44) in the form

$$\sigma^2(\theta) = (\theta/\theta_0)^{5/3}. \tag{2.46}$$

Putting Eq. (2.46) into Eq. (2.38) gives the decrease of the Strehl ratio as a function of θ

$$R \propto \exp[-(\theta/\theta_0)^{5/3}]. \tag{2.47}$$

Hence θ_0 is the angular distance from the guide source over which the Strehl ratio of a compensated image is divided by a factor e.

The dependence of turbulence-related parameters on wavelength λ and zenith angle γ is summarized in Table 2.1 for convenience.

References

Fried, D. L. (1965) Statistics of a geometrical representation of wave-front distortion. *J. Opt. Soc. Am.*, **55**, 1427–35.

Fried, D. L. (1982) Anisoplanatism in adaptive optics. *J. Opt. Soc. Am.* **72**, 52–61.

Fried, D. L. (1990) Time-delay-induced mean-square error in adaptive optics. *J. Opt. Soc. Am. A* **7**, 1224–5.

Fried, D. L. (1994) Atmospheric turbulence optical effects: understanding the adaptive-optics implications. In: *Adaptive Optics for Astronomy*, eds D. M. Alloin, J.-M. Mariotti, NATO-ASI Series, 423, pp. 25–57. Kluwer Academic Publ., Dordrecht.

Greenwood, D. P. (1977) Bandwidth specification for adaptive optics systems. *J. Opt. Soc. Am.* **67**, 390–3.

Hudgin, R. J. (1977) Wave-front compensation error due to finite corrector-element size. *J. Opt. Soc. Am.* **67**, 393–5.

Noll, R. J. (1976) Zernike polynomials and atmospheric turbulence. *J. Opt. Soc. Am.* **66**, 207–11.

Roddier, F. (1981) The effects of atmospheric turbulence in optical astronomy. *Progress in Optics* **19**, 281–376.

Roddier, F. (1989) Optical propagation and image formation through the turbulent atmosphere. In: *Diffraction-Limited Imaging with Very Large Telescopes*, eds D. M. Alloin, J.-M. Mariotti, NATO-ASI Series, 274, pp. 33–52. Kluwer Academic Publ., Dordrecht.

Roddier, F., Gilli, J. M. and Vernin, J. (1982a) On the isoplanatic patch size in stellar speckle interferometry. *J. Optics* (Paris) **13**, 63–70.

Roddier, F., Gilli, J. M. and Lund, G. (1982b) On the origin of speckle boiling and its effects in stellar speckle interferometry *J. Optics* (Paris) **13**, 263–271.

Roddier, F. and Roddier, C. (1986) National Optical Astronomy Observatories (NOAO) Infrared Adaptive Optics Program II. Modeling atmospheric effects in adaptive optics systems for astronomical telescopes. In *Advanced Technology Telescopes III*, ed. L. D. Barr, Proc. SPIE vol 628, pp. 298–304.

Further references

Greenwood, D. P. and Fried, D. L. (1976) Power spectra requirements for wave-front-compensative systems. *J. Opt. Soc. Am.* **66**, 193–206.

Troxel, S. E., Welsh, B. Roggemann, M. C. (1995) Anisoplanatism effects on signal-to-noise ratio performance of adative optical systems. *J. Opt. Soc. Am. A* **12**, 570.

Tyler, G. A. (1984) Turbulence-induced adaptive-optics performance degradation: evaluation in the time domain. *J. Opt. Soc. Am. A* **1**, 251–62.

Part two

The design of an adaptive optics system

3

Theoretical aspects

FRANÇOIS RODDIER

Institute for Astronomy, University of Hawaii, USA

3.1 The principles of adaptive optics

In this chapter, we consider AO systems in general, mostly regardless of any practical implementation. An AO system basically consists of three main components, a wave-front corrector, a wave-front sensor, and a control system. They operate in a closed feedback loop. The wave-front corrector first compensates for the distortions of the incoming wave fronts. Then part of the light is diverted toward the wave-front sensor to estimate the residual aberrations which remain to be compensated. The control system uses the wave-front sensor signals to update the control signals applied to the wave-front corrector. As the incoming wave-front evolves, these operations are repeated indefinitely.

A key aspect of adaptive optics is the need for a 'guide' source to sense the wave front. Bright point sources work best. Fortunately, nature provides astronomers with many point sources in the sky, in the form of stars. However, they are quite faint. With current systems, observations are limited to the vicinity of the brightest stars, that is a few percent of the sky. Wave-front sensing is also possible with extended, but preferably small sources, provided they are bright enough. This includes not only solar system objects such as asteroids, or satellites of the main planets, but also a few galaxy cores, and small nebulosities. A whole chapter of this book is devoted to the problem of solar observations (Chapter 10). Another to the creation of artificial guide sources with laser beacons (Chapter 12).

Let us consider the wave-front corrector. It has a finite number P of parameters, or actuators one can control. Similarly, the wave-front sensor provides only a finite number M of measurements. For AO systems operating with bright sources, M can be large, and is always taken larger than P. The system is said to be overdetermined, and a least square solution is chosen for the control parameters. In this case, the accuracy of the compensation is mainly limited by the mirror. For astronomical observations with faint natural guide stars, where only a few photons are available, M is necessarily limited. Taking $M = P$ obviously requires a careful match of the wave-front sensor sensitivity to the particular aberrations the wave-front corrector can compensate. One could even envisage underdetermined systems for which $M < P$, and choose the control parameters that minimize the correction (minimum norm solution). In this case, the accuracy of the compensation would be essentially limited by the sensor. In practice wave-front correctors are not perfect. Actuators often introduce wave-front distortions which have to be corrected. To sense these distortions, one needs to have at least as many measurements as actuators, that is $M \geqslant P$.

In general, there is no one-to-one relationship between the sensor signals and the actuators. Acting on one actuator modifies all the sensor signals. In a closed feedback loop, the signals are small, and the wave-front sensor response can be considered as linear. The response of the sensor to each actuator is then described by a $P \times M$ matrix called the interaction matrix. A singular value decomposition of the interaction matrix gives a set of singular values. The number N of non-zero singular values is called the number of degrees of freedom of the system. It is the number of linearly independent parameters one can control to compensate the wave-front distortions. This number largely determines the compensation performance of an AO system. In this chapter, we determine what is the best possible performance that a theoretically ideal AO system with N degrees of freedom can achieve. One can then define the efficiency of a real system by comparison with this theoretical model.

3.2 Modal wave-front representation

Atmospheric turbulence produces randomly distorted wave fronts. The wave-front corrector attempts to compensate the distortions with an N-parameter fit. In some cases a good fit may be obtained. In other cases the fit will be poorer. The residual wave-front error is clearly random. The most efficient AO system is the one which *on the average* produces the smallest error. The solution to this optimization problem can be found by expanding the atmospheric wave fronts in a series of orthogonal functions. Optical physicists usually expand wave-front distortions in terms of functions called Zernike modes. Because of their simple analytic expressions, we will also use Zernike expansions. However, there is an infinite number of other possible expansions. Only one has statistically independent coefficients. It is called a Karhunen–Loève expansion. At the end of this section, we show that among all possible AO systems with N degrees of freedom, the most efficient one is the one which fully compensates the first N Karhunen–Loève modes.

3.2.1 Zernike expansion

A review of the properties of the Zernike modes relevant to the description of atmospherically distorted wave-fronts can be found in a paper by Noll (1976). For a circular aperture without obstruction, using polar coordinates (r, α), the Zernike modes are defined by

$$Z_n^m(r, \alpha) = \sqrt{n+1}\, R_n^m \begin{cases} \sqrt{2}\cos(m\alpha) \\ \sqrt{2}\sin(m\alpha)\,, \\ 1\ (m = 0) \end{cases} \tag{3.1}$$

where

$$R_n^m = \sum_{s=0}^{(n-m)/2} \frac{(-1)^s (n-s)!}{s![(n+m)/2 - s]![(n-m)/2 - s]!} r^{n-2s} \tag{3.2}$$

are the Zernike polynomials. The index n is called the radial degree, the index m the azimuthal frequency. Table 3.1 shows the first 15 Zernike modes $Z_n^m(r, \alpha)$, where j is a usual ordering number. Note that even j values correspond to cosine terms.

The Zernike modes are orthogonal over a circle of unit radius. Expressing Z as a function of the vector $\mathbf{r}(r, \alpha)$,

$$\int W(\mathbf{r}) Z_j(\mathbf{r}) Z_k(\mathbf{r})\, d\mathbf{r} = \delta_{jk}, \tag{3.3}$$

where δ_{jk} is the Kronecker symbol equal to one if $j = k$, and equal to zero if $j \neq k$. The weighting function $W(\mathbf{r})$ is given by

$$W(\mathbf{r}) = \begin{cases} 1/\pi & (r \leqslant 1) \\ 0 & (r > 1). \end{cases} \tag{3.4}$$

Any wave-front phase distortion $\varphi(\mathbf{r})$ over a circular aperture of unit radius can be expanded as a sum of Zernike modes

$$\varphi(\mathbf{r}) = \sum_j a_j Z_j(\mathbf{r}). \tag{3.5}$$

The sum is over an infinite number of terms. The coefficients a_j of the expansion are given by

$$a_j = \int W(\mathbf{r}) Z_j(\mathbf{r}) \varphi(\mathbf{r})\, d\mathbf{r}. \tag{3.6}$$

Here we are dealing with random wave-front aberrations. The coefficients a_j are random and we are interested in their statistical properties. Their covariance is given by

$$\langle a_j a_k \rangle = \left\langle \int W(\mathbf{r}) Z_j(\mathbf{r}) \varphi(\mathbf{r})\, d\mathbf{r} \int W(\mathbf{r}') Z_k(\mathbf{r}') \varphi(\mathbf{r}')\, d\mathbf{r}' \right\rangle \tag{3.7}$$

Table 3.1. *Expression of the first 15 Zernike modes*

$n\downarrow$ $m\rightarrow$	0	1	2	3	4
0	$Z_1 = 1$ Piston				
1		$Z_2 = 2r\cos\theta$ $Z_3 = 2r\sin\theta$ Tip/tilt			
2	$Z_4 = \sqrt{3}(2r^2 - 1)$ Defocus		$Z_5 = \sqrt{6}r^2\sin 2\theta$ $Z_6 = \sqrt{6}r^2\cos 2\theta$ Astigmatism (3rd order)		
3		$Z_7 = \sqrt{8}(3r^3 - 2r)\sin\theta$ $Z_8 = \sqrt{8}(3r^3 - 2r)\cos\theta$ Coma		$Z_9 = \sqrt{8}r^3\sin 3\theta$ $Z_{10} = \sqrt{8}r^3\cos 3\theta$ Trefoil	
4	$Z_{11} = \sqrt{5}(6r^4 - 6r^2 + 1)$ Spherical		$Z_{12} = \sqrt{10}(10r^4 - 3r^2)\cos 2\theta$ $Z_{13} = \sqrt{10}(10r^4 - 3r^2)\sin 2\theta$ Astigmatism (5th order)		$Z_{14} = \sqrt{10}r^4\cos 4\theta$ $Z_{15} = \sqrt{10}r^4\sin 4\theta$ Ashtray

which can be rewritten as a double integral

$$\langle a_j a_k \rangle = \int\int W(\mathbf{r})Z_j(\mathbf{r})W(\mathbf{r}')Z_k(\mathbf{r}')\langle\varphi(\mathbf{r})\varphi(\mathbf{r}')\rangle \, d\mathbf{r} \, d\mathbf{r}' \qquad (3.8)$$

or, with $\mathbf{r}' = \mathbf{r} + \rho$,

$$\langle a_j a_k \rangle = \int \langle\varphi(\mathbf{r})\varphi(\mathbf{r} + \rho)\rangle \int W(\mathbf{r})Z_j(\mathbf{r})W(\mathbf{r} + \rho)Z_k(\mathbf{r} + \rho) \, d\mathbf{r} \, d\rho. \qquad (3.9)$$

This integral is best calculated by using Parseval's theorem. Assuming stationarity, the covariance $\langle\varphi(\mathbf{r})\varphi(\mathbf{r} + \rho)\rangle$ is a function of ρ only. Its Fourier transform is the power spectrum $\Phi(\boldsymbol{\kappa})$ of the random wave-front phase $\varphi(\mathbf{r})$. The second integral is a convolution. Its Fourier transform is the product of the Fourier transforms of the convolution factors. Let $Q_j(\boldsymbol{\kappa})$ be the Fourier transform of $W(\mathbf{r})Z_j(\mathbf{r})$. Eq. (3.9) can be written

$$\langle a_j a_k \rangle = \int \Phi(\boldsymbol{\kappa})Q_j(\boldsymbol{\kappa})Q_k(\boldsymbol{\kappa}) \, d\boldsymbol{\kappa}. \qquad (3.10)$$

The power spectrum $\Phi(\boldsymbol{\kappa})$ of the wave-front phase can be obtained from its structure function (Tatarski 1961). With the above notations (Roddier 1981)

$$\Phi(\boldsymbol{\kappa}) = 7.2 \times 10^{-3}(D/r_0)^{5/3}|\boldsymbol{\kappa}|^{-11/3}. \qquad (3.11)$$

The functions $Q_j(\boldsymbol{\kappa})$ are given by (Noll 1976)

$$Q_n^m(f, \alpha) = \sqrt{n+1}\frac{J_{n+1}(2\pi\kappa)}{\pi\kappa}\begin{cases} (-1)^{(n-m)/2}\sqrt{2}\cos m\alpha \\ (-1)^{(n-m)/2}\sqrt{2}\sin m\alpha \\ (-1)^{n/2} \quad (m = 0), \end{cases} \qquad (3.12)$$

where κ and α are the modulus and argument of $\boldsymbol{\kappa}$. The function J_n is a Bessel function of order n. Putting Eq. (3.11) and Eq. (3.12) into Eq. (3.10), shows that only Zernike terms with the same azimuthal frequency m are correlated. Moreover, cosine terms are uncorrelated with sine terms. The covariance takes the form

$$\langle a_j a_k \rangle = c_{jk}(D/r_0)^{5/3}. \qquad (3.13)$$

The coefficients c_{jk} are functions of the radial degrees n and n' of the two Zernike terms, and of their common azimuthal frequency m. An analytic expression for the c_{jk} was first derived by Noll (1976). For n or $n' \neq 0$,

$$c_{jk} = 7.2 \times 10^{-3}\sqrt{(n+1)(n'+1)}(-1)^{(n+n'-2m)/2}\pi^{8/3} \qquad (3.14)$$

$$\times \frac{\Gamma(14/3)\Gamma[(n+n'-5/3)/2]}{\Gamma[(n-n'+17/3)/2]\Gamma[(n'-n+17/3)/2]\Gamma[(n+n'+23/3/2]},$$

where Γ is Euler's Gamma function. For $n = n' = 0$, the integral in Eq. (3.10) diverges. The variance of the piston term is infinite because Eq. (3.11) assumes turbulence with an infinite outerscale. Values for the first c_{jk} have been published by Wang and Markey (1978), and N. Roddier (1990). For $j > 1$, the c_{jj} are all finite and positive. They decrease monotonically as j increases. For the cross terms, tip and tilt are anticorrelated with their corresponding coma terms; the coefficients are $c_{2,8} = c_{3,7} = -1.41 \times 10^{-2}$. Defocus is anticorrelated with spherical aberration, and third order astigmatism with fifth order; the coefficients are $c_{4,11} = c_{5,13} = c_{6,12} = -3.87 \times 10^{-3}$.

The mean square wave-front phase fluctuation is defined as

$$\sigma^2 = \left\langle \int W(\mathbf{r})\varphi^2(\mathbf{r})\,d\mathbf{r} \right\rangle. \tag{3.15}$$

Putting Eq. (3.5) into Eq. (3.15) and rearranging the order of summation gives

$$\sigma^2 = \sum_j \sum_k \langle a_j a_k \rangle \int W(\mathbf{r}) Z_j(\mathbf{r}) Z_k(\mathbf{r})\,d\mathbf{r} \tag{3.16}$$

or, taking Eq. (3.3) into account,

$$\sigma^2 = \sum_j \langle a_j^2 \rangle. \tag{3.17}$$

If we include the $j = 1$ piston term, this quantity is infinite. However, we are only interested in the deviation from the mean surface. The mean square deviation, as defined by Eq. (2.14), is given by

$$\sigma_1^2 = \sum_{j=2}^{\infty} \langle a_j^2 \rangle \tag{3.18}$$

which is finite. Taking Eq. (3.13) into account, gives

$$\sigma_1^2 = \sum_{j=2}^{\infty} c_{jj}(D/r_0)^{5/3}. \tag{3.19}$$

One can now imagine a theoretical AO system that would compensate Zernike modes. Since the low order terms have the highest variance, one wants to compensate them first. Hence, a system with N degrees-of-freedom would do best by compensating the first N modes (other than piston). The variance of the residual wave-front distortion will simply be

$$\sigma_{N+1}^2 = \sum_{j=N+2}^{\infty} c_{jj}(D/r_0)^{5/3}. \tag{3.20}$$

A table giving the sum of the c_{jj} from $N+2$ to infinity can be found in the paper by Noll (1976), for N ranging from 0 to 20. In Fig. 3.1, we have plotted as a function of $N+1$ the variance of the compensated wave front as a fraction of the variance of the uncompensated wave front, that is the ratio $\sigma_{N+1}^2/\sigma_1^2$. It shows the number of parameters one has to control to reduce

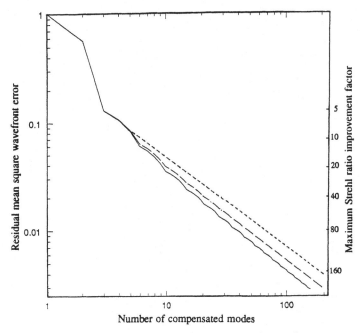

Fig. 3.1. Residual mean square wave-front distortion $\sigma_{N+1}^2/\sigma_1^2$ as a function of the number N of compensated modes. *Full line*: Karhunen–Loève modes. *Dashed line*: Zernike modes. *Dotted line*: zonal compensation.

the rms amplitude of the wave-front distortion by a given amount. For instance, a perfect compensation of the first 50 Zernike terms divides the wave-front rms amplitude by about a factor 10. According to Noll (1976), the variance decreases as $N^{-\sqrt{3}/2} = N^{-0.87}$ for large N values. This is slightly faster than the $N^{-5/6} = N^{-0.83}$ decrease law for zonal compensation (see Section 2.2).

3.2.2 Karhunen–Loève expansion

An even faster decrease can be obtained if our theoretical AO system compensates for statistically independent modes, that is Karhunen–Loève (K–L) modes. K–L modes can be expressed in terms of the Zernike modes by diagonalization of the Zernike covariance matrix. Let us consider the infinite column vector

$$\mathbf{A} = \begin{bmatrix} a_1 \\ a_2 \\ a_3 \\ \vdots \end{bmatrix}. \tag{3.21}$$

The covariance matrix of the coefficients a_j can be written

$$E(\mathbf{AA^t}) = E\left(\begin{bmatrix} a_1 \\ a_2 \\ a_3 \\ \vdots \end{bmatrix} [a_1 a_2 a_3]\right) = \begin{bmatrix} E(a_1 a_1) & E(a_1 a_2) & \dots \\ E(a_2 a_1) & E(a_2 a_2) & \dots \\ \vdots & \vdots & \ddots \end{bmatrix} \qquad (3.22)$$

where $E()$ denotes ensemble averages. Since the covariance matrix is hermitian, it can be diagonalized. That is there is always a unitary matrix $\mathbf{U}$ such that $\mathbf{U}E(\mathbf{AA^t})\mathbf{U^t}$ is diagonal. The coefficients b_j of the K–L expansion are the components of the column vector $\mathbf{B} = \mathbf{UA}$. Indeed, their covariance matrix is

$$E(\mathbf{BB^t}) = E(\mathbf{UAA^tU^t}) = \mathbf{U}E(\mathbf{AA^t})\mathbf{U^t}. \qquad (3.23)$$

Since this matrix is diagonal, the coefficients b_j are statistically independent. The coefficients a_j can be expressed as a function of the coefficients b_j by the inverse relation $\mathbf{A} = \mathbf{U^t B}$, or

$$a_j = \sum_k u_{jk} b_k. \qquad (3.24)$$

Putting Eq. (3.24) into Eq. (3.5) gives

$$\varphi(\mathbf{r}) = \sum_j \sum_k u_{jk} b_k Z_j(\mathbf{r}) = \sum_k b_k \sum_j u_{jk} Z_j(\mathbf{r}) \qquad (3.25)$$

which is a development of the wave-front phase distortion $\varphi(\mathbf{r})$ in terms of the K–L modes

$$K_k(\mathbf{r}) = \sum_j u_{jk} Z_j(\mathbf{r}). \qquad (3.26)$$

Equation (3.26) expresses the K–L modes as a development in terms of Zernike modes. Since $\mathbf{U}$ is a unitary matrix, the scalar products are conserved and the K–L modes are orthogonal. The first coefficients u_{jk} of the development are given by Wang and Markey (1978). Each K–L function is a linear combination of Zernike modes with the same azimuthal frequency m. One can therefore also characterize a K–L mode by its azimuthal frequency. The radial degree of a K–L mode can be defined as the degree of the lowest order Zernike term in the expansion. By doing so, one produces a one-to-one relationship between K–L and Zernike modes. Low order K–L modes are very similar to the related Zernike modes. For instance K–L 'tip/tilt' is a tip/tilt with 3% of negative coma, plus 0.2% of the fifth degree term, and so on. K–L 'defocus' has 18% of negative spherical aberration, plus 2% of the sixth degree term, and so on. As the order increases the difference between a Zernike mode and its related K–L mode becomes more and more pronounced. All the above summations are over an infinite number of terms. In practice a good approximation is obtained by summing over a large number of terms. Equation (3.25) can be used to simulate an atmospherically distorted wave front on a computer (see Section 7.5.2).

Let us now consider a theoretical AO system that would perfectly compensate the N first K–L terms and nothing beyond. The residual mean square wave-front distortion will be

$$\sigma'^2_{N+1} = \sum_{j=N+2}^{\infty} \langle b_j^2 \rangle. \tag{3.27}$$

Figure 3.1 shows σ'^2_{N+1}/σ^2 as a function of $N+1$. Note that it decreases faster than σ^2_{N+1}/σ^2. Hence, for a given number N of degrees of freedom, it is more efficient to compensate K–L modes than Zernike modes. It can be shown that a perfect compensation of the first N K–L modes is the best compensation one can possibly achieve with N degrees of freedom. Indeed, for a linear system, the residual wave-front distortion $e(\mathbf{r})$ is a weighted sum of the residual distortions $e_j(\mathbf{r})$ that the system would leave for each K–L mode

$$e(\mathbf{r}) = \sum_j b_j e_j(\mathbf{r}) \tag{3.28}$$

and the mean square residual wave-front variance is

$$\sigma_e^2 = \int W(\mathbf{r}) \langle e^2(\mathbf{r}) \rangle \, d\mathbf{r}. \tag{3.29}$$

Putting Eq. (3.28) into Eq. (3.29) gives

$$\sigma_e^2 = \int W(\mathbf{r}) \sum_j \sum_k \langle b_j b_k \rangle e_j(\mathbf{r}) e_k(\mathbf{r}) \, d\mathbf{r} \tag{3.30}$$

or, the coefficients b_j being uncorrelated,

$$\sigma_e^2 = \sum_j \langle b_j^2 \rangle \int W(\mathbf{r}) e_j^2(\mathbf{r}) \, d\mathbf{r}. \tag{3.31}$$

Since all the terms in this sum are positive, it can only be minimized by minimizing each term. With N degrees of freedom, the minimum error is obtained by cancelling the N terms which have the largest weight, that is the first N K–L terms.

3.3 Ideal compensation performance

We now know how to best reduce the mean square wave-front distortion with a given number N of degrees of freedom. How well do we improve the image? To answer this question, one must first define some criterion of image quality. Astronomers are used to defining image quality in terms of the full width at half maximum (fwhm) of a point source image. This is an acceptable criterion as long as the image profile is independent of seeing quality. In Chapter 2 we saw that the profile of an uncompensated, seeing-limited point source image is nearly Gaussian and indeed independent of seeing conditions. Only the width of the profile changes. However, we have also seen that a compensated image consists of a narrow diffraction-limited core surrounded by a halo of light scattered by the uncompensated residual wave-front errors. The ratio of the amount of light in the core to that in the halo varies with the degree of compensation. Depending on this ratio, the image fwhm may fall in the narrow core, or in halo. Hence, the criterion loses its significance.

A good criterion would be, of course, the ratio of the light in the core to that in the halo. For historical reasons, one instead uses a directly related criterion called the Strehl ratio, which is the ratio of the maximum intensity in the compensated image to that in a perfect diffraction-limited image (see Section 2.3). Compared to the image fwhm, the Strehl ratio is more sensitive to residual wave-fronts errors. It is therefore a better test of image quality. If one insists on measuring image quality in terms of a width, one can also define an equivalent width as the width of a uniformly illuminated disk with the same flux, and the same central intensity. This quantity has sometimes been referred to as the Strehl width (Roddier, *et al.* 1991). Depending on the application, other criteria could be used. Examples are the diameter of the circle which contains 50% of the energy, or − for spectroscopy − the width of a slit that would let through this amount. Here, we will mainly discuss image quality in terms of the Strehl ratio, the most widely used criterion.

As shown in Section 2.3, and with the definition Eq. (3.4) of the weighting function W, the optical transfer function for long exposures images can be written as

$$G(\mathbf{f}) = \pi \int W(\mathbf{r}) W(\mathbf{r} + 2\lambda \mathbf{f}/D)$$

$$\times \exp\{-\tfrac{1}{2}\langle[\varphi(\mathbf{r}) - \varphi(\mathbf{r} + 2\lambda \mathbf{f}/D)]^2\rangle\} \, d\mathbf{r}$$

$$= \pi \int W(\mathbf{r}) W(\mathbf{r} + 2\lambda \mathbf{f}/D)$$

$$\times \exp\{\langle\varphi(\mathbf{r})\varphi(\mathbf{r} + 2\lambda \mathbf{f}/D)\rangle - \langle|\varphi(\mathbf{r})|^2\rangle\} \, d\mathbf{r}. \tag{3.32}$$

Putting a K–L expansion of $\varphi(\mathbf{r})$ into Eq. (3.32) gives

$$G(\mathbf{f}) = \pi \int W(\mathbf{r})W(\mathbf{r} + 2\lambda\mathbf{f}/D)$$

$$\times \exp\left\{ \sum_{j=N+2}^{\infty} \langle |b_j|^2 \rangle [K_j(\mathbf{r})K_j(\mathbf{r} + 2\lambda\mathbf{f}/D) - |K_j(\mathbf{r})|^2] \right\} d\mathbf{r} \quad (3.33)$$

Here we assume that the first N K–L modes (other than piston) are perfectly compensated, and the sum is extended over the uncompensated K–L modes. These can be estimated numerically using truncated series of Zernike modes. One can then estimate $G(\mathbf{f})$ with a truncated series of K–L modes (Wang and Markey 1978). Another possibility is to estimate Eq. (3.32) directly by averaging randomly drawn phase terms (N. Roddier 1990). The Strehl ratio is obtained by integrating $G(\mathbf{f})$ over all frequencies

$$R = \frac{\displaystyle\int G(\mathbf{f})\,d\mathbf{f}}{\displaystyle\int T(\mathbf{f})\,d\mathbf{f}} \quad (3.34)$$

where

$$T(\mathbf{f}) = \pi \int W(\mathbf{r})W(\mathbf{r} + 2\lambda\mathbf{f}/D)\,d\mathbf{r} \quad (3.35)$$

is the transfer function of the diffraction-limited telescope. A drawback of the Strehl ratio is that it is relative to a given telescope. Instead of using the telescope transfer function $T(\mathbf{f})$ for normalization, one can use the atmospheric transfer function $A(\mathbf{f})$ (Chapter 2, Eq. (2.35)). This was done by Fried (1966) and Wang and Markey (1978), who calculate the ratio

$$\mathscr{R} = \frac{\displaystyle\int G(\mathbf{f})\,d\mathbf{f}}{\displaystyle\int A(\mathbf{f})\,d\mathbf{f}}. \quad (3.36)$$

$\mathscr{R}$ is the gain in resolution over that of a pure seeing-limited image (no diffraction), and is referred to as Fried's normalized resolution. The width of a point source image varies as the inverse square root of it.

Figure 3.2 shows a log–log plot of this normalized resolution as a function of D/r_0 for different degrees of compensation. Under given seeing conditions, it gives the normalized resolution as a function of the telescope diameter. For a given telescope, it gives the normalized resolution as a function of seeing, the largest D/r_0 values corresponding to the worst seeing. For a given telescope under given seeing conditions, it also gives the normalized resolution as a function of wavelength (see arrows on top of the figure). Each curve is drawn

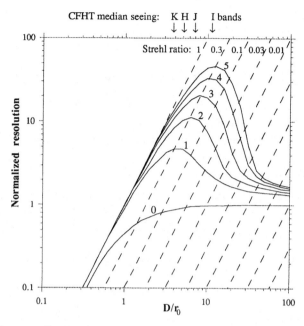

Fig. 3.2. Fried's normalized resolution as a function of D/r_0. Curve number n is for a perfect compensation of the first $N = (n+1)(n+2)/2$ K–L modes. Dashed lines are lines of equal Strehl ratio. Arrows on the top point to typical D/r_0 values for I-, J-, H-, and K-band observations with the CFH telescope.

for a perfect compensation of the first N K–L modes, where N is the number of the equivalent Zernike modes up to the polynomial degree indicated on the curve. Curve 0 shows the normalized resolution of uncompensated images. When D/r_0 is small, it is limited by diffraction. As D/r_0 increases, it tends asymptotically toward the seeing limit which has been set equal to unity. The normalized resolution of fully compensated images (Strehl ratio equal to unity) is represented by a straight line of slope 2. It grows as the square of the telescope diameter (the width of a diffraction-limited image decreases as the inverse of the telescope diameter). Curve 1 is for a perfect compensation of the K–L tip/tilt modes which we have seen to be quite close to pure tip/tilt modes. Curve 2 includes compensation of the K–L 'defocus' and 'astigmatism' modes, and so on.

Except for curve 0, all the curves go to a maximum at which the normalized resolution is the highest. The maximum occurs at $D/r_0 \approx 2.7\sqrt{N}$. At this point a Strehl ratio of $R \approx 0.3$ is obtained. The gain in angular resolution brought about by the compensation also goes to a maximum at about the same point. At this point the AO system is the most effective. Although Fig. 3.2 applies to K–L modes, similar curves can be drawn for Zernike-mode or zonal correction. In all cases a maximum gain is observed at which the compensated image has a Strehl ratio of ≈ 0.3. For perfect zonal correction, the maximum occurs at $D/r_0 = 2.3\sqrt{N}$ and the gain on the central intensity is by a factor 1.6 N (Roddier 1998).

With a given system under given seeing conditions, one can always choose the operating point by choosing the wavelength. Although the best possible image (highest Strehl ratio) will always be obtained at the longest wavelength, one will often select a shorter wavelength, and operate close to the maximum efficiency point, because at this point the larger gain in resolution brings a larger amount of new information. One can also change the operating point by changing the size of the telescope aperture. The existence of a maximum implies that the resolution can be actually increased by stopping down the telescope aperture. Although this may sound awkward to most astronomers, solar astronomers are familiar with it. This is because they use short exposures which are not affected by wave-front tip/tilt errors. In this case the resolution is given by curve 1 which does go to a maximum. The best images of the solar granulation have often been obtained with a reduced aperture. With adaptive optics, one might also occasionally wish to stop down the aperture, or use the same AO system on a smaller telescope. This is the case if one wants to observe at shorter wavelengths. On stellar sources, the gain in central intensity can be sufficient to balance the loss of photons to the point where the exposure time does not have to be increased.

One may note the high gain in resolution brought about by a simple tip/tilt compensation. For a stellar image, the gain in central intensity reaches a factor 5 at $D/r_0 = 4$. It has motivated the development of simple image stabilizers on many telescopes. For various reasons, the actual gain has often been much lower. In the visible, even under exceptional seeing conditions ($r_0 = 20$ cm), the maximum gain can only be achieved with small telescopes ($D < 1$ m). In the infrared, where r_0 easily reaches one meter, the maximum gain can be obtained with a 4-m telescope. However, at these wavelengths thermal background becomes important. One often takes only short exposures that can be easily recentered and co-added without requiring adaptive compensation. Also the gain brought about by image stabilization can be lower than that indicated in Fig. 3.2, because the outerscale may no longer be much larger than the telescope diameter (in Fig. 3.2 it is supposed to be infinite). In addition, the poor optical quality of many infrared telescopes has been a limiting factor. To date, the best results have been obtained with 2-m class optical telescopes used in the very near infrared (I, J, and H bands). A gain of 4.7 has been experimentally demonstrated with the CFH telescope stopped down to 1 meter (Graves *et al.* 1992a; Graves *et al.* 1992b).

The performance of actual adaptive optics systems follow curves similar to that in Fig. 3.2, albeit with smaller gains. Therefore, one can express their performance in terms of the number $N(K-L)$ of K–L modes one must compensate to produce similar results. This number has been referred to as the 'order of compensation', and the highest degree n of the corresponding Zernike terms as the 'degree of compensation' (Roddier 1994). The ratio of the order of compensation $N(K-L)$ to the actual number $N(\text{actual})$ of degrees of freedom of the system is a measure of the compensation efficiency

$$\eta_c = \frac{N(K-L)}{N(\text{actual})}.$$
(3.37)

By definition of the K–L modes, this number is necessarily below unity. A more practical definition of the efficiency of an AO system can be given in terms of the number $N(\text{zonal})$ of degrees of freedom of a perfect zonal compensation system which has the same performance, that is

$$q_c = \frac{N(\text{zonal})}{N(\text{actual})}$$
(3.38)

As indicated above, perfect zonal correction yields a maximum gain of $1.6N(\text{zonal})$. Therefore the efficiency q_c of a real system is simply

$$q_c = 0.63 \frac{(\text{Maximum gain in central intensity})}{(\text{Actual number of degrees of freedom})}$$
(3.39)

Eq. (3.39) provides a practical means to estimate the efficiency of an AO system, by simply measuring the maximum gain one can possibly achieve when operating at various wavelengths (various r_0 values). To do this, long exposure images must be recorded first with the feedback loop open, while static voltages are applied to the deformable mirror to compensate any static aberration in the system. Telescope jitter must also be avoided, otherwise the gain observed when closing the loop will be overestimated. Curvature-based AO systems (Chapter 9) have a typical efficiency $q_c \approx 50\%$. Shack–Hartmann systems built for the European Southern Observatory (Chapter 8) have a lower efficiency (approx 30%). Other higher order systems have even lower efficiencies (see Roddier 1998).

Aliasing errors due to coarse wave-front sampling, and matching errors between the sensor and the deformable mirror are the main causes of efficiency reduction. During the design process, the efficiency of a system can be estimated from computer simulations (see Chapter 7). Once the efficiency is known, Fig. 3.2 can be used to estimate the system performance. For instance, to obtain a Strehl ratio of 0.8, the number of degrees of freedom required is $\eta_c^{-1}(D/r_0)^2$. A Strehl ratio of 0.3 requires seven times less degrees of freedom.

It gives a maximum gain in resolution and will often be acceptable. The closer one wants to approach the diffraction limit, the costlier it becomes not only in terms of hardware, but also in terms of number of photons required to sense the wave front. To express this more quantitatively, it requires seven times the number of subapertures to improve the Strehl ratio from 0.3 to 0.8. As we shall see, one also has to run the feedback loop $\sqrt{7}$ faster. Hence $7\sqrt{7} = 18.5$ times more photons are needed from the reference source which must be 3 magnitudes brighter. When natural guide stars are used, this is a high penalty. It shows the importance of developing high efficiency systems for astronomy.

By taking the Fourier transform of Eq. (3.33), one obtains a theoretical image profile. Such a profile is shown in Fig. 3.3. The calculation was done assuming a perfect compensation of the Zernike modes of degree three or less. The ratio D/r_0 was taken to be eight, which gives the maximum gain in central intensity. The horizontal scale is in arcseconds for a 3.6-m aperture at $\lambda = 1.2$ μm. At this wavelength, $r_0 = 45$ cm. It corresponds to $r_0 = 15.7$ cm at

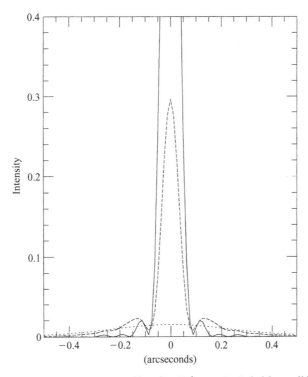

Fig. 3.3. Theoretical stellar image profiles for $D/r_0 = 8$. *Solid line*: diffraction-limited image. *Dotted line*: uncompensated image. *Dashed line*: with perfect compensation of all the Zernike modes of degree 3 or less. The horizontal scale is for a 3.6-m telescope observing at $\lambda = 1.2$ μm.

$\lambda = 0.5$ µm, that is a 0.64″ uncompensated image fwhm, which is typical for a good astronomical site. The intensity scale is normalized to unity at the maximum of the diffraction-limited image (upper solid line). The dashed line is the partially compensated image. It is worth noting that, although the Strehl ratio of this partially compensated image is only 0.3, its fwhm is very close to the diffraction limit. Only the wings are higher. Such images can easily be further improved by deconvolution.

3.4 Temporal and angular dependence of the Zernike modes

So far we have described the performance of an ideal AO system with an infinite bandwidth, observing in the direction of the guide source. We have determined the minimum number of degrees of freedom required to achieve a given Strehl ratio in that direction. We now need to determine the bandwidth requirements for the AO system, and to estimate how the compensation performance degrades when observing away from the the guide source. To do this, we consider again an AO system in which a finite number of Zernike modes are compensated. To determine how fast the Zernike coefficients evolve with time, we derive an analytic expression for the temporal power spectrum of each Zernike term, and then compute a time delay error. The same formalism is then used to estimate how fast the Zernike coefficients vary with the distance to the guide source, and to determine the size of the isoplanatic area.

3.4.1 Temporal power spectra

We consider first a single atmospheric layer with frozen-in turbulence propagating at the wind velocity $\mathbf{v}$. The phase distortion at time t is $\varphi(\mathbf{r} - \mathbf{v}t)$. According to Eq. (3.6), the value of the Zernike coefficient a_j at time t is

$$a_j(\mathbf{v}t) = \int W(\mathbf{r})Z_j(\mathbf{r})\varphi(\mathbf{r} - \mathbf{v}t)\,d\mathbf{r}. \qquad (3.40)$$

In other words $a_j(\mathbf{r})$ is given by a convolution product

$$a_j(\mathbf{r}) = \varphi(\mathbf{r}) * W(\mathbf{r})Z_j(\mathbf{r}). \qquad (3.41)$$

Hence, the spatial power spectrum $\Phi_j(\boldsymbol{\kappa})$ of $a_j(\mathbf{r})$ is related to the power spectrum $\Phi(\boldsymbol{\kappa})$ of $\varphi(\mathbf{r})$ by

$$\Phi_j(\boldsymbol{\kappa}) = \Phi(\boldsymbol{\kappa})|Q_j(\boldsymbol{\kappa})|^2, \qquad (3.42)$$

where $|Q_j(\boldsymbol{\kappa})|^2$ is the square modulus of the Fourier transform of $W(\mathbf{r})Z_j(\mathbf{r})$. Putting Eq. (3.11) and Eq. (3.12) into Eq. (3.42) gives

$$\Phi_j(\boldsymbol{\kappa}) = 7.2 \times 10^{-3}\pi^{-2}(D/r_0)^{5/3}(n+1)\kappa^{-17/3}J_{n+1}^2(2\pi\kappa) \begin{cases} 2\cos^2(m\alpha) \\ 2\sin^2(m\alpha) \\ 1 \ (m=0) \end{cases}.$$

$$(3.43)$$

The spatial covariance $B_j(\rho)$ of $a_j(\mathbf{r})$ is defined as the ensemble average

$$B_j(\rho) = \langle a_j(\mathbf{r})a_j(\mathbf{r}+\rho)\rangle. \tag{3.44}$$

According to the Wiener–Kinchin theorem, it is the inverse Fourier transform of the power spectrum $\Phi_j(\boldsymbol{\kappa})$. Let ξ and η be the components of the vector ρ. The Wiener–Kinchin theorem states that

$$B_j(\xi, \eta) = \int \Phi_j(\kappa_x, \kappa_y)\exp[2i\pi(\xi\kappa_x + \eta\kappa_y)]\, d\kappa_x\, d\kappa_y, \tag{3.45}$$

where κ_x and κ_y are the components of the vector $\boldsymbol{\kappa}$ conjugate to ξ and η.

Let $s_j(t) = a_j(\mathbf{v}t)$ describe the time evolution of the Zernike coefficient a_j. The temporal covariance $C_j(\tau)$ of $s_j(t)$ is given by

$$C_j(\tau) = \langle s_j(t)s_j(t+\tau)\rangle$$

$$= \langle a_j(\mathbf{v}t)a_j(\mathbf{v}t+\mathbf{v}\tau)\rangle$$

$$= B_j(\mathbf{v}\tau). \tag{3.46}$$

Let us choose the ξ component in the direction of propagation of the wind. Putting Eq. (3.45) into Eq. (3.46) gives

$$C_j(\tau) = \int\int \Phi_j(\kappa_x, \kappa_y)\exp(2i\pi\upsilon\kappa_x\tau)\, d\kappa_x\, d\kappa_y$$

$$= \int \exp(2i\pi\upsilon\kappa_x\tau)d\kappa_x \int \Phi_j(\kappa_x, \kappa_y)\, d\kappa_y, \tag{3.47}$$

where $\upsilon = |\mathbf{v}|$ is the wind speed. Introducing the temporal frequency $\nu = \upsilon\kappa_x$, gives

$$C_j(\tau) = \frac{1}{\upsilon}\int \exp(2i\pi\nu\tau)\int \Phi_j\left(\frac{\nu}{\upsilon}, \kappa_y\right)\, d\kappa_y\, d\nu. \tag{3.48}$$

The temporal frequency spectrum of $s_j(t)$ is the one-dimensional Fourier transform of the temporal covariance $C_j(\tau)$. According to Eq. (3.48), it is simply given by

$$F_j(\nu) = \frac{1}{\upsilon}\int \Phi_j\left(\frac{\nu}{\upsilon}, \kappa_y\right)\, d\kappa_y. \tag{3.49}$$

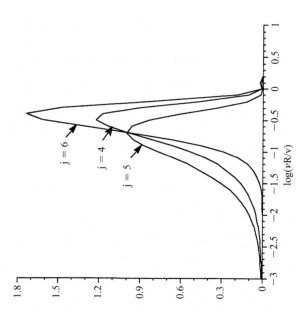
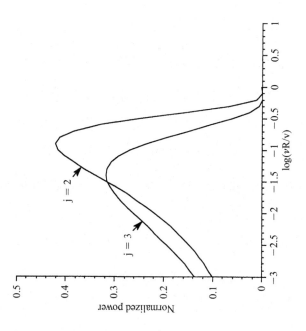

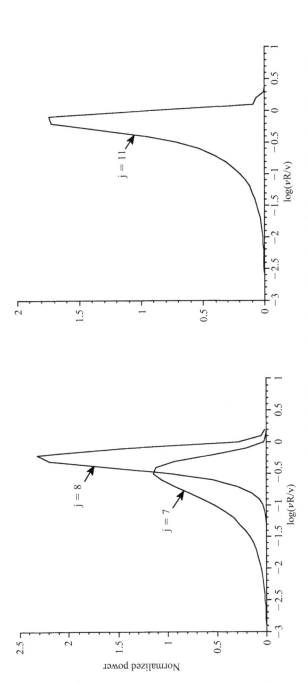

Fig. 3.4. Plots of the normalized power $\nu F_j(\nu)/\int F_j(\nu)\,d\nu$ as a function of $\log(\nu R/v)$ for tip/tilt $(j = 2, 3)$, defocus $(j = 4)$, astigmatism $(j = 5, 6)$, coma $(j = 7, 8)$, and spherical aberration $(j = 11)$.

Putting Eq. (3.43) into Eq. (3.49) gives

$$F_j(v) = \frac{1}{v} \int d\kappa_y \left[\left(\frac{v}{v} \right)^2 + \kappa_y^2 \right]^{-17/6} \left| J_{n+1} \left[2\pi \sqrt{ \left(\frac{v}{v} \right)^2 + \kappa_y^2 } \right] \right|^2 \begin{cases} 2\cos^2(m\alpha) \\ 2\sin^2(m\alpha) \\ 1 \ (m=0) \end{cases} .$$

(3.50)

If the wind propagates along a direction making an angle α_0 with the axes, one simply replaces α by $\alpha - \alpha_0$ in Eq. (3.50). Figure 3.4 shows normalized spectra $vF_j(v)/\int F_j(v)dv$ as a function of $\log(v)$ for a few Zernike terms. This display has the advantage that a change in the wind speed amounts to a simple translation of the curves along a horizontal axis, whereas the area under the curve still represents the total energy or variance (here normalized to unity). In Fig. 3.4, the frequency v is expressed in v/R units, where R is the aperture radius. It is a non-dimensional quantity. By adding the sine and cosine spectra one obtains the total power associated with a given type of aberration. Equation (3.50) shows that it is independent of the wind direction.

Let us now consider the case where several turbulent layers contribute to image degradation. The total wave-front distortion is the sum of the distortions produced by each layer. Since these distortions are statistically independent, the spectrum of the sum is the sum of the spectra. In a $\log(v)$ plot, the normalized spectrum is a sum of normalized spectra each shifted by an amount proportional to the log of the layer wind speed, and weighted by the layer contribution. In other words, for each type of aberration, the normalized spectrum is that of a single layer convolved with the distribution of turbulence expressed as a function of the log of the wind speed. Experimentally observed spectra are consistent with theoretical ones. Attempts to deconvolve them with the theoretical spectrum for a single layer have produced information on the distribution of turbulence in the atmosphere (Roddier *et al.* 1993).

3.4.2 *Time delay error*

Knowing how fast the Zernike coefficients change with time, one can now estimate the bandwidth requirements for the control system. To do this we need a model describing how the control system works. Since the wave-front sensor operates in a close loop, it sees only the residual wave-front error after a correction has been applied by the wave-front compensator. Using the wave-front sensor signals the control system computes how the voltages currently applied to the compensator must be modified to provide an improved compensation. New corrections are applied at each iteration. In other words, the currently applied compensation is a sum of all the corrections applied during all the previous iterations. Such a control system is called an integrator. Since we have decided to discuss here only a theoretically ideal system, we will assume that the delay between the read-out of the sensor signals and the application of the correction is negligibly small, that is we will model the control system as a pure integrator.

The control system is schematically described in Fig. 3.5. The input $x(t)$ is any Zernike coefficient of the uncompensated wave front. The wave-front compensator subtracts a quantity $y(t)$ from it, so that the residual error $e(t)$ is

$$e(t) = x(t) - y(t). \tag{3.51}$$

Using capital letters for the Laplace transform, one has

$$Y(p) = E(p)G(p), \tag{3.52}$$

where $G(p)$ is the transfer function of the control system, also called *open loop transfer function*. Putting Eq. (3.52) into the Laplace transform of Eq. (3.51) gives

$$E(p) = X(p) - E(p)G(p). \tag{3.53}$$

The ratio of the error term $E(p)$ over the input term $X(p)$ is called the *closed loop error transfer function*. It is given by

$$\frac{E(p)}{X(p)} = \frac{1}{1 + G(p)}. \tag{3.54}$$

It should not be confused with the *closed loop output transfer function*, which is the ratio of the output term $Y(p)$ over the input term $X(p)$ and is given by

$$\frac{Y(p)}{X(p)} = \frac{G(p)}{1 + G(p)}. \tag{3.55}$$

The open loop transfer function of a pure integrator varies as $1/p$, hence

$$G(p) = \frac{g}{p} \tag{3.56}$$

where g is the loop gain. Putting Eq. (3.56) into Eq. (3.54) and replacing p with $2i\pi\nu$ gives the closed-loop error transfer function as a function of the time frequency ν

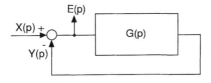

Fig. 3.5. Schematic diagram of the control system. $X(p)$, $Y(p)$, and $E(p)$ are the Laplace transforms of the control system input $x(t)$, output $y(t)$, and residual error $e(t)$. $G(p)$ is the open loop transfer function.

$$T(\nu) = \frac{i\nu}{\nu_c + i\nu} \tag{3.57}$$

where $\nu_c = g/2\pi$. The power spectrum of the residual error $e(t)$ is simply the power spectrum of the input wave front $x(t)$ (discussed in the last section) multiplied by the squared modulus of the error transfer function:

$$|T(\nu)|^2 = \frac{\nu^2}{\nu_c^2 + \nu^2}. \tag{3.58}$$

The quantity ν_c is the frequency at which the variance of the residual wavefront error is half the variance of the input wave front. It is called the 3 dB closed-loop bandwidth of the control system. On a log–log scale, $|T(\nu)|^2$ grows linearly with a slope of 6 dB/octave until ν becomes of the order of ν_c. Then it saturates to a value equal to unity. In real systems there is often an overshoot due to the delay in the computation that we have neglected here.

The variance of the residual error due to the finite system bandwidth is obtained by integrating the error power spectrum over all frequencies. Figure 3.6 shows this residual variance as a function of the degree of the Zernike polynomial, for a system with a 3 dB closed loop bandwidth equal to $\nu_c = v/R$. The residual variance is that of the sum of all the Zernike terms with the

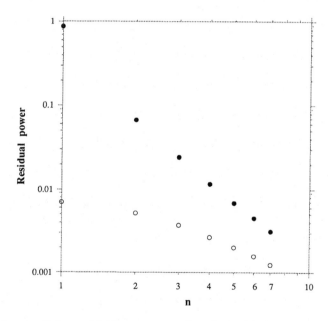

Fig. 3.6. Variance of the residual error as a function of polynomial degree n. The variance is that of the sum of all the Zernike terms of the same degree n. It is expressed as a fraction of the total variance of the uncompensated wave front. *Full circles:* before compensation. *Open circles:* after compensation with a system bandwidth $\nu_c = v/R$.

same polynomial degree n. It is expressed as a fraction of the total variance of the uncompensated wave-front. Although the tip/tilt terms are the slowest modes (Fig. 3.4), their amplitude is so large that they still contribute the most to image degradation. This was first emphasized by Conan *et al.* (1995).

Figure 3.7 shows the variance of the residual error as a function of the system bandwidth v_c expressed in v/R units. The variance is that of the sum of all the compensated terms. These go up to a maximum degree n ranging from 1 to 5. The variance of the sum of the remaining uncompensated terms is shown as horizontal lines. It allows one to give specifications for the bandwidth requirement. Clearly, one wants the residual error variance to be smaller than or at most equal to that of the uncompensated wave fronts. The point at which they are equal is indicated by a full circle on Fig. 3.7. The abcissa of these points gives a minimum allowable bandwidth v_0. To express this quantity in Hertz one needs an estimate of the wind velocities in turbulent layers. These fluctuate randomly with statistitics depending on the site being considered. Mean values are of the order of 10 m/s with variations typically ranging from 3 m/s up to 30 m/s or more. Since one wants the system bandwidth to exceed

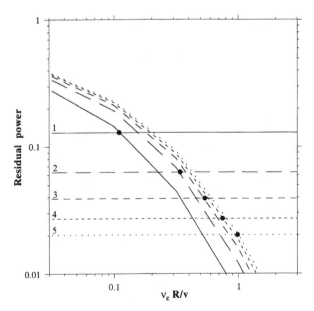

Fig. 3.7. Variance of the residual error of the compensated terms as a function of the system bandwidth v_c expressed in v/R units. Lines with decreasing dash length are for a compensation of all the Zernike terms up to degree 1 (full line), 2, 3, 4, and 5. Horizontal lines show the residual errors due to the uncompensated terms. The abscissa of the cross over points (black circles) gives the minimum allowable bandwidth v_0. Variances are expressed as a fraction of the uncompensated wave-front total variance.

Table 3.2. *Minimum allowable bandwidth* v_0

n	1	2	3	4	5
$v_0 R / v$	0.11	0.34	0.54	0.75	0.99
v_0 (Hz)[a]	1.9	5.4	8.7	12.4	16.7
τ_0 (ms)	84	30	18	13	9.5

[a] For a 3.6-m telescope with a maximum wind speed of 30 m/s.

the minimum allowable bandwidth in most weather conditions, we take here 30 m/s as a high wind speed value. The minimum bandwidths associated with each degree of compensation (up to $n = 5$) are shown in Table 3.2, together with the characteristic integration time $\tau_0 = 1/(2\pi v_0)$. The reader should be reminded that these are for the integrator only. For the approximation of a pure integrator to be valid, the bandwidth of any other part of the control system should be at least an order of magnitude larger. Significantly higher bandwidths may also be necessary to allow for the hysteresis of piezoelectric actuators.

3.4.3 *Isoplanatic patch size*

Let us consider first a single atmospheric turbulent layer at an altitude h above ground. We assume that observations are made at a zenith distance γ with a guide source at an angular distance $\boldsymbol{\theta}$ from the object. The two light beams will cross the turbulent layer at a distance $\rho = \boldsymbol{\theta} h / R \cos(\gamma)$ expressed in pupil radius units. The mean square difference between the Zernike coefficients is

$$e_j^2(\theta) = \langle |a_j(\mathbf{r}) - a_j(\mathbf{r} + \boldsymbol{\theta} h / R \cos(\gamma))|^2 \rangle. \tag{3.59}$$

It depends only on the modulus θ of the vector $\boldsymbol{\theta}$. Expanding the above expression, and introducing the covariance $B_j(\rho)$ of $a_j(\mathbf{r})$ defined by Eq. (3.44) gives

$$e_j^2(\theta) = 2[B_j(0) - B_j(\boldsymbol{\theta} h / R \cos(\gamma))]. \tag{3.60}$$

The covariance $B_j(\rho)$ can be calculated from Eq. (3.43) and Eq. (3.45).

It is convenient to normalize the mean square error by dividing it with the variance $\langle a_j^2 \rangle = B_j(0)$ of the Zernike coefficient. One obtains

$$\frac{e_j^2(\theta)}{\langle a_j^2 \rangle} = 2[1 - \Gamma_j(\boldsymbol{\theta} h / R \cos(\gamma))] \tag{3.61}$$

where $\Gamma_j(\rho) = B_j(\rho)/B_j(0)$ is the correlation coefficient of the Zernike terms for the two beams. Note that if the correlation coefficient becomes less than 0.5, then the normalized error becomes larger than unity and the AO system

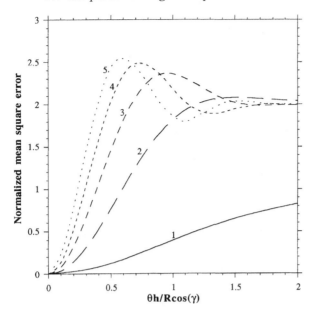

Fig. 3.8. Mean square error on a Zernike coefficient as a function of the angular distance θ expressed in $R\cos(\gamma)/h$ units (cosine and sine terms have been added). The error depends only on the degree n of the Zernike polynomial, indicated near each curve. The error is normalized as a fraction of the Zernike coefficient variance.

degrades the wave front instead of improving it. Figure 3.8 shows the normalized mean square error $e_j^2(\theta)/\langle a_j^2 \rangle$ for Zernike terms of degree $n = 1$ to $n = 5$, as a function of the angular distance θ. For small values of θ, the error grows approximately as θ^2, not as $\theta^{5/3}$ as one would expect from Eq. (2.46). However, the 5/3 power law becomes valid as $n \to \infty$ (Roddier *et al.* 1993). When several atmospheric layers are present, these power laws are still valid, but the altitude h must be replaced by an effective altitude $\overline{h}$ (Roddier *et al.* 1993).

Figure 3.9 shows the total mean square error for all the compensated terms up to a polynomial degree (n) ranging from 1 to 5. The variance of the sum of the remaining uncompensated terms is shown as a horizontal line for each n. Again, the system performance will not be significantly degraded if the residual error on the compensated terms is much smaller than or at most equal to that due to the uncompensated wave-front modes. The point at which they are equal is indicated by a full circle on Fig. 3.9. The abcissa of these points gives a maximum angular distance θ_0 we can use as a measure of the isoplanatic patch size. To express this quantity in arcseconds, one needs to know the effective altitude of the turbulent layers. Measurements of the isoplanatic patch size have recently been made at the Mauna Kea Observatory, both with the University of

Table 3.3. *Isoplanatic angle* θ_0

n	1	2	3	4	5
$\theta_0 h / R \cos(\gamma)$	0.56	0.28	0.17	0.13	0.10
θ_0 (arcsec)[a]	208	104	64	48	36

[a] For observations at zenith with a 3.6-m telescope and an effective turbulence height of 1 km.

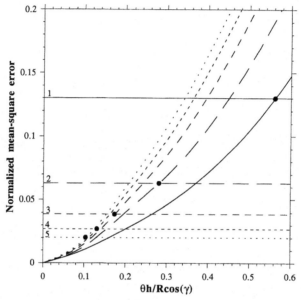

Fig. 3.9. Total mean square error on the compensated terms as a function of the angular distance θ expressed in $R\cos(\gamma)/h$ units. Lines with decreasing dash length are for a compensation of all the Zernike terms up to degree 1 (full line), 2, 3, 4, and 5. Horizontal lines shows the residual error due to the uncompensated terms. The abscissa of the cross over points (black circles) gives the isoplanatic patch size θ_0. Error is expressed as a fraction of the uncompensated wave-front total variance.

Hawaii AO system and with the AO bonette of the Canada–France–Hawaii telescope. Results show that it is larger than originally expected, and typically represented by taking $\overline{h} = 1$ km. The corresponding isoplanatic angles associated with each degree of compensation (up to $n = 5$) are shown in Table 3.3.

3.5 Sensor noise limitation

As for any measurement, the wave-front sensor measurements are affected by sensor noise which produces errors in the wave-front estimation. The noise properties of sensors and light detectors used for adaptive optics are discussed in Chapter 5. Here we limit ourselves to a general discussion of the effect of sensor noise on the residual wave-front error, and establish a minimum brightness requirement – or limiting magnitude – for a stellar type guide source. To do this, we will use again the criterion that the error on the wave-front estimation due to sensor noise should not exceed the error due to the uncompensated wave-front modes.

Since sensor noise depends on the particular sensor choice, it is difficult to establish general results, which are independent of the implementation. Instead, we consider a particular system, chosen for its simplicity, and discuss the general implications of the result. Our chosen system consists of a segmented mirror controlled by a Shack–Hartmann sensor. In such a system, an image of the telescope entrance aperture is formed on a lenslet array. Each lenslet forms an image of the guide source on a detector array. A distorted wave front will shift the position of these images, compared to that of an undistorted wave front. Each shift gives a measure of the local wave-front slope averaged over the lenslet area. Signals from the detector array are used to estimate these slopes. If each lenslet is optically conjugate to a segment of the mirror, one can tilt the conjugated mirror segment to compensate the wave-front slope. Piston motions must be calculated and applied to maintain continuity of the wave front. The number M of measurements (two per subaperture) being less than the number P of actuators (three per subaperture), one has indeed to use of a minimum norm (smoothest) solution (see Section 3.1). Although such systems have been built, it is not advisable to build them because of the sensitivity of these systems to uncontrolled imperfections in the deformable mirror. However, their properties can be easily derived analytically, and used as an example of the general behavior of AO systems.

Since this chapter is about theoretically ideal systems, we consider here an ideal detector array, able to detect each photon impact and measure its position with a perfect accuracy, that is we consider only the fundamental source of noise produced by the quantum nature of photodetection. The probability of finding a photon at a given location is proportional to the intensity at that location. Hence, the probability distribution of photon impacts is the intensity distribution in the image. The sensor seeks to determine the center of the intensity distribution produced by each subaperture. A single photon event gives the center location with a mean square error equal to the variance of the intensity distribution, that is the width θ_b of a subimage. For a single photon event, the mean square angular error $\langle\theta^2\rangle$ on local wave-front slopes is of the order of θ_b^2. If the guide source provides n_{ph} independent photon events the mean square error is n_{ph} times smaller, that is $\langle\theta^2\rangle = \theta_b^2/n_{ph}$. Over the subaperture size d, an error θ on the slope angle produces an error $\delta = \theta d$ on the optical path, with variance

$$\langle \delta^2 \rangle = \frac{\theta_b^2 d^2}{n_{ph}}. \tag{3.62}$$

If we assume that each subaperture is larger than r_0 (at the sensor wavelength λ), then each subimage is blurred and has an angular width $\theta_b \simeq \lambda/r_0$ (see Section 2.3). Assuming that the guide source provides p photons per unit area on the telescope aperture, then $n_{ph} = pd^2$, and Eq. (3.62) becomes

$$\langle \delta^2 \rangle = \frac{\lambda^2}{p r_0^2}. \tag{3.63}$$

Although Eqs. (3.62) and (3.63) have been established for a particular sensor (Shack–Hartmann), they also apply to other sensors, albeit with different numerical coefficients all of the order of unity (see Chapter 5). They are therefore quite general.

The ratio of the mean square error on the reconstructed wave-front surface over $\langle \delta^2 \rangle$ is called the error gain. Detailed calculations (Fried 1977; Hudgin 1977; Noll 1978) show that the error gain for a Shack–Hartmann sensor grows only slowly, as the logarithm of the number of subapertures, and is typically of the order of unity. For the purposes of our calculation we will assume that $\langle \delta^2 \rangle$, as given by Eq. (3.63), also represents the mean square wave-front error. This error should not exceed the so-called fitting error due to the finite number of actively controlled elements. For a segmented mirror, the fitting error is given by Eq. (2.17), or in terms of optical path fluctuations,

$$\langle \delta_{fit}^2 \rangle = 0.134 \left(\frac{\lambda}{2\pi} \right)^2 (d/r_0)^{5/3}. \tag{3.64}$$

Note that, since r_0 varies as $\lambda^{6/5}$, this expression is wavelength independent. Hence both r_0 and λ can be evaluated at the sensor wavelength. The minimum number p of photons per cm^2 is obtained by equating Eq. (3.63) and Eq. (3.64). This gives

$$p = \frac{4\pi^2}{0.134} d^{-5/3} r_0^{-1/3}, \tag{3.65}$$

where r_0 is estimated at the sensor wavelength.

Astronomers express the brightness of a star in stellar magnitudes. For a given wavelength λ, the magnitude m is defined as

$$m = -2.5 \log \left(\frac{F}{F_0} \right), \tag{3.66}$$

where F/F_0 is the ratio of the observed stellar flux over the flux given by a magnitude zero star with spectral type AO. For the sensor wavelength we take $\lambda = 0.63$ μm, the value at which silicon detectors have a maximum sensitivity. At this wavelength, the flux of a magnitude zero AO star is 2.5×10^{-12} W cm^{-2} per micron bandwidth (Johnson 1966), and the photon energy is $hc/\lambda = 3.15 \times 10^{-19}$ joules. The corresponding photon flux is therefore

$$F_0 = \frac{2.5 \times 10^{-12}}{3.15 \times 10^{-19}} \approx 8 \times 10^6 \text{ photons s}^{-1} \text{ cm}^{-2} \text{ μm}^{-1}. \qquad (3.67)$$

Putting this value into Eq. (3.66) gives a photon flux

$$F = 8 \times 10^6 \times 10^{-0.4m} \text{ photons s}^{-1} \text{ cm}^{-2} \text{ μm}^{-1}. \qquad (3.68)$$

The number p of photons detected per cm^2 is

$$p = 8 \times 10^{3-0.4m} \tau \eta \int q(\lambda)\, d\lambda \qquad (3.69)$$

where τ is the integration time in seconds, η is the transmission coefficient of the system, and $q(\lambda)$ is the detector quantum efficiency. The integral is over the detector bandwidth, and is now expressed in nanometers. Equating Eq. (3.65) and Eq. (3.69) gives

$$10^{-0.4m} = \frac{3.68 \times 10^{-2}}{\tau \eta \int q(\lambda)\, d\lambda} d^{-5/3} r_0^{-1/3}. \qquad (3.70)$$

For the integration time, we take the characteristic integration time for tip/tilt correction over a subaperture of diameter d as given by Table 3.2, that is $1/\tau = 1.382(v/d)$. This gives

$$10^{-0.4m} = \frac{5.1 \times 10^{-2}}{\eta \int q(\lambda)\, d\lambda} v d^{-8/3} r_0^{-1/3}. \qquad (3.71)$$

For a numerical application, we assume the use of the best photon counting detectors now available, avalanche photodiodes (APDs) with a maximum quantum efficiency of 70% at 0.63 μm, and $\int q(\lambda)\, d\lambda = 300$ nm. We take $\eta = 0.4$, a typical value for actual AO systems. We assume $v = 3 \times 10^3$ cm/sec, and $r_0 = 20$ cm (at 0.63 μm), values that are typical for the Mauna Kea Observatory. These give

$$10^{-0.4m} = 0.47 d^{-8/3} \qquad (3.72)$$

or

$$m = 0.82 + 6.67 \log(d_{\text{cm}}) \qquad (3.73)$$

where d_{cm} is d expressed in centimeters. Expression (3.73) is particularly simple because the result depends only on the size d of a subaperture, and is independent of the telescope size. In Section 2.3, we saw that a Strehl ratio of 0.37 can be achieved with a segment size $d = 3.3 r_0$. This corresponds to a 1 radian rms error on the wave front. Assuming $r_0 = 20$ cm at 0.63 μm, gives

$r_0 = 20(\lambda/0.63)^{6/5}$ cm at wavelength λ, hence a corresponding subaperture size

$$d = 66(\lambda/0.63)^{6/5} \text{ cm.} \tag{3.74}$$

Putting Eq. (3.74) into Eq. (3.73) gives a limiting magnitude as a function of wavelength

$$m = 14.6 + 8\log(\lambda_{\mu m}). \tag{3.75}$$

Table 3.4 gives this maximum magnitude for a number of standard spectral bands. These are limiting magnitudes that a high performance system can practically reach today. It should be emphasized that it is not an absolute limit. For instance, the use of detectors more sensitive than silicon at longer wavelengths could still increase these numbers.

Adaptive optics systems based on natural guide stars are only effective within an isoplanatic patch distance of a suitable guide source. In general, this represents only a fraction of the sky. This fraction has sometimes been taken as a measure of the system effectiveness. It represents the probability of finding a guide star brighter than the limiting magnitude within an isoplanatic patch distance of an arbitrary object. This probability can be estimated from star counts (see for instance Bahcall and Soneira 1981). Figure 3.10 shows equal probability contours in a magnitude versus distance plot. Contours are for a 30° Galactic longitude. A 50% probability contour is also given for the Galactic pole.

For the maximum distance we take again a 1 radian rms error criterion, and use the isoplanatic angle given by Eq. (2.45) of Section 2.4. For actual systems, more accurate values are given in Table 3.3. However, Eq. (2.45) has the advantage of being system independent, and will give us smaller, conservative values. We take again $r_0 = 20(\lambda/0.63)^{6/5}$ cm, $\overline{h} = 1$ km, and assume observations at zenith ($\gamma = 0$). The corresponding angular distance θ_0 is given in arcseconds by

$$\theta_0 = 22.5''\lambda_{\mu m}^{6/5} \tag{3.76}$$

This is the maximum distance over which the compensation is still considered acceptable. For each spectral band, Table 3.4 shows both the maximum guide star magnitude, and the maximum distance to the guide star. Figure 3.10 shows the corresponding operating points having these two values as coordinates. At these points, the isoplanicity and sensor noise errors add quadratically each contributing to one square radian. The result is a loss by a factor $e^2 = 7.4$ in Strehl ratio compared to the performance with a bright on-axis star. Taking this as the limit beyond which image improvement is no longer significant, Figure 3.10 shows that one can still observe a significant image improvement over

Table 3.4. *Guide star maximum magnitude and angular distance*

Image spectral band	R	I	J	H	K
Wavelength (for imaging)	0.65	0.85	1.22	1.65	2.2
Maximum guide star mag (at 0.63 μm)	13.1	14.0	15.2	16.3	17.3
Maximum angular distance (arcsec)	13.4	18.6	28.6	41.1	58.1

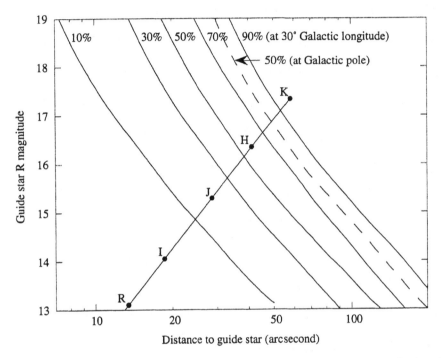

Fig. 3.10. Probability of finding a guide source brighter than a given magnitude within a given distance. Contours are for a 30° Galactic longitude. A 50% probability contour is given for the Galactic pole. Black dots indicate the guide star maximum distance and magnitude for the standard spectral bands R, I, J, H, and K.

almost the full sky in the K band, and over more than 10% of the sky in the J band. Only down to the visible, the sky coverage becomes quite low. This result is roughly independent of the size of the telescope being used. Methods to improve the sky coverage at shorter wavelengths are discussed in Part 4.

References

Bahcall, J. N. and Soneira, R. M. (1981) The distribution of stars to V = 16th magnitude near the north galactic pole: normalization, clustering properties and counts in various bands. *Astrophys. J.* **246**, 122–35.

Conan, J.-M., Rousset, G. and Madec P.-Y. (1995) Wave-front temporal spectra in high resolution imaging through turbulence. *J. Opt. Soc. Am. A* **12**, 1559–70.

Fried, D. L. (1966) Optical resolution through a randomly inhomogeneous medium for very long and very short exposures. *J. Opt. Soc. Am* **56**, 1372–9.

Fried, D. L. (1977) Least-square fitting a wave-front distortion estimate to an array of phase difference measurements. *J. Opt. Soc. Am.* **67**, 370–5.

Graves, J. E., Roddier, F., MacKenna, D. and Northcott, M. (1992a) Latest results from the University of Hawaii Prototype Adaptive Optics System. In: Proc. *Laser Guide Star Adaptive Optics Workshop*, ed. R. Q. Fugate, Vol. 2, pp. 511–21. SOR, Phillips Lab/LITE, Kirtland AFB, New Mexico.

Graves, J. E., MacKenna, D., Northcott, M. and Roddier, F. (1992b) Recent results of the UH Adaptive Optics System. In: *Adaptive Optics for Large telescopes*, Tech. Digest OSA/AF Conf., Lahaina (Maui), August 17–21, 1992.

Hudgin, R. H. (1977) Wave-front reconstruction for compensated imaging. *J. Opt. Soc. Am.* **67**, 375–8.

Johnson, H. L. (1966) Astronomical measurements in the infrared. *Ann. Rev. Astron. Astrophys.* **4**, 201.

Noll, R. J. (1976) Zernike polynomials and atmospheric turbulence. *J. Opt. Soc. Am.* **66**, 207–11.

Noll, R. J. (1978) Phase estimates from slope-type wave-front sensors. *J. Opt. Soc. Am.* **68**, 139–40.

Roddier, F. (1981) The effects of atmospheric turbulence in optical astronomy. *Progress in Optics* **19**, 281–376.

Roddier, F. (1994) The problematic of adaptive optics design. In: *Adaptive Optics for Astronomy*, eds D. M. Alloin, J.-M. Mariotti, (NATO-ASI Series), 423, pp. 89–111. Kluwer Academic Publ., Dordrecht.

Roddier, F. (1998) Maximum gain and efficiency of adaptive optics systems. *Pub. Astr. Soc. Pac.* **110**, 837–40.

Roddier, F., Northcott, M. and Graves, J. E. (1991) A simple low-order adaptive optics system for near-infrared applications. *Pub. Astr. Soc. Pac.* **103**, 131–49.

Roddier, F., Northcott, M. J., Graves, J. E., McKenna, D. L. and Roddier, D. (1993) One-dimensional spectra of turbulence-induced Zernike aberrations: time-delay and isoplanicity error in partial adaptive compensation. *J. Opt. Soc. Am.* **10**, 957–65.

Roddier, N. (1990) Atmospheric wave-front simulation using Zernike polynomials. *Opt. Eng.* **29**, 1174–80.

Tatarski, V. I. (1961) *Wave Propagation in a Turbulent Medium*. Dover, New York.

Wang, J. Y. and Markey, J. K. (1978) Modal compensation of atmospheric turbulence phase distortion. *J. Opt. Soc. Am.* **68**, 78–87.

Further references

Chassat, F. (1989) Calcul du domaine d'isoplanétisme d'un système d'optique adaptative fonctionnant à travers la turbulence atmosphérique. *J. Optics* (Paris) **20**, 13–23.

Valley, G. C. and Wandzura, S. M. (1979) Spatial correlation of phase-expansion coefficients for propagation through atmospheric turbulence. *J. Opt. Soc. Am.* **69**, 712–7.

4

Wave-front compensation devices

MARC SÉCHAUD

Office National d'Études et de Recherches Aérospatiales (ONERA) Châtillon, France

4.1 Introduction

Image quality can be degraded by both phase and amplitude distortions of the optical wavefront across a telescope aperture. However, as shown in Chapter 2, the effect of phase fluctuations is predominant. AO systems are designed to provide a real time compensation of these fluctuations by means of phase correctors. Such devices introduce an optical phase shift φ by producing an optical path difference δ. The phase shift is

$$\varphi = \frac{2\pi}{\lambda}\delta. \tag{4.1}$$

The quantity δ is the variation of the optical path ne

$$\delta = \Delta(ne), \tag{4.2}$$

where n and e are respectively the refractive index and the geometrical path spatial distribution of the corrector. Geometrical path differences Δe can be introduced by deforming a mirror surface. Index spatial differences Δn can be produced by birefringent electro-optical materials. To date, deformable mirrors are preferably used because they are well suited to astronomical adaptive optics. They provide short response times, large wavelength-independent optical path differences, with a high uniform reflectivity that is insensitive to polarization, properties that are not commonly shared by birefringent materials.

Since the early 1970s, and with the initial impetus given by defense-oriented research, a wide variety of deformable mirrors have been developed. The performance requirements of deformable mirrors vary according to applications which include high energy laser focusing, compensated imagery through atmospheric turbulence, and laser cavity control. Compared with astronomical applications, requirements related to defense applications are often more demanding, such as the need for a cooled reflective surface to support a high energy laser or the need for a larger number of actuators and a faster response time for high resolution imaging in the visible. On the other hand some requirements can be relaxed for systems operating at longer wavelengths, or with monochromatic light.

The characteristics for a deformable mirror are dictated by the statistical spatial and temporal properties of the phase fluctuations and the required degree of correction. For astronomical applications they are given in Chapter 3. The number of actuators is proportional $(D/r_0)^2$, where D is the telescope diameter, and r_0 is Fried's diameter. It ranges from at least 2 (tip/tilt correction) to several hundreds, depending on the wavelength of the observations and the brightness of the available wave-front reference sources. The required stroke is proportional to $\lambda(D/r_0)^{5/6}$. It is practically wavelength independent, and of the order of at least several microns. The required optical quality (root mean square surface error) varies in proportion to the wavelength of observations. It is of the order of a few tens of nanometers. The required actuator response time is proportional to the ratio $r_0/\overline{v}$, and is of the order of at least a few milliseconds (see Chapter 3). It increases as the degree of correction decreases.

Several types of deformable mirrors have been studied. A wide variety of effects have been proposed to deform the mirror, such as the magnetostrictive, electromagnetic, hydraulic effects (Hansen 1975; Pearson 1976; Freeman and Pearson 1982; Eitel and Thompson 1986; Neal *et al.* 1991; Tyson 1991; Ribak 1994). In most operational deformable mirrors, the actuators use the ferro-electric effect, in the piezoelectric or electrostrictive form. These actuators are described in Section 4.2.1. The decisive advantages of ferroelectric actuators are high packing density, efficient electro-mechanical interaction with the face-plate, low power dissipation, fast response time, high accuracy, and high stability. Ferroelectric actuators have been used to build both segmented or continuous facesheet deformable mirrors such as monolithic, discrete actu-ators, and bimorph mirrors, presented in Sections 4.2.2–4.2.5. Other deform-able mirrors, like electrostatic actuator membranes, adaptive secondaries, and liquid crystal mirrors, have been studied but have not yet been used in operational systems. They are presented in Section 4.3.

Section 4.4 deals with the spatial correction efficiency which ultimately limits the performance of deformable mirrors for adaptive optics. Beam steering mirrors used to achieve full tip/tilt corrections are presented in Section 4.5. Section 4.6 addresses the intensity fluctuation compensation issue. Finally, Section 4.7 is dedicated to customer users who have to choose a deformable mirror and specify its characteristics.

4.2 Deformable mirrors with ferroelectric actuators

4.2.1 Ferroelectric actuators

4.2.1.1 The piezoelectric effect

An electric field applied to a permanently polarized piezoelectric ceramic induces a deformation of the crystal lattice and produces a strain proportional to the electric field (Herbert 1982). Lead zirconate titanate $Pb(Zr, Ti)O_3$, commonly referred to as PZT, exhibits the strongest piezoelectric effect. The poling is created by applying an intense field to the ceramic, aligning the previously randomly oriented dipoles parallel to the field. In the case of a disk actuator, the effect of a longitudinal electric field E is to change the relative thickness $\Delta e/e$ to (see Fig. 4.1)

$$\frac{\Delta e}{e} = d_{33}E, \tag{4.3}$$

where d_{33}, the longitudinal piezoelectric coefficient, refers to a field parallel to the poling axis P which is the axis of deformation. Introducing the voltage $V = Ee$

$$\Delta e = d_{33}V, \tag{4.4}$$

showing that the change in thickness is thickness independent. Values of d_{33}

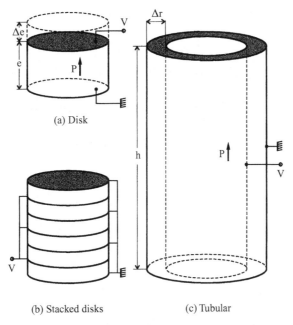

(a) Disk

(b) Stacked disks   (c) Tubular

Fig. 4.1. PZT actuators (a) disk, (b) stacked disks, (c) tubular.

are typically between 0.3 and 0.8 μm/kV. To obtain a stroke of several microns with voltages of a few hundred volts, (compatible with solid state electronics) N disks can be stacked and electrically connected in parallel dividing the voltage by N (see Fig. 4.1).

Tubular actuators are used as well. The electric field is transversally and radially applied. As shown in Fig. 4.1 the axial relative deformation is

$$\frac{\Delta h}{h} = d_{31} E, \tag{4.5}$$

so that

$$\Delta h = d_{31} V \frac{h}{\Delta r}, \tag{4.6}$$

where d_{31} is a transverse piezoelectric coefficient which refers to a voltage perpendicular to the poling axis P. The value of d_{31} is roughly $3/8$ of d_{33} and of opposite sign, Δr is the thickness of the shell and h its height.

For a given voltage, the maximum electric field E_{max} which can be applied is theoretically limited to the depolarization field of the material, but practically limited to a lower value to reduce hysteresis. The minimum thickness e or Δr is equal to V/E_{max}. Finally, the maximum displacements produced by stacked actuators and a cylindrical actuator of the same height h are respectively

$$\Delta e = h E_{max} d_{33}, \tag{4.7}$$

and

$$\Delta e = h E_{max} d_{31}, \tag{4.8}$$

showing that stacked actuators provide the larger stroke for a given height.

Initially, stacked actuator generation was mechanically preloaded to avoid any interface-breaking element. Improvement of the technological bonding techniques allowed suppression of the preload, decreasing the aging and simplifying the actuator realization. PZT wafers are now bonded to form a block which is diced to make the actuators (Aldrich 1980).

4.2.1.2 Electrostrictive effect

The electrostrictive effect generates a relative deformation $\Delta e/e$ which is proportional to the square of the applied electric field E (Uchino *et al.* 1980; Uchino *et al.* 1981; Uchino 1986; Eyraud *et al.* 1988) so that

$$\Delta e/e = a E^2 = a \left(\frac{V}{e}\right)^2, \tag{4.9}$$

where a is the electrostriction coefficient. Note that with electrostrictive material, the change in thickness is thickness dependent. For a given applied electric field, the lower the value of V, the thinner e. In piezoelectric ceramics the deformation induced by an electric field is due to the superposition of both electrostrictive and piezoelectric effects. Lead magnesium niobate $Pb(Mg_{1/3}Nb_{2/3})O_3$, commonly referred to as PMN, is a pure electrostrictive material which has been extensively studied. Other compositions have been studied, such as PMN : PT with substitution of $PbTiO_3$, and Ba : PZT with partial substitution of Pb by Sr − Ba (Eyraud *et al.* 1988; Galvagni 1990; Galvagni and Rawal 1991; Blackwood *et al.* 1991). The electrostrictive effect does not require a remanent polarization in the ceramic, and it produces a more stable device with less aging compared with PZT. But a major drawback is that the response depends on the temperature, due to a Curie point around 0 °C (Ealey 1991). Hysteresis and strain sensitivities to field increase as the temperature decreases from 25 °C to the Curie point. The strain may decrease by a factor 2 at −5 °C. Hysteresis may be negligible at 25 °C but is about 10% at 8 °C (Blackwood *et al.* 1991). However, the electrostrictive effect allows processes to be used such as coating, that would depole PZT material if performed above its Curie point of about 200 °C.

A polarization field may linearize the response of electrostrictive materials and increase their deformation: local sensitivities as high as 2 μm/kV have been obtained, with a polarization electric field of the order of 750 V/mm. With non-polarized material, the non-linear response limits the voltage range to only positive or only negative values.

4.2.1.3 Fabrication processes

Ferroelectric materials are produced by sintering, generally through solid processing, or liquid processing by coprecipitation to get a higher density and improved piezoelectric coefficients (Eyraud *et al.* 1988). Classical stacked actuators consist of adhesive bounded disks of about 1 mm thickness, leading to actuators of several centimeters. But actuators could also be developed with techniques currently used for capacitor fabrication (Ealey and Davis 1990; Galvagni and Rawal 1991), leading to multilayer cofired actuators with a structure consisting of multiple thin layers (125 to 250 μm stroke). Strokes larger than 10 μm for an applied voltage of 150 V are only limited by the interlaminar shear strength of the multilayer structure. However this delicate fabrication process requires specific plants. Note that it is well suited to provide the thin wafers needed to produce electrostrictive actuators. The electric field required being typically 600 V/mm, this leads to a thickness as small as 250 μm for a 150 V applied voltage.

4.2.1.4 Hysteresis

Piezoelectric materials generally exhibit hysteresis which increases as the applied electric field approaches the depolarization field (typically 2 kV/mm). From a physical standpoint, an hysteresis cycle characterizes the behavior of polarization and strain versus electric field. From an experimental standpoint, it can be shown that the relation between strain and polarization or charge is more linear, suggesting that charge (i.e. current) should be used to drive ferroelectric actuators, instead of voltage (Newcomb and Flinn 1982; Eyraud *et al.* 1988). From a control standpoint, hysteresis is a phase lag which does not depend on the frequency (Madec, personal communication; Kibblewhite *et al.* 1994). The hysteresis cycle is characterized by the response stroke versus alternating applied voltage. During the cycle, the strokes for the zero voltages differ. The ratio of the stroke difference for zero voltage ΔS over the difference between the maximum and the minimum strokes $(S_{max} - S_{min})$ gives the amount of relative hysteresis H_{rel}. The phase lag $\Delta\varphi$ can be expressed as

$$\Delta\varphi = \sin(H_{rel}) = \sin\left(\frac{\Delta S}{S_{max} - S_{min}}\right). \qquad (4.10)$$

Typical values of H_{rel} range from less than 1% to more than 10%. It increases with the sensitivity (stroke/voltage) for PZT materials. It depends on the temperature for electrostrictive materials. Phase lag lower than 5° for a 10% full stroke at the temporal sampling frequency is considered negligible (Madec, personal communication).

A wide variety of materials has been studied and used to date. The performance of available actuators satisfy AO requirements. But in the future, new materials may be found to make their use still easier with a linear and non-hysteretic higher sensitivity (Eyraud *et al.* 1996).

4.2.1.5 Power supply

The function of the power supply is to deliver output analog high voltage signals to the actuators from the input digital low voltage signals delivered by the control computer. The power supply comprises a stabilized high voltage generator and high voltage amplifiers. The value of the required voltage is deduced from the sensitivity of the mirror multiplied by the required stroke. The voltage applied to contiguous actuators is always limited to less than the maximum voltage by a protection control because in the long term it would endanger the actuators coupled by high stresses through the coupling by face-plate, especially during the test phase of the adaptive optics system. The high voltage generator is characterized by the maximum delivered current, which depends on the spectral characteristics of the required correction. Below we estimate the variance of this current.

The current required to control a piezoelectric actuator is given by

$$i = C_t \, \mathrm{d}V/\mathrm{d}t, \tag{4.11}$$

where C_t is the capacitance of the actuator plus that of its connection wire. Although the power dissipation may be low, the capacitive load results in a high instantaneous current at a high frequencies which, with the high voltage, produces large reactive power. The capacitance C_a of the free actuator is

$$C_a = \epsilon \epsilon_0 S/e, \tag{4.12}$$

where S is the surface of the electrodes (half the sum of the electrodes plus the ground electrode for a monolithic piezoelectric mirror, or the total electrode surface for a stacked actuators mirror), e is the thickness of the capacitance, ϵ and ϵ_0 are the relative and vacuum permittivity respectively. Typically, ϵ is about 1300 for PZT and 12 000 for PMN; ϵ_0 is equal to 8.85×10^{-12} F/m. The capacitance of the connection wire, typically 100 pF/m, is generally negligible.

The control voltage V is proportional to the stroke, that is the optical path difference δ. The temporal Fourier spectrum of the current i is proportional to the product of the temporal frequency ν and the temporal Fourier spectrum of δ. The spectral density of the current Φ_i is therefore proportional to $\nu^2 \Phi_\delta$, where Φ_δ is the spectral density of δ. For atmospheric compensation the spectral density of δ is given by Kolmogorov's law (see Chapter 2). Finally, the current fluctuation variance required for the actuator is given by

$$\sigma_i^2 = \frac{C_t^2}{K^2} \int \nu^2 \Phi_\delta(\nu) \, \mathrm{d}\nu, \tag{4.13}$$

where K is the sensitivity (stroke/voltage) of the actuator. K lies from a few μm/kV to a few tens of μm/kV.

4.2.2 Segmented mirrors

Segmented mirrors consist of a juxtaposition of elementary mirrors which are individually controlled as depicted in Fig. 4.2. A typical segmented mirror is shown in Fig. 4.3. In the early deformable mirror developments, segmented mirrors were considered as a simple and low risk concept (Smith 1977; Freeman and Pearson 1982). Segmented mirrors equip the AO system of the Sac Peak Solar Telescope (Acton 1992) and the AO prototype system developed by Durham University for the 4.2-m William Herschel Telescope (Busher *et al.* 1995). Their scalable structures are well suited to provide a large number

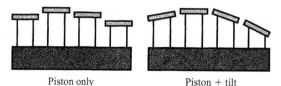

Piston only   Piston + tilt

Fig. 4.2. Segmented mirror.

Fig. 4.3. A 512-element segmented mirror from ThermoTrex Corporation. (Courtesy D. Sandler.)

of actuators, and mirrors with up to 1500 actuators have been built (Hardy 1989; Hulburd and Sandler 1990; Hulburd *et al.* 1991).

The main advantage of segmented mirrors is that they use a set of identical and easily repairable elementary mirrors distributed over a square or hexagonal array (Malakhov *et al.* 1984). Elementary mirrors being quite mechanically independent, the mechanical design study is minimized. But low weight, stiff segments are needed to reduce the accelerated mirror mass, to increase the bounce frequency of the mass/actuator spring system, and to avoid wing beating effect.

The main drawback of segmented mirrors is their high fitting error compared with a continuous facesheet deformable mirror with the same number of actuators. If the elementary mirror is activated by a sole actuator, the motion is limited to piston. To get the same mirror fitting error as a continuous surface deformable mirror, roughly four to eight times more piston mirrors are needed (Hudgin 1977). An additional drawback is the edge diffraction effect induced by gaps between segments.

To avoid having an excessive number of elementary mirrors to control (the computing power being proportional to the square of the number of degrees-of-freedom) the solution is to use piston tip/tilt elementary mirrors. A tubular PZT actuator (see Fig. 4.1) is activated by three independent electrodes deposited on the external surface of the tube, the interior electrode being grounded. Then the problem turns out to be the control of the piston mode of each mirror to insure the wavefront continuity from an elementary mirror to its neighbors for white light operation. This could be done with an additional internal servo-loop, using dedicated sensors, but with an increase of the complexity (Hulburd *et al.* 1991). With piston and tip/tilt control over each segment, the fitting error is the same as that of a continuous facesheet mirror with 2/3 as many actuators (Sandler *et al.* 1994).

To minimize the fitting error, continuous facesheet deformable mirrors are the most widely used and many different structures have been developed.

4.2.3 *Monolithic piezoelectric mirrors*

Developed in the mid-1970s, the monolithic piezoelectric mirror (MPM) was the first deformable mirror installed on a ground-based telescope in 1982, and used for operational space surveys with the Compensated Imaging System of the Air Force AMOS station located in Maui, on top of Mt Haleakala (Hardy *et al.* 1977; Greenwood and Primmerman 1992).

MPM structure is shown in Fig. 4.4. Figure 4.5 shows a 345-actuator MPM. A thin reflecting glass plate is bonded to the upper face of a monolithic piezoelectric disk. A set of actuators is defined on the upper surface by an electrode network. The electrical addressing leads go through holes drilled into the disk (Feinleib *et al.* 1974; Hudgin and Lipson 1975), or are deposited on the upper face of the disk to increase the stroke (Séchaud *et al.* 1987; Séchaud and Madec 1987; Madec *et al.* 1989), the lower face forming the common ground electrode. A voltage applied to an electrode induces a local smooth deformation at the mirror surface and none at the bottom surface if the block thickness is greater than the spacing between electrodes.

Compared with discrete actuator deformable mirrors, the structure of a MPM is very compact and exhibits good optical flatness due to very small mechanical strains between the PZT disk and the glass face-plate. Analytical models of the mirror provide a coarse optimization of the main parameters, that is the face-plate thickness, of the order of 1 mm, the disk thickness, of the order

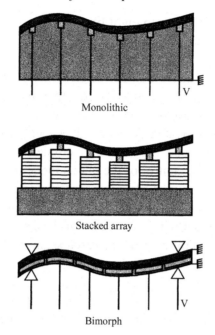

Monolithic

Stacked array

Bimorph

Fig. 4.4. Continuous facesheet piezoelectric deformable mirrors.

Fig. 4.5. A 345-actuator monolithic piezoelectric mirror (MPM) from ITEK (Courtesy M. Ealey).

of 15 mm, and the electrode size going from 3 mm to 1 cm. Fine structural analysis of the mirror requires a 3-D modeling coupling the mechanical and piezoelectric equations (Favre 1989). Simulations show that the sensitivity of a MPM is about 70% that of a free piezoelectric actuator displacement, due to the opposite piezoelectric transverse effect. The maximum voltages being settled by interelectrode breakdown voltage, the main drawback of a MPM is a small stroke lower than 2 μm. This value limits the application of MPMs to the compensation of turbulence on small size telescopes, typically 1-m class.

4.2.4 Deformable mirrors with discrete actuators

4.2.4.1 Principle

Deformable mirrors (DM) with discrete actuators have been the most widely used. Particularly, they were used on the first operational astronomical AO systems, COME-ON and COME-ON PLUS, on the 3.6-m ESO telescope at La Silla, Chile (see Chapter 8, Rousset *et al.* 1990; Rousset *et al.* 1993). DMs with discrete actuators are now also used in astronomical AO systems at Mount Wilson, Lick, Keck, Calar Alto, and Palomar observatories (Shelton and Baliunas 1993; Olivier 1994; Wizinovitch *et al.* 1994; Wirth *et al.* 1995; Dekany 1996). They have long been used in systems developed for defense applications at AMOS by ITEK and MIT Lincoln Laboratory, and at the Starfire Optical Range by the Phillips Laboratory (Primmerman *et al.* 1991; Fugate *et al.* 1991; Hardy 1993).

Their structure consists of a reflective glass facesheet deformed by an array of discrete axial push–pull actuators mounted on a rigid support (see Fig. 4.4). A large variety of mirrors have been developed to date. The main characteristiscs are the number of actuators, the spacing between them, their stroke and voltage. The first deformable mirrors with continuous facesheets developed for turbulence compensation were dedicated to focus high energy infrared lasers (10.6 μm CO_2, 3.8 μm DF or 1.3 μm iodine lasers). The second generation consisted of uncooled mirrors with improved optical quality for near-infrared and visible wavelengths, to be used in astronomical and defense-related compensated imaging. Such mirrors are now referred to as 'stacked actuator mirrors' or 'SAMs'. SAMs of the first generation had a few tens of actuators with actuator spacing of 20–30 mm, and withstanding voltages higher than 1.5 kV. They were followed by high packing density, low voltage actuators (Everson *et al.* 1981), PZT or PMN, typically 250, with 7–8 mm actuator spacing and low voltages, typically 400 V driving signals. SAMs with several thousands of PMN actuators were then built (Ealey 1993). The piezoelectric actuators are no longer

discrete and individually assembled, which requires many adjustments, but
ferroelectric wafers are bonded together and treated to isolate the different
actuators (Ealey and Wheeler 1989; Jagourel and Gaffard 1989; Ealey and Davis
1990; Lillard and Schell 1994) (see Fig. 4.6 and Fig. 4.7).

4.2.4.2 Dynamic behavior

Facesheet deformable mirrors mainly consist of two parts which are mechanically in series: a plate and an array of actuators. The fundamental resonant
frequency of the mirror is given by the lowest resonant frequency of the plate
and of the actuators.

The dynamic equation of the deformation W of a plate is

$$S_p \nabla^2 W - \rho_p t_p \left(\frac{\nu_p}{2\pi} \right)^2 W = 0, \qquad (4.14)$$

where S_p, ρ_p, t_p and ν_p are respectively the stiffness, the mass density, the
thickness, and the characteristic frequency of the plate and where ∇^2 denotes
the two-dimensional Laplacian. The stiffness of a clamped plate of radius R
with a central load is given by (Timoshenko and Woinowsky-Krieger 1959)

$$S_p = \frac{E_p t_p^3}{12R^2(1 - \sigma_p^2)}, \qquad (4.15)$$

Fig. 4.6. A 341-actuator deformable mirror (SELECT) from LITTON-ITEK (Courtesy
M. Ealey).

Fig. 4.7. A 249-actuator deformable mirror (SAM) from CILAS (Courtesy P. Jagourel).

where E_p and σ_p are the Young modulus and the Poisson coefficient of the plate material. This leads to a resonant frequency for part of the plate clamped to a distance of the order of the actuator spacing r_s given by

$$\nu_p = c \frac{t_p}{r_s^2} \sqrt{\frac{E_p}{\rho_p(1 - \sigma_p^2)}},$$ (4.16)

where c is a constant nearly equal to 1.6 (Taranenko 1981).

The stiffness of the actuators S_a depends on the Young modulus E_a, the surface S of a section and the height h of the actuator

$$S_a = \frac{E_a S}{h}.$$ (4.17)

The lowest compression resonant frequency for a clamped-free actuator is

$$\nu_{ac} = \frac{1}{4}\sqrt{\frac{S_a}{m}} = \frac{1}{4h}\sqrt{\frac{E_a}{\rho_a}},$$ (4.18)

where m is the mass of the actuator and ρ_a is the actuator mass density, showing that the resonant frequency is inversely proportional to the height of the actuator.

The ratio v_p/v_{ac} is typically equal to $4t_p h/r_s^2$ (Ealey 1991). For large h, the lowest resonant frequency is that of the actuators. As h decreases, it increases to that of the plate (as far as beam and shear effects can be neglected). Typical values are generally higher than several tens of kHz.

The lowest bending resonant frequency is related to the compression frequency

$$v_{ab} = v_{ac}\frac{s}{h},\qquad(4.19)$$

where s is the lateral size of the actuator. The theoretical lowest resonant frequency of the mirror is generally the bending frequency, but it is experimentally found that the lowest resonant frequency is the axial frequency because the bending modes are not excited. It should be noted that if the stiffness of the actuator is larger than the stiffness of the plate, the deformation of the plate may be 20–30% smaller than the free deformation of the actuator because of high mechanical coupling.

4.2.5 Bimorph mirrors

4.2.5.1 Principle

Although bimorph mirrors were envisaged a long time ago (Kokorowski 1979; Steinhaus and Lipson 1979), the actual use of a bimorph mirror in adaptive optics was first demonstrated in 1994 in a system developed at the University of Hawaii for astronomical applications (Roddier *et al.* 1994). A bimorph mirror now equips the PUEO user AO system of the Canada–France–Hawaii Telescope (Lai *et al.* 1995). Bimorph mirrors are also under test for the SUBARU AO system (Takami *et al.* 1995) and for the Anglo-Australian Telescope AO system (Bryant *et al.* 1995).

A bimorph mirror consists of two piezoelectric ceramic wafers which are bonded together and oppositely polarized, parallel to their axis. An array of electrodes is deposited between the two wafers. The front and bottom surfaces are grounded (see Figs 4.4 and 4.8), (Kokorowski 1979; Steinhaus and Lipson 1979; Roddier 1988; Vorontsov *et al.* 1989; Jagourel *et al.* 1990). When a voltage is applied to an electrode, a wafer contracts locally and laterally as the other wafer expands, inducing a bending. It is difficult to find high density PZT materials providing a surface roughness appropriate for direct coating to form the reflective surface. One solution is to cover the front of the wafer by an optically polished thin glass plate (Lipson *et al.* 1994). A drawback of this

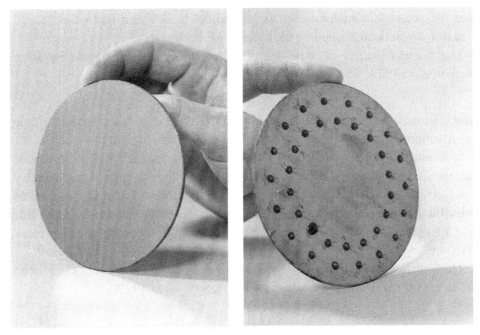

Fig. 4.8. A 36-actuator bimorph mirror (BIM) from CILAS. *Left*: front. *Right*: back (Courtesy P. Jagourel).

solution is a sensitivity to temperature changes. It could be reduced by covering the back of the wafer with a symmetrical plate, but at the expense of a reduced stroke. A very efficient solution is to cover the face of the bimorph with a mirror replica (Jagourel *et al.* 1990).

The relative change in length induced on an electrode of size l is given by

$$\frac{\Delta l}{l} = \frac{V d_{31}}{t}, \tag{4.20}$$

where d_{31} is the transverse piezoelectric coefficient and t is the thickness of the wafer. Neglecting the stiffness of the wafers and three-dimensional effects, the radius of curvature becomes

$$R = \frac{tl}{2\Delta l} = \frac{t^2}{2V d_{31}}. \tag{4.21}$$

For a spherical deformation over the diameter d, the bimorph sensitivity S_b expressed as the ratio stroke/voltage is

$$S_b = \frac{d^2}{8RV} = \frac{d^2}{4t^2} d_{31}. \tag{4.22}$$

Taking typical values $d = 40$ mm, $t = 1$ mm, $d_{31} = 0.2$ μm/kV, it is found that $S_b = 80$ μm/kV. This should be compared with the longitudinal sensitivity of a free piezoelectric actuator which is around 0.3 μm/kV and that of a stacked actuator which is multiplied by the number of its elements.

The static equation of state for an ideal bimorph mirror has the form (Kokorowski 1979; Roddier 1988)

$$\nabla^2(\nabla^2 W + AV) = 0, \tag{4.23}$$

where ∇^2 denotes the two-dimensional Laplacian, $W(x, y)$ is the mirror surface deformation, $V(x, y)$ is the voltage distribution on the wafer, and $A = 8d_{31}/t^2$.

It should be pointed out that the equilibrium is reached when the mirror surface is the solution of a Poisson equation with appropriate boundary conditions. Radial tilts at the edge provide the boundary conditions required to solve the Poisson equation. A simple way to control these tilts is to use an extra ring of electrodes and to limit the pupil to the inner part of their surfaces (Jagourel *et al.* 1990).

The expression of the spatial spectrum of the displacement $\tilde{W}(k)$ is related to the spatial spectrum of the voltage $\tilde{V}(k)$ (Kokorowski 1979)

$$\tilde{W}(k) = \tilde{V}(k)\left[\frac{8d_{31}}{t^2 k^2} - bd_{31}\right], \tag{4.24}$$

where b is a coefficient which depends on the material and lies between 0.4 and 0.6 (Jagourel *et al.* 1990). This expression shows that the spectrum of the bending deformation decreases as k^{-2}, which is very close to the $k^{-11/6}$ decrease of the Kolmogorov spectrum of the phase fluctuations (Roddier 1992).

Besides this displacement, the applied voltage also causes opposite changes of the thickness of each wafer (Kokorowski 1979). Even if this effect cancels for the entire bimorph, it still produces a displacement of the top and bottom surfaces, which is added to the displacement caused by bending. From Eq. (4.23) the deformation due to thickness changes over an electrode may be written $W_t(x, y) = -bV(x, y)d_{31}$. This effect can be compared with the pure bending deformation W_b over an electrode of diameter d given by Eq. (4.22). These effects are opposite and the resulting displacement is locally zero when the wafer diameter d becomes of the order the thickness t. To have a good bimorph efficiency, that is $W_b \gg W_t$, a good criterion is to have an electrode diameter t at least four times larger than the wafer thickness. Since the ratio of the diameter of the whole wafer to its thickness is limited by polishing considerations, the number of electrodes is limited by the bimorph diameter-to-thickness ratio. Typically a few tens of electrodes are used. Hence, bimorph mirrors are best suited for low-order compensation systems.

It should be pointed out that, owing to the k^{-2} spectrum dependence, bimorph mirrors have a sufficient stroke at low spatial frequencies to compensate for turbulence induced tip/tilt errors. Thanks to their light weight, larger but slower telescope tracking errors can be compensated by mounting them on a tip/tilt platform. This avoids the use of a separate tip/tilt mirror and reduces the number of reflective surfaces.

4.2.5.2 Dynamic behavior

The resonant frequency of a free supported circular plate is of the same form as Eq. (4.16), with c of the order of 0.8 instead of 1.6 for a clamped plate. The main resonant frequency is typically of the order of several kHz, which is lower than that of a deformable mirror with displacement actuators. For a given t_p/r_s ratio, the resonant frequency varies as $1/r_s$. For a large number of actuators, it may be lower than the required AO bandwidth, and the deformation may be in a dynamic regime where control is more complex. Faint modes have been observed at frequencies as low as a few hundreds of Hertz due to the support.

4.3 Deformable mirrors with non-ferroelectric actuators

4.3.1 Membrane mirrors

4.3.1.1 Principle

A membrane mirror consists of a reflective membrane, stretched over a ring and deformed by means of electrostatic forces in a partial vacuum chamber, as depicted in Fig. 4.9 (Yellin 1976; Grosso and Yellin 1977; Merkle *et al.* 1982; Centamore and Wirth 1991; Bonaccini *et al.* 1991; Takami and Iye 1994). Compared to a continuous facesheet deformable mirror, a membrane mirror has no inertia and no hysteresis.

The local curvature is proportional to the square of the voltage, which is applied between a network of conducting pads and the membrane. To linearize the response, a bias voltage V_0 is added to the signal voltage V_S. This voltage may be applied to a window coated with a transparent electrode, with the

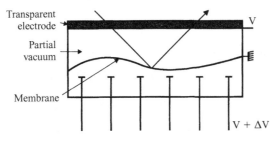

Fig. 4.9. Membrane mirror.

membrane grounded and the pads containing the voltage $V_S + V_0$. With no signal V_S, the membrane stays flat. Possible drawbacks of the coated window are ghost reflections and a limitation of the spectral range. Another solution is to apply the bias voltage to the membrane and to compensate for the bias curvature of the membrane with a concave anti-reflection coated lens window.

With an external pressure $P(x, y)$ applied to the circular membrane and with no viscous damping, the static equation of state of an ideal membrane is (Morse 1948)

$$\nabla^2 W(x, y) = -\frac{P(x, y)}{T(x, y)}, \tag{4.25}$$

where W is the deformation of the membrane and T is the stress/length ratio. This ideal model assumes a perfect elastic membrane, a linear behavior and neglects the boundary effects. As for bimorph mirrors, the equilibrium is reached when the membrane surface is the solution of a Poisson equation with appropriate boundary conditions. In other words, the effect of applying a local pressure is to change the local curvature of the membrane. The stress applied with an electrostatic deflector is (Morse 1948)

$$F = AP = \frac{A\epsilon_0}{2}\frac{V^2}{l_e^2},$$

where A is the active area, $\epsilon_0 = 8.85 \times 10^{-12}$ F/m, V and l_e are the voltage and the distance between an electrode and the membrane. If the bias voltage is applied to the membrane, the peak deflection inside the pad radius is (Morse 1948)

$$W_p = \frac{\epsilon_0}{4T}\left[\frac{V_0 + V_S}{l_e}\right]^2. \tag{4.27}$$

If the bias voltage is applied to the window, the peak deflection becomes

$$W_p = \frac{\epsilon_0}{4T}\left[\frac{(V_0 + V_S)^2}{l_e^2} - \frac{V_0^2}{l_w^2}\right], \tag{4.28}$$

where l_w is the distance between the window and the membrane. The membrane is typically a 0.5–2.5 µm thick metallic or polymer foil. The higher value of V/l is limited by electric discharge threshold. Typical values are some 200 V for V and 100 µm for l.

In vacuum, the membrane surface is unstable, whereas at ambient air pressure the damping significantly reduces the stroke. It is also sensitive to environmental acoustic perturbations. To control the membrane transient behavior and its acoustical sensitivity, the air damping must be optimized. The interior of the membrane assembly is evacuated to about several torrs. Stable linear response with a 25 mm active diameter membrane leads to a sensitivity of the order 10–20 μm/kV.

4.3.1.2 Dynamic behavior

The lowest characteristic frequency of a membrane is given approximately by (Grosso and Yellin 1977)

$$\nu_0 = \frac{0.76}{D} \sqrt{\frac{T}{\sigma}}, \tag{4.29}$$

where D is the membrane diameter, T is the membrane tension, σ is the mass/area ratio. It is interesting to note that the fundamental resonant frequency is independent of the thickness of the membrane.

From experimental results (Grosso and Yellin 1977), it appears that at low pressure, the response of the membrane is undamped, while at higher pressure it is overdamped. A typical optimized pressure cavity is of the order of few torrs with an operating frequency higher than 5 kHz for a 50 mm diameter membrane and a gap of about 60 μm (Grosso and Yellin 1977).

4.3.2 Deformable secondary mirrors

An adaptive secondary mirror is a concept recently proposed to both eliminate the optical components required to conjugate a deformable mirror at a reimaged pupil and minimize thermal emission (Salinari *et al.* 1993; Martin and Anderson 1995; Bruns *et al.* 1996). When compared to a deformable mirror, the major difference is that the resonant frequency of an adaptive secondary mirror may be lower than the AOs bandwidth. This is due to a larger interactuator spacing (see Eq. (4.16)). The deformations of the mirror, which are no longer in a quasi-static regime, should be described in a dynamic one. From a control standpoint (see Section 6.4.1.6) the mirror transfer function cannot be considered as constant since it exhibits an overshoot at the resonant frequency. Although the system is still linear, the control matrix is a function of the frequency, thus inducing a more complex control. Such a wave-front corrector is under development for the 6.5-m MMT (see Section 13.5.1.2). It consists of a 2-mm thick convex mirror, 640 mm in diameter, supported on 320 force actuators (voice coil), about 25 mm apart, mounted in a rigid glass substrate. The mirror deformation is controlled with a wave-front sensor. Capacitive position sensors control the complex dynamics through a 10 kHz internal feedback loop. Adaptive secondary mirrors are also envisaged to equip the 6.5-m Magellan telescopes and the 8.4 m Large Binocular Telescope.

4.3.3 Liquid crystal devices

Several emerging technologies developed for display components offer alter-
natives to deformable mirrors. Among them, liquid crystal devices (LCDs) are
particularly attractive because of their low cost, large number of correcting
elements, low power consumption, and compact size with no moving parts.

Liquid crystals (LC) refer to a state of matter intermediate between solid and
liquid (De Gennes 1975). The fundamental optical property of LCs is their
birefringence. It is higher than that of electro-optical crystals, due to the
process of orientation of anisotropic molecules (e.g. long thin ones) and is of
the order of 0.2 for a refractive index of 1.5. This process also explains their
relatively long response time.

LCs are classified in nematic and smectic crystals, depending on the long-
distance ordering of the centers of gravity of their molecules. Nematic and
smectic crystals differ in their electrical behavior. Ferroelectricity is the most
interesting phenomenon for a variety of smectic crystals. Nematic and ferroelectric LCs have been studied for adaptive optics applications (Riehl *et al.*
1988; Vorontsov *et al.* 1989; Bonaccini *et al.* 1991). Only nematic crystals
provide continuous index control, compared with the binary modulation given
by ferroelectric crystals. However, the response time of nematic crystals is
longer than that of ferroelectric crystals. Their rise time is related to the forced
alignment of molecules by the applied electric field, and it is of the order of
10 ms. The relaxation time is longer, of the order of 100 ms. But the decay time
may be forced to reach the rise time. One solution is to use two excitation
frequencies (Wu 1985).

Nematic LCs are uniaxial. In the so-called electrically controlled birefringence configuration, the extraordinary index of a thin LC film can be modulated, producing an optical path variation with polarized light (Soref and
Rafuse 1972). Modified devices which can operate with non-polarized beams
have been proposed (Bonaccini *et al.* 1994; Love *et al.* 1996a). The liquid
crystal film, of the order of 5–10 μm, is sandwiched between two pieces of
optical quality glass to ensure uniform cell thickness. A crucial part of the
fabrication is to fix the orientation of the molecules by a proper hooking of the
LCs to the cell surfaces.

The typical elementary active area, of the order of 1 mm, is significantly
smaller than mechanical actuator spacing. Each pixel has its own individual
electrode and is controlled via silicon integrated circuit chips. The typical
control voltage is of the order of 1 V. It introduces a piston-like correction
and the fitting error is similar to that of a segmented mirror. But compared
with a segmented mirror, it is easier to use the larger number of pixels

required to achieve the same fitting error as a continuous facesheet deformable mirror.

Demonstration of phase correction has been achieved (Love *et al.* 1994; Love and Restaino 1995). Response frequency for a stroke of 1 µm is about 10 Hz, and is far from the AO requirements for turbulence effects to be fully compensated. It is mandatory to overcome the actual bandwidth limitation before LCDs can replace deformable mirrors. Another drawback is the potentially limited spectral range. But LCDs seem particularly well suited for high spatial resolution compensation of slowly evolving wave fronts like instrument aberrations in the so-called active optics systems.

4.4 Spatial correction efficiency

A main feature of a deformable mirror is its mechanical efficiency to fit the wave-front perturbations with the best accuracy. Karhunen–Loève polynomials are the eigenmodes of turbulent wavefronts. If a low order mode like tip/tilt is perfectly compensated from a spatial point of view with a plano mirror, residual errors occur in compensation of higher order modes because actual deformable mirrors' eigenmodes are not Karhunen–Loève modes (see Chapter 3).

To first order, the deformable mirror filters frequencies higher than the cut-off Nyquist frequency are equal to the ratio $1/(2r_s)$ where r_s is the actuator spacing. A more accurate estimation requires a more precise description of the mirror spatial mechanical response: for instance, a point-like shape will be less efficient than the one of a continuous facesheet deformable mirror with a smoothed profile. In a preliminary design phase, analytical computations give a good estimate of the mechanical mirror behavior (Timoshenko and Woinowski-Krieger 1959; Roarke and Young 1982). Finite element analysis is only useful in the design phase to determine detailed specifications and to optimize the spatial mirror mechanical response.

4.4.1 Optimization of the mirror influence function

The lowest mechanical resonant frequency of a deformable mirror has to be sufficiently higher than the sampling frequency to introduce a tolerable phase lag in the AO servo loop. In other words, the mechanical deformations are in a quasi-static regime and the deformations of the mirrors are small enough to assume the linearity of the deformation. The elementary deformation produced by one activated actuator, the others only acting as springs, is called the influence function $D_i(x, y)$. When all the actuators are activated with an amplitude A_i, the resulting deformation $D(x, y)$ is therefore given by

$$D(x, y) = \sum_{i=1}^{N} A_i D_i(x, y). \qquad (4.30)$$

Some authors have proposed a model of a Gaussian-like shape influence function (Hudgin 1977; Garcia and Brooks 1978; Taranenko *et al.* 1981). This shape may approximately fit the experimental data between two actuators, but it has no physical origin. Beyond some distance, of the order of twice r_s the Gaussian shape does not represent actual profiles which can take negative values beyond the location of the first ring of actuators. Approximate formulas giving the influence function can be found (Roarke and Young 1982; Ealey 1991). They are close to the simple expression of the normalized deformation $W(r/R)$ of a clamped circular face-plate of radius R under the force applied by a point-like central actuator (Roarke and Young 1982)

$$W\left(\frac{r}{R}\right) = 1 - \left(\frac{r}{R}\right)^2 + 2\left(\frac{r}{R}\right)^2 \log\left(\frac{r}{R}\right). \qquad (4.31)$$

From a mechanical standpoint, the fundamental parameter is the mechanical coupling, that is the value of the normalized deformation at the location of the nearest actuator. The influence function is generally nearly axisymmetrical and independent of the actuator position. The influence function of a central load supported by four points in a square pattern is nearly the same as that of a load applied to a circular area encircling the same square. This is even truer with an hexagonal array. Beyond the first ring, asymmetry may arise due to a low mechanical coupling, square arrays being more sensitive than hexagonal arrays. Concentric cylindrical sections are also used with bimorph mirrors: this structure matches the structure of Zernike polynomials more closely (Centamore and Wirth 1991).

It should be noted that the face-plate may be free, i.e. only supported by the array of actuators, or it may be clamped on an external ring. The advantage of a free face-plate is that the deformation at the edge is controlled in a better way than with a clamped face-plate. The drawback is that an additional ring of actuators outside the pupil must be controlled. This requires a larger computing power, the number of operations to perform varying as the square of the number of actuators. With a clamped plate, the external actuators have a non-symmetrical influence function and the pupil has to be limited to the center of the external actuators to minimize the mirror fitting error. The distance between these actuators and the edge of the plate has to be at least 1.5 r_s (Madec 1989). However the support acts as a reference to control the polishing, making it easier. For a clamped plate, even if all the actuators provide the same displacement, the surface is not perfectly flat but rippled. The amplitude of the ripples depends on the stiffness of the plate and decreases with higher mechanical coupling.

More generally, the mirror fitting error depends on the mechanical spatial response of the mirror and on the shape of the wave front to be corrected. Using the same formalism as that of Section 6.3, the mechanical response of the mirror is determined by the mechanical interaction matrix denoted D_{mec}. If $|V\rangle$ is the control voltage vector, the corresponding wave-front correction $|\varphi\rangle$ introduced by the deformable mirror is

$$|\varphi\rangle = D_{mec}|V\rangle. \tag{4.32}$$

A column vector of the matrix D_{mec} is the correction vector corresponding to a unity voltage applied to one of the deformable mirror actuators, that is the influence function of this actuator. These influence functions can be theoretically estimated by finite element analysis or experimentally measured (Garcia and Brooks 1978).

For a given correction vector $|\varphi\rangle$, the control vector minimizing the norm of the mirror fitting error is $D_{mec}^{*}|\varphi\rangle$ where

$$D_{mec}^{*} = (D_{mec}^{t} D_{mec})^{-1} D_{mec}^{t}. \tag{4.33}$$

Superscript t denotes the transposition operator.

The mirror fitting error vector is given by

$$|\epsilon\rangle = (I - D_{mec} D_{mec}^{*})|\varphi\rangle \tag{4.34}$$

where I is the identity matrix. The squared norm of the error may be written

$$\|\epsilon\|^{2} = \mathrm{trace}|\epsilon\rangle\langle\epsilon| = \mathrm{trace}\, M|\varphi\rangle\langle\varphi|M^{t} \tag{4.35}$$

where $M = I - D_{mec} D_{mec}^{*}$. The instantaneous turbulent phase to be compensated may be expanded in a series of Zernike polynomials, and the turbulent phase vector may be written

$$|\varphi(t)_{\mathrm{tur}}\rangle = Z|a(t)\rangle \tag{4.36}$$

where Z is the column matrix of Zernike polynomials and $|a(t)\rangle$ the vector of the expansion coefficients. It is then straightforward to show that the variance of the mirror fitting error in compensating for turbulent wave fronts is

$$\|\epsilon_{\mathrm{tur}}\|^{2} = \overline{\|\epsilon_{\mathrm{tur}}(t)\|^{2}} = \mathrm{trace}\, MZC_{Z,\mathrm{tur}}Z^{t}M^{t} \tag{4.37}$$

where the overline $\overline{(\quad)}$ denotes the mean value with respect to time and $C_{Z,\mathrm{tur}}$ is the covariance matrix of the Zernike polynomials

$$C_{Z,\mathrm{tur}} = \overline{|a(t)\rangle\langle a(t)|}. \tag{4.38}$$

If a Karhunen–Loève expansion is used instead of a Zernike one, then the covariance matrix becomes diagonal, and Eq. (4.35) becomes a sum of independent terms, the fitting errors associated to each Karhunen–Loève mode [see Eq. (3.31)].

Several authors have proposed expressions for the fitting error as a function of the Fried diameter r_0 and the actuator spacing r_s of the form (Hudgin 1977; Pearson and Hansen 1977; Greenwood 1979; Tyson and Byrne 1980; Winocur 1982; Belsher and Fried 1983; Sandler *et al.* 1984)

$$\|\epsilon_{\text{tur}}\|^2 = \mu \left(\frac{r_s}{r_0} \right)^{5/3}.$$

$$(4.39)$$

The value of μ depends on the influence functions and ranges from 0.15 (piston) to 1.26 (Gaussian).

4.4.2 Optical quality

To achieve a sufficient mirror optical quality, accurate mechanical modelling is required but the experience of mechanical and optical engineers is also crucial. For instance, with SAMs, the thermal effects due to the mismatch of thermal coefficients between the actuator materials, the plate and the support have to be minimized by a good choice of materials but also by design tricks. After building, any deformation with typical size smaller than the actuator spacing has to be avoided because it cannot be compensated by the mirror itself. Changes in the sensitivity of the actuators due to aging are compensated in closed-loop operation. The deformation, mainly defocus, due to thermal effects resulting from a difference between the operating temperature and the temperature during the polishing may be compensated as well. Furthermore, during the polishing a high mechanical coupling minimizes the ripples.

In a SAM, the actuator forces must be applied normally to the mirror surface, otherwise moments are introduced in the mirror. The locally induced deformation cannot be compensated. Furthermore, if the stiffness of the actuators is not sufficiently high, bending frequencies significantly lower than axial frequencies may be excited in discrete actuator deformable mirrors (see Section 4.2.4.2). In practice, the design of the mechanical interface between the actuator and the plate is a key point to reduce print-through effects and to obtain a high optical quality. These problems are avoided with bimorph mirrors.

4.5 Tip/tilt mirrors

As shown in Chapter 3, tip and tilt corrections require the largest stroke to be corrected, typically a few tens of microns peak-to-peak. Residual telescope tracking error also needs to be compensated.

It is difficult to obtain such strokes with deformable mirrors because they require high voltages. But they are easily produced by dedicated flat steering mirrors, or possibly by a two-axis tilt secondary mirror. Ferroelectric, electro-magnetic and linear voice coil type actuators have been used, generally arranged in push–pull pairs (Germann and Braccio 1990; Loney 1990, Marth *et al.* 1991; Gaffard *et al.* 1994; Bruns *et al.* 1996). The mechanical mounting design has to minimize the piston excitation, especially for mirrors used in interferometry. The only limitation of steering mirrors comes from the require-ment of an accurate correction at high temporal frequencies which increases with the number of spatial corrected modes (see Chapter 3). Tip/tilt mirrors generally provide a medium or low bandwidth pointing and deformable mirrors compensate for the residual high bandwidth pointing and also for the dynamic deformations of the steering mirror.

4.6 Intensity fluctuation compensation

Amplitude fluctuations are generally small and their effect on image degrada-tion remains limited (see Chapter 2). Their correction is not crucial, except for detection of exo-solar planets (Angel 1994; Stahl and Sandler 1995; Love and Gourlay 1996b). Nevertheless, spatial intensity compensator concepts have been proposed (Casasent 1977; Fisher and Warde 1983; Fisher 1985). Any simple passive spatial intensity corrector, e.g. with a liquid crystal device, dims the light and is not generally well suited for astronomical applications. Active correctors, which amplify light using non-linear optical processes are still under study (Yariv 1978; Yariv and Koch 1982; Ikeda *et al.* 1984; Reintjes 1988). Passive correctors using phase compensators like deformable mirrors may be considered as well. The basic idea is that intensity fluctuations result from the propagation of phase fluctuations induced by high altitude layers (see Chapter 2) and may be compensated by appropriate phase correctors conjugate with these layers. Because high altitude layers are also at the origin of the anisoplanatism, such correctors may increase the field of correction of adaptive optics system as well (Mc Call and Passner 1978).

4.7 How to specify a deformable mirror

From a user standpoint, the practical question is how to specify a deformable mirror. Table 4.1 summarizes the characteristics of a deformable mirror which have to be addressed, with a few comments. For astronomical applications, aperture size and optical quality are important considerations. The aperture diameter determines the size of the instrument and should be

Table 4.1. *Required characteristics for a deformable mirror*

Actuator characteristics

Number of actuators	Depends on wavelength of observations, desired Strehl ratio, availability of suitable guide sources, and cost. Free face-plate mirrors and bimorph mirrors require a ring of actuators outside the pupil area.
Actuator spacing	$\geq$ 6 mm for mirrors commercially available to date. Ultimately limited by the stroke. The smaller the spacing, the more compact the optical system will be.
Actuator geometry	Should match the sensor sampling geometry: square or hexagonal (SH sensors); annular (curvature sensors).

Mechanical characteristics

Actuator stroke	Depends on the telescope diameter and the worse seeing one expects to compensate. Should include typically 10% additional stroke to correct for the deformable mirror static aberrations. *Stacked actuator mirrors:* expressed in microns, and independent of aberration mode. *Bimorph mirrors:* depends on aberration mode. Given by the minimum radius of curvature. See Eq. (9.4)
Fitting error	Average accuracy with which random atmospheric wave-fronts are compensated. Depends on the number of actuators, and the mirror ability to fit the atmospheric Karhunen–Loève modes. Calculated from the mirror influence functions (Eq. (4.37)).
Actuator mechanical coupling	For stacked actuator mirrors only. Typically 15%. Trade-off between stroke and influence function.
Lowest mechanical frequency	Typically a few kHz. Should be much larger than the servo bandwidth. The effect should be estimated from a servo model (see Chapter 6).
Phase lag	Typically < 5% at 1 kHz depending on sampling frequency. The effect should be estimated from a servo model (see Chapter 6).
Actuator hysteresis	Typically < 5% (full stroke). The effect should be estimated from computer simulations (see Chapter 7).
Probability of failure	*Stacked actuators:* typically $\leq$ 0.01 $\times$ actuator number/year. Inquire about actuator replacement possibility (cost/delay).

Optical characteristics

Pupil diameter	Depends on the number of actuators or the actuator spacing. Should be as small as possible since it determines the overall system size.

continues

Table 4.1. (*cont.*)

Optical characteristics (cont.)	
Optical quality	*Static aberrations:* should typically require $\leqslant$ 10% of the mirror stroke to be compensated. *Closed-loop residuals:* should not significantly affect the closed loop point spread function. Typically 0.03 µm rms. *Surface roughness, scratches and digs:* as for other optical components. Typical roughness: 1 nm rms.
Spectral reflectivity	As high as possible over both sensing and imaging bandwidths. Protected silver coating recommended for usual applications.
Power supply	
Number of channels	Equal to the number of controlled actuators.
Input signal	Typically ± 10 V from digital-to-analog converter.
Cut-off frequency	Typically $>$ 1 kHz (first order filter). The effect should be estimated from a servo model (see Chapter 6).
Phase shift	Typically $>$ 5° at 100 Hz. The effect should be estimated from a servo model (see Chapter 6).
Signal-to-noise ratio	Typically $>$ 1000. Effect on Strehl ratio should be negligible.
Controls	Amplifier offset and gain adjustments.
Output monitoring	Typically 10% of the output signal for each channel.
Thermal dissipation	To be minimized (typically $<$ 100 W).
Environmental conditions	
Temperature range (air)	*Functional:* Typically -10 to 25 °C. *Operational:* Typically 0–15 °C.
Temperature gradient (air)	Typically 0.4–0.7 °C/h.

minimized. The level and the structure of the light scattered around a star affects the detection of nearby faint objects. When comparing the different available mirror structures, the other question is how to choose a well-adapted technology. Owing to its mechanical structure, the bimorph mirror presently seems to be the cheapest wave-front corrector device, providing the best optical quality, allowing compensation of turbulence tip/tilt wave-front disturbances and thus avoiding the use of a dedicated additional mirror. It seems only limited by its maximum number of actuators, which means it is best suited to low order compensation. For high order compensation a SAM may be the most efficient solution.

4.8 Conclusion

Wave-front compensation through deformable mirrors is now an operational reality. Certainly, the technology is mature. Future trends may be to have cheaper, perhaps non-mechanical, and smaller components to reduce the size of the optical system, using the benefit of developments non-specific to adaptive optics. Long-term perspective may arise from non-linear optics approaches to overcome the key issue of anisoplanatism and the present wide angle field of view limitation.

Acknowledgements

I would like to express my deep gratitude, for innumerable fruitful discussions during more than 10 years, to Professor François Roddier (University of Hawaii) and to my colleagues Gérard Rousset and Pierre-Yves Madec (ONERA) on adaptive optics systems, to Pascal Jagourel and Jean-Paul Gaffard (CILAS) on deformable mirrors, and to Professor Lucien Eyraud and Paul Eyraud (INSA Lyon) on ferroelectric materials. I thank Gordon Love (USAF Phillips Laboratory) for his kind support in the field of liquid crystal devices. I am grateful to Caroline Dessenne and Laurent Mugnier (ONERA) and David Sandler (ThermoTrex) for their detailed reading of my contribution and help in preparing the document. And many more besides.

References

Acton, D. S. (1992) Status of the Lockheed 19-Segment Solar Adaptive Optics System. In: *Real Time and Post Facto Solar Image Correction*. Proc. Thirteenth National Solar Observatory Sacramento Peak Summer Shop Series 13, Sunspot (New Mexico), 15–18 September 1992, pp. 1–5.

Aldrich, R. E. (1980) Requirements for piezoelectric materials for deformable mirrors. *Ferroelectrics* 27, 19–25.

Angel, J. R. P. (1994) Ground-based imaging of extrasolar planets using adaptive optics. *Nature* (London) 368, 203–7.

Belsher, J. F. and Fried, D. L. (1983) Adaptive optics mirror fitting error. Optical Sciences Co, Placentia, CA, Report TR-521.

Blackwood, G. H., Davis, P. A. and Ealey, M. A. (1991) Characterization of PMN:BA electrostictive plates and SELECT actuators at low temperature. In: *Active and Adaptive Optical Components*. Proc. SPIE 1543, pp. 422–9.

Bonaccini, D., Brusa, G., Esposito, S., Salinari, P., Stefanini, P. and Biliotto, V. (1991) Adaptive optics wavefront corrector using addressable liquid crystal retarders II. In: *Active and Adaptive Optical Components*. Proc. SPIE 1543, pp. 133–43.

Bonaccini, D., Esposito, S. and Brusa, G. (1994) Adaptive optics with liquid crystal phase screens. In: *Adaptive Optics in Astronomy*. Proc. SPIE 2201, pp. 1155–8.

Bruns, D., Barrett, T., Brusa, G., Biasi, R. and Gallieni, D. (1996) Adaptive Secondary Development. In: *Adaptive Optics* OSA Conf. Maui, Hawaii, July 8–12, 1996 Tech. Digest Series 13, pp. 302–4.

Bryant, J. J., O'Byrne, J. W., Minard, R. A. and Fekete, P. W. (1995) Low order adaptive optics at the Anglo-Australian telescope. In: *Adaptive Optics*, OSA/ESO Conf., Garching-bei-München (Germany), October 2–6, 1995. Tech. Digest Series 13, pp. 23–8.

Busher, D. F., Doel, A. P., Andrews, N., Dunlop, C., Morris, P. W., Myers, R. M. *et al.* (1995) Novel adaptive optics with the Durham University Electra system. In: *Adaptive Optics*, OSA/ESO Conf., Garching-bei-München (Germany), October 2–6, 1995. Tech. Digest Series 23, pp. 63–8.

Casasent, D. (1977) Spatial light modulators *Proc. IEEE* **65** No. 1, pp. 143–57.

Centamore, R. M. and Wirth, A. (1991) High bias membrane mirror. In: *Active and Adaptive Optical Components*. Proc. SPIE 1543, pp. 128–32.

De Gennes, P. G. (1975) *The Physics of Liquid Crystals*. Clarendon Press, Oxford.

Dekany, G. R. (1996) The Palomar adaptive optics system. In: *Adaptive Optics*, OSA Conf., Maui, Hawaii, July 8–12, 1996. Tech. Digest Series 13, pp. 40–2.

Ealey, M. A. and Wheeler, C. E. (1989) Modular adaptive optics. In: *Active Telescope Systems*. Proc. SPIE 1114, pp 134–44.

Ealey, M. A. and Davis, P. A. (1990) Standard SELECT Electrostrictive PMN actuators for active and adaptive components. *Opt. Eng.* 29, 1373–82.

Ealey, M. A. (1991) Active and adaptive optical components: the technology and future trends. In: *Active and Adaptive Optical Components*. Proc. SPIE 1543, pp. 2–34.

Ealey, M. A. (1993) Low Voltage SELECT Deformable Mirrors. In: *Smart Structures and Materials 1993: Active and Adaptive Optical Components and Systems II*. Proc. SPIE 1920, pp 91–102.

Eitel, F. and Thompson, C. (1986) Liquid Cooled Deformable Mirror with Close Packed Actuators. U.S. Patent No. 4, 844, 603.

Everson, J. H., Aldrich, R. E. and Albertinetti, N. P. (1981) Discrete actuator deformable mirror. *Opt. Engr.* 20, 316–19.

Eyraud, L., Eyraud, P., Gonnard, P. and Troccaz, M. (1988) Matériaux électrostrictifs pour actuateurs. *Revue Phys. Appl.* 23, 879–89.

Eyraud, L., Eyraud, P., Audigier, D. and Claudel, B. (1996) Influence of the fluoride ion on the piezoelectric properties of a PZT ceramics. *Ferroelectric* 175, 241–50.

Favre, M. (1989) Étude du miroir monolithique. SINAPTEC report, ONERA contract 738040.

Feinleib, J., Lipson, S. G. and Cone, P. F. (1974) Monolithic piezoelectric mirror for wavefront correction. *Appl. Phys. Lett.* 25, 311–15.

Fisher, A. D. and Warde, C. (1983) Technique for real-time high-resolution adaptive phase compensation. *Opt. Lett.* **8**, 353–5.

Fisher, A. D. (1985) Self-referenced high resolution adaptive wavefront estimation and compensation. Proc. SPIE 551, pp. 102–12.

Freeman, R. H. and Pearson, J. E. (1982) Deformable mirrors for all seasons and reasons. *Appl. Opt.* **21**, 580–88.

Fugate, R. Q., Fried, D. L., Ameer, G. A., Boeke, B. R., Browne, S. L., Roberts, P. H., *et al.* (1991) Measurement of atmospheric wave-front distorsion using scattering light from a laser guide star. *Nature* **353**, 144–6.

Gaffard, J. P. Jagourel, P. and Gigan, P. (1994) Adaptive optics: description of available components at Laserdot. In: *Adaptive Optics in Astronomy*. Proc. SPIE 2201, pp. 688–702.

Galvagni, J. (1990) Electrostrictive actuators and their use in optical applications. *Opt. Engr.* **29**, 1389–91.

Galvagni, J. and Rawal, B. (1991) A comparison of piezoelectric and electrostrictive actuator stacks. In: *Active and Adaptive Optical Components*. Proc. SPIE 1543, pp. 296–300.

Garcia, H. R. and Brooks, L. D. (1978) Characterization techniques for deformable metal mirror. In: *Adaptive Optical Components*. Proc. SPIE 141, pp. 74–81.

Germann, L. and Braccio, J. (1990). Fine-steering mirror technology supports 10 nano-radian systems. *Opt. Engr.* **29**, 1351–9.

Greenwood, D. P. (1979) Mutual coherence function of a wavefront corrected by zonal adaptive optics. *J. Opt. Soc. Am.* **69**, 549–54.

Greenwood, D. P. and Primmerman, C. A. (1992) Adaptive optics research at Lincoln Laboratory. *Lincoln Lab. Journal* **5**, 3–24.

Grosso, R. P. and Yellin, M. (1977) The membrane mirror as an adaptive optical element. *J. Opt. Soc. Am.* **67**, 399–406.

Hansen, S. (1975) Final report, Contract N60921-75-C-0067 (June 1975) (available from NATAC Tech. Info. Serv.).

Hardy, J. W., Lefebvre, J. E. and Koliopoulos, C. L. (1977) Real time atmospheric compensation. *J. Opt. Soc. Am.* **67**, 360–9.

Hardy, J. W. (1989) Instrumental limitations in adaptive optics for astronomy. Proc. SPIE 1114, pp. 2–13.

Hardy, J. W. (1993) Twenty years of active and adaptive optics. In: *Active and Adaptive Optics*, ed. F. Merkle, ICO 16 Satellite Conf., ESO Conf. Proc. 48, pp. 29–34. ESO, Garching.

Herbert, J. M. (1982) *Ferroelectrics Transducers and Sensors*. Gorden and Breach Science Publ., New York.

Hudgin, R. and Lipson, S. (1975) Analysis of monolithic piezoelectric mirror. *J. Appl. Phys.* **46**, 510–12.

Hudgin, R. (1977) Wave-front compensation error due to finite corrector-element size. *J. Opt. Soc. Am.* **67**, 393–5.

Hulburd, B. and Sandler, D. (1990) Segmented mirrors for atmospheric compensation. *Opt. Eng.* **29**, 1186–90.

Hulburd, B., Barrett, T., Cuellar, L. and Sandler, D. (1991) High band-width, long stroke segmented mirror for atmospheric compensation. In: *Active and Adaptive Optical Components*. Proc. SPIE, 1543, pp. 64–75.

Ikeda, O., Takehara, M. and Sato, T. (1984) High-performance image-transmission system through a turbulent medium using multi reflectors and adaptive focusing combined with four wave-mixing. *J. Opt. Soc. Am.* A **1**, 176–9.

Jagourel, P. and Gaffard, J. P. (1989) Adaptive optics components in Laserdot. In: *Active and Adaptive Optical Components*. Proc. SPIE 1543, pp. 76–87.

Jagourel, P., Madec, P.-Y. and Séchaud, M. (1990) Adaptive optics: a bimorph mirror for wavefront correction. In: *Adaptive Optics and Optical Structures*. Proc. SPIE 1271, pp. 160–71.

Kibblewhite, E., Smutko, M. F. and Fang Shi (1994) The effect of hysteresis on the performance of deformable mirrors and methods of its compensation. In: *Adaptive Optics in Astronomy*. Proc. SPIE 2201, pp. 754–61.

Kokorowski, S. A. (1979) Analysis of adaptive optical elements made from piezoelectric bimorphs. *J. Opt. Soc. Am.* **69**, 181–7.

Lai, O., Arsenault, R., Rigaut, F., Salmon, D., Thomas, J., Gigan, P., *et al.* (1995) CFHT Adaptive optics integration and characterization. In: *Adaptive Optics*, OSA/ESO Conf., Garching-bei-München (Germany), October 2–6, 1995. Tech. Digest Series 23, pp. 491–6.

Lillard, R. L. and Schell, J. D. (1994) High-performance deformable mirror for wavefront correction. In: *Adaptive Optics in Astronomy*. Proc. SPIE 2201, pp. 740–53.

Lipson, S. G., Ribak, E. N. and Schwartz, C. (1994) Bimorph deformable mirror design. In: *Adaptive Optics in Astronomy*. Proc. SPIE 2201, pp. 703–14.

Loney, G. C. (1990) Design of a small-aperture steering mirror for high bandwidth acquisition tracking. *Opt. Engr.* **29**, 1360–5.

Love, G. D., Major, J. V. and Purvis, A. (1994) Liquid-crystal prisms for tip-tilt adaptive optics. *Opt. Lett.* **19**, 1170–2.

Love, G. D. and Restaino, S. R. (1995) High quality liquid crystal spatial light modulators for adaptive optics. In: *Adaptive Optics*, OSA Conf., Garching-bei-München (Germany), October 2–6, 1995. Tech. Digest Series 23, pp. 223–5.

Love, G. D., Restaino, S. R., Carreras, R. C., Loos, G. C., Morrison, R. V., *et al.* (1996a) Polarization-insensitive 127-segment liquid crystal wave-front corrector. Proc. OSA Conf. 13, pp. 288–90.

Love, G. D. and Gourlay, J., (1996b) Intensity-only modulation for atmospheric scintillation correction by liquid-crystal spatial light modulators. *Opt. Lett.* **21**, 1496–8.

Mc Call, S. L. and Passner, A. (1978) In: *Adaptive Optics and Short Wavelength Sources* 6, Physics of Quantum Electronics, Addison-Wesley Pub. New York.

Madec, P. Y., Séchaud, M., Rousset, G., Michau, V. and Fontanella, J. C. (1989) Optical adaptive systems: recent results at ONERA. In: *Active Telescope Systems*. Proc. SPIE 1114, pp. 43–64.

Malakhov, M. N., Matyukhin, V. F. and Pilepskii, B. V. (1984) Estimating the parameters of a multielement mirror of an adaptive optical system. *Sov. J. Opt. Technol.* **51**, 141–4.

Marth, H., Donat, M. and Pohlhammer C. (1991) Latest experience in design of piezoelectric driven fine steering mirrors. In: *Active and Adaptive Optical Components*. Proc. SPIE 1543, pp. 248–61.

Martin, H. M. and Anderson, D. S (1995) Steps toward optical fabrication of an adaptive secondary mirror. In: *Adaptive Optics*, OSA/ESO Conf., Garching-bei-München (Germany), October 2–6, 1995, Tech. Digest Series 23, pp. 401–6.

Merkle, F., Freischlad, F. and Reichmann H.-L. (1982) Development of an active optical mirror for astronomical applications. In: *Advanced Technology Optical Telescopes*. Proc. SPIE 332, pp. 260–8.

Morse, P. M. (1948) Vibration and Sound, 2nd edn, McGraw-Hill Publ., New York.

Neal, D. R., McMillin, P. L. and Michie, R. B. (1991) An astigmatic unstable resonator with an intracavity deformable mirror. In: *Active and Adaptive Optical Systems.* Proc. SPIE 1542, pp. 449–58.

Newcomb, C. V. and Flinn, I. (1982) Improving the linearity of piezoelectric ceramic actuators. *Electr. Letters* **18**, 442–3.

Olivier, S. S., An, J., Avicola, K., Bissinger, H. D., Brake, J. M., Fiedman, H. W., *et al.* (1994) Performance of laser guide star adaptive optics at Lick Observatory. In: *Adaptive Optics in Astronomy.* Proc. SPIE 2201, pp. 1110–20.

Pearson, J. E. (1976) Compensation of propagation distortion using optical adaptive techniques (COAT). *Opt. Eng.* **15**, pp. 151–7.

Pearson, J. E. and Hansen, S. (1977) Experimental studies of a deformable mirror adaptive optical system. *J. Opt. Soc. Am.* **67**, 325–33.

Primmerman, C. A., Murphy, D. V., Page, D. A., Zollars, B. G. and Barclays, H. T. (1991) Compensation of atmospheric optical distortion using a synthetic beacon. *Nature* **353**, 141–3.

Reintjes, J. F. (1988) Nonlinear and adaptive techniques control laser wavefronts. *Laser Focus* **24** (12), 63–78.

Ribak, E. N. (1994) Deformable mirrors In: *Adaptive Optics for Astronomy* Series C: Mathematical and Physical Sciences, Vol. 423, Kluwer Academic Pub., Dordrecht.

Riehl, J., Taufflieb, E. and Séchaud, M. (1988) *Modulateurs spatiaux de phase: applications à l'optique adaptative.* ONERA Technical Report 65/7177 SY, Nov. 1988.

Roarke, R. J. and Young, W. C. (1982) *Formulas for Stress and Strain.* McGraw Hill Publ., New York.

Roddier, F. (1988) Curvature sensing and compensation: a new concept in adaptive optics. *Appl. Opt.* **27**, 1223–5.

Roddier, F. (1992) Towards lower cost adaptive optics systems. In: *Adaptive Optics for Large Telescopes.* Tech. Digest OSA meeting, Lahaina (Maui, Hawaii) Aug. 17–21, 1992, pp. 173–5.

Roddier, F., Anuskiewicz, J., Graves, J. E., Northcott, M. J. and Roddier, C. (1994) Adaptive optics at the University of Hawaii I: Current performance at the telescope. In: *Adaptive Optics in Astronomy.* Proc. SPIE 2201, pp. 2–9.

Rousset, G., Fontanella, J. C., Kern, P., Gigan, P., Rigaut, F., Léna, P., *et al.* (1990) First diffraction-limited astronomical images with adaptive optics. *Astron. Astrophys.* **230**, 29–32.

Rousset, G., Beuzit, J. L., Hubin, N., Gendron, E., Boyer, C., Madec, P. Y., *et al.* (1993) The COME-ON PLUS adaptive optics system: results and performance. In: *Active and Adaptive Optics*, ed. F. Merkle, ICO-16 Satellite Conf., ESO Conf. Workshop Proc. 48, pp. 65–70.

Salinari, P., Del Vecchio, C. and Billiotti, V. (1993) A study of an adaptive secondary mirror. In: *Active and Adaptative Optics*, ed. F. Merkle, ICO-16 Conf., ESO Conf. Proc. 353, pp. 247–53.

Sandler, D. G., Stagat, R. and White, W. (1984) Mission Research Corporation MRC-N-635, (April 1984).

Sandler, D. G., Cuellar, L., Lefebvre, M., Barrett, R., Arnold, R., Johnson, P., *et al.* (1994) Shearing interferometry for laser-guide-star atmospheric correction at large D/r_0. *J. Opt. Soc. Am.* A **11**, 858–73.

Séchaud, M., Eyraud, M. and Thiriot, A. (1987a) Miroir déformable à densité de moteurs élevée et faible temps temps de réponse. French Patent No 87 06 404., 6 May 1987.

Séchaud, M. and Madec, P.-Y. (1987b) Système optique de correction par réseau actif pour télescope. French Patent No 87 06 407, 6 May 1987.

Shelton, J. C. and Baliunas, S. L. (1993) Results of Adaptive Optics at Mount Wilson Observatory. In: *Smart Structures and Materials 1993: Active and Adaptive Optical Components and Systems II*. Proc. SPIE, 1920, p. 371.

Smith, D. C. (1977) High-Power Laser Propagation: Thermal Blooming. Proc. IEEE 65.

Soref, R. A. and Rafuse, M. J. (1972) Electrically controlled birefringence of thin nematic films. *J. Appl. Phys.* **43**, 2029.

Stahl, S. S. and Sandler, D. G. (1995) Optimization and performance of adaptive optics for imaging extrasolar planets. *Astrophys. J. Letter* **454**, 153–6.

Steinhaus, E. and Lipson, S. G. (1979) Bimorph piezoelectric flexible mirror. *J. Opt. Soc. Am.* **69**, 478–81.

Takami, H. and Iye, M. (1994). Membrane deformable mirror for Subaru adaptive optics. In: *Adaptive Optics in Astronomy*. Proc. SPIE 2201, pp. 762–67.

Takami, H., Iye, M., Takato, N. and Hayamo, Y. (1995) Subaru adaptive optics program. In: *Adaptive Optics*, OSA/ESO Conf., Garching-bei-München (Germany), October 2–6, 1995. *Tech. Digest Series* 23, pp. 43–8.

Taranenko, V. G., Koshelev, G. P. and Romanyuk, N. S. (1981) Local deformation of solid mirrors and their frequency dependence. *Sov. J. Opt. Technol.* **48**, 650–2.

Timoshenko, S. and Woinowsky-Krieger, S. (1959) Theory of plates and shells, 2nd edn, McGraw Hill, New York.

Tyson, R. K. and Byrne, D. M. (1980) The effect of wavefront sensor characteristics and spatiotemporal coupling on the correcting capability of a deformable mirror. In: *Active Optical Devices and Applications*. Proc. SPIE 228, pp. 21–5.

Tyson, R. K. (1991) *Principles of Adaptive Optics*. Academic Press, Inc. Harcourt Brace Jovanovich, Pub.

Uchino, K. (1986) Electrostrictive actuators: materials and applications. *Ceramic Bulletin* **65**, 647–52.

Uchino, K., Cross, L. E. and Nomura, S. (1980) Inverse hysteresis of field induced elastic deformation in the solid solution 90 mol% $Pb(Mg_{1/3}Nb_{2/3})O_3 - 10$ mol% $PbTiO_3$. *J. Mater. Sci.* **15**, 2643.

Uchino, K., Nomura, S., Cross, L., Newnham, R. and Jang, S. (1981) Review of the electrostrictive effect in perovskites and its transducer applications. *J. Mat. Sci.* **16**, 569–78.

Vorontsov, M. A., Katulin, V. A. and Naumov, A. F. (1989) Wavefront control by an optical-feedback interferometer. *Opt. Commun.* **71**, 35–8.

Winocur, J. (1982) Modal compensation of atmospheric turbulence induced wave front aberrations. *Appl. Opt.* **21**, 433–8.

Wirth, A., Landers, F., Trvalik, B., Navetta, J. and Bruno, T. (1995) A laser guide star atmospheric compensation system for the 3.5 m Calar Alto telescope. In: *Adaptive Optics*, OSA/ESO Conf., Garching-bei-München (Germany), October 2–6, 1995. *Tech. Digest Series* 23, pp. 8–9.

Wizinovitch, P. L., Nelson, J. E., Mast, T. S. and Glecker, A. D. (1994) W. M. Keck Observatory adaptive optics program. In: *Adaptive Optics in Astronomy*. Proc. SPIE 2201, pp. 22–3.

Wu, S. T. (1985) Toward high speed modulation by nematic liquid crystals. Proc. SPIE 567, p. 74.

Yariv, A. (1978) Compensation for optical propagation distortion through phase adaptation by nonlinear techniques (PANT). In: *Adaptive Optics and Short*

Wavelength Sources 6, Physics of Quantum Electronics, Addison-Wesley Pub. New York.

Yariv, A. and Koch, T. L. (1982) One-way coherent imaging through a distorting medium using four-wave mixing. *Opt. Lett.* **7**, 113–15.

Yellin, M. (1976) Using membrane mirrors in adaptive optics. In: *Imaging Through the Atmosphere*. Proc. SPIE 75, p. 97.

5

Wave-front sensors

GÉRARD ROUSSET

Office National d'Études et de Recherches Aérospatiales (ONERA), France

5.1 Introduction

The wave-front sensor (WFS) is one of the basic elements of an adaptive optics system. The requirements are to sense the wave front (WF) with enough spatial resolution and enough speed for real time compensation of atmospheric seeing. The importance of WF sensing is better understood by considering the image formation theory (see Chapter 2). The optical transfer function (OTF) can be derived from the knowledge of the WF in the pupil plane of the instrument. For incoherent light, the OTF $\tilde{S}(\mathbf{r}/\lambda)$ is given by the autocorrelation of the field in the pupil

$$\tilde{S}(\mathbf{r}/\lambda) = P(\mathbf{r})\exp[i\varphi(\mathbf{r})] * P(\mathbf{r})\exp[i\varphi(\mathbf{r})], \qquad (5.1)$$

where $\mathbf{r}$ is the position vector in the pupil plane, λ the observing wavelength, $\mathbf{r}/\lambda$ the angular spatial frequency and $P(\mathbf{r})$ the pupil transmission function (1 inside the aperture, 0 outside). The quantity $\varphi(\mathbf{r})$ is the WF phase and represents the phase shift at wavelength λ introduced by refractive index fluctuations in the atmosphere and by telescope aberrations. Writing the complex field in Eq. (5.1) as $\exp[i\varphi(\mathbf{r})]$ is called the near-field approximation, as it neglects amplitude fluctuations produced by Fresnel diffraction in the upper layers of the atmosphere (Roddier 1981) (see also Chapter 2). Therefore, the OTF can be fully characterized by the measurement of φ. This is of great value when imaging objects through the atmosphere using real time compensation by AO or by *post facto* compensation as in speckle interferometry. The possibility of coupling a WFS to a speckle camera was pointed out by Fontanella (1985).

What are the first requirements a WFS must fulfil in astronomy?

- Measurement quality: the sensitivity and the accuracy are usually specified in terms of fractions of waves λ/X where X may be between 10 and 20 for high image quality (Maréchal and Françon 1970).

- Limiting magnitude: the sensor must work on faint objects. It requires the use of detectors with high quantum efficiency and low noise.
- Incoherent sources: it must work with white light and on extended sources.

5.2 How can the wave front be sensed at optical wavelengths?

It is not possible to directly measure the WF phase at optical wavelengths, as today no existing detector responds at the temporal frequencies involved. In fact, the available optical detectors measure the intensity of the light. Generally, indirect methods must be used to translate information related to the phase into intensity signals to be processed, a well-known technique in optics is interferometry.

5.2.1 Focal plane techniques

A first technique which seems obvious is to derive the phase from the intensity distribution in the focal plane. Indeed, this distribution for a point source and a monochromatic beam is the point-spread function (PSF) which is the Fourier transform of the OTF directly related to φ by Eq. (5.1). For an extended object, it is the convolution of the object by the PSF. The phase estimation is an inverse problem, so-called 'phase retrieval' (Fienup 1982), which is not at all obvious. In general, there is not a unique solution, but multiple measurements and/or *a priori* constraints can be used to ensure the uniqueness. The Gershberg–Saxton iterative algorithm is the basic principle of the inversion, using the relations between PSF, OTF and φ (Gonsalves 1976). New developments have been recently made in order to use a number of intensity distributions encoded by known aberrations: the so-called 'phase diversity' technique (Paxman and Fienup 1988; Paxman *et al.* 1992). The method is illustrated in Fig. 5.1. The phase aberrations are estimated from two simultaneously recorded images. One image is the conventional focal plane image degraded by the unknown WF phase disturbances. The additional image of the same object (diversity image) is formed after reflexion by a beam splitter on a second detector array defocused by a known small amount. The goal of the inversion algorithm is to identify the combination of object and WF phase which are consistent with the data, using the relations between images, PSFs, the known defocus (phase diversity), φ, and the object. This technique has proved to work for extended objects (Paxman and Fienup 1988; Restaino 1992; Kendrick *et al.* 1994). The main drawbacks of the method, when applied in AO for astronomy, are the requirement of a narrow spectral band and the computing burden to reach the convergence of the solution (Kendrick *et al.* 1994).

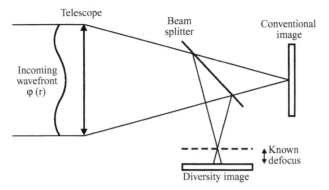

Fig. 5.1. Principle of the phase diversity technique. Adapted from Paxman *et al.* (1992).

Another related method is the pioneering multidither technique which consists in modulating the corrective phase at frequencies much higher than the turbulence frequencies and in finding after synchronous demodulation the phase that maximizes an intensity criterion, e.g. the on-axis intensity. An example is the so-called coherent optical adaptive technique (COAT), developed for atmospheric turbulence compensation in laser beam propagation in the 1970s (O'Meara 1977). The main drawbacks of this method are the need for bright sources (Von der Luhe 1987) and the limited number of channels that can be implemented within the finite bandwidth of a multidither mirror. Therefore, this technique does not seem to be applicable to astronomy.

The major advantage of the focal plane techniques is the direct access to the WF phase from the intensity distribution instead of the phase derivatives, as with the methods presented below. Therefore, no phase reconstruction is required (see Section 5.4). Many developments are still needed to implement such focal plane techiques in AO. Today, phase diversity has been applied to experimental data using only a limited number of degrees of freedom in the WF (Paxman and Fienup 1988; Kendrick *et al.* 1994; Lloyd-Hart *et al.* 1992). No detailed analysis of the noise is available. In particular, the noise limitation in terms of the WF spatial resolution which can be achieved by this technique is not known.

5.2.2 Pupil plane techniques

The most popular techniques of WF sensing in AO are derived from the methods used in optical testing. There are two classes of methods based on either interferometry or geometrical optics concepts. Techniques of the first class use the principle of light beam superposition to form interference fringes coding the phase differences between the two beams, while techniques of the second class use the property that light rays are orthogonal to the WF (Born and Wolf 1980).

In the optical shop, the interferometer is a well-known instrument. The Twyman–Green and Mach–Zehnder interferometers, for instance, are used to measure the aberrations of mirrors or transmissive optical elements (Born and Wolf 1980). The principle is to form an interference pattern between the beam coming from the test object and the beam coming from the reference mirror. In AO, a plane wave reference beam is not available, therefore the beam has to be self-referenced. This is the case for the Smartt point-diffraction interferometer: the reference is generated from a spatially filtered sample of the object beam (Underwood *et al*. 1982). The spatial filtering is made by a transmissive or a reflective pinhole. The main disadvantages of this scheme are the need for a large spatial coherence in the incoming beam and the unequal intensities between the test beam and the reference (only a small part of the incoming light is diffracted by the pinhole) limiting the sensitivity of the method. Angel (1994) proposed a similar scheme with a Mach–Zehnder interferometer in which the spatial filter is progressively reduced in diameter as the loop is closed. With all these interferometers, the phase can be directly determined from the recorded fringe patterns. As for the focal plane technique, this can be an advantage when compared to the devices usually used in AO.

A very powerful approach is the shearing interferometer which makes use of the principle of self-referencing. In a shearing interferometer, the beam is amplitude-divided into two beams which are mutually displaced and super-imposed to produce an interference pattern. An important property of these interferometers is their ability to work with partially coherent light. Several methods for producing sheared WFs are known: rotational shearing, radial shearing and lateral shearing (Armitage and Lohmann 1965; Wyant 1974). We will restrict our discussion to the lateral shearing interferometer (see Section 5.3.1), the most widely used WFSs in AO 15 years ago.

The second class of methods based on geometrical optics concepts are also well known for optical testing. The Foucault test uses a knife edge near focus blocking half the return beam. Any deviation of the light rays from their undisturbed path results in a fluctuation of the amount of light crossing the focal plane (Born and Wolf 1980). The aberrations of the test object are analyzed by recording the intensity distribution in a conjugate pupil plane after the focus. This method has been adapted for fluid flow visualization and is the so-called Schlieren technique. Adapted spatial filters have been proposed instead of the knife edge using the variable transparencies of liquid crystals (Von der Luhe 1988). The main drawbacks are the loss of half of the light and non-linearity problems. But there is no requirement about coherence. Another technique is the Hartmann test. This technique, used for testing telescope mirrors, employs an opaque mask with holes, in front of the optical element under test. Each hole defines a light ray. The analysis is made after the focal plane of the tested mirror in order to record an array of spots. With a proper calibration, the position of each spot is a direct measurement of the local WF tilt experienced by each ray. This technique has been modified by Shack who placed lenses in place of the holes (Shack and Platt 1971). Later, it was adapted to AO by using lenslet arrays allowing 100% light efficiency (Schmutz *et al.* 1979; Fontanella 1985). This technique is widely used today. It is discussed in Section 5.3.2.

The above-mentioned methods (shearing interferometer, Shack–Hartmann sensor) lead to the determination of the angle of arrival of the rays, i.e. the local slope of the WF, that is to say its spatial first derivative (the gradient). Another technique, recently developed, measures the second derivative of the phase, or more precisely its Laplacian (Roddier 1988). This technique is called curvature sensing and is presented in Section 5.3.3. To summarize, the WFSs used in AO do not measure the WF directly, but its gradient or Laplacian. It is therefore necessary to 'reconstruct' the WF from the measurements by appropriate algorithms, i.e. a kind of spatial integration (see Section 5.4). A matrix multiplication must be performed in real time. On the contrary, for the techniques estimating the phase directly, e.g. Smartt interferometer or phase diversity, no matrix multiplication is required in principle and each phase pixel may correspond to a deformable mirror or liquid crystal device actuator. In practice, a matrix multiplication is usually helpful in managing the coupling effects between neighboring actuators.

5.3 The three main wave-front sensors in adaptive optics

5.3.1 The lateral shearing interferometer

5.3.1.1 Principle and signal analysis

The lateral shearing interferometer (LSI) is the most commonly-used interferometer in AO (Hardy *et al.* 1977; Greenwood and Primmerman 1992; Sandler *et al.* 1994). The LSI combines the WF with a shifted version of itself to form interferences. As shown in Fig. 5.2, a shearing device splits the incoming WF into two components and shifts one of them. The two WFs are mutually displaced by a distance **s**, the so-called shear. They interfere in their overlap area. By their position, the interference fringes are a measurement of

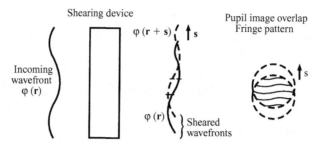

Fig. 5.2. Principle of the lateral shearing interferometer.

the phase difference over the shear distance in the shear direction. In a pupil image plane, the resulting intensity is simply given by

$$I(\mathbf{r}) = \tfrac{1}{2}|\exp[i\varphi(\mathbf{r})] + \exp[i\varphi(\mathbf{r} + \mathbf{s})]|^2$$

$$= 1 + \cos[\varphi(\mathbf{r}) - \varphi(\mathbf{r} + \mathbf{s})]. \qquad (5.2)$$

In principle, the LSI measures the phase differences for a shear $\mathbf{s}$ in the pupil. Let us notice that Eq. (5.2) is a chromatic expression since the phase is inversely proportional to the wavelength. Indeed, the phase is given by

$$\varphi(\mathbf{r}) = \frac{2\pi}{\lambda}\delta(\mathbf{r}), \qquad (5.3)$$

where $\delta(\mathbf{r})$ is the optical path difference (OPD) induced by the atmospheric turbulence, nearly independent of λ (see Chapter 2). If the shear distance is reduced and the WF deformation is small, the phase difference can be expanded in a Taylor series (Wyant 1974; Koliopoulos 1980), e.g. for a shear along the x-direction, we have

$$\varphi(\mathbf{r}) - \varphi(\mathbf{r} + \mathbf{s}) = s\frac{\partial\varphi}{\partial x}(\mathbf{r}) + \epsilon(s), \qquad (5.4)$$

where $\epsilon(s)$ represents the higher order terms of the expansion. In first approximation for signal analysis $\epsilon(s)$ is neglected. Considering now lateral shearing interferometers which use gratings to produce shear, we obtain nearly achromatic fringes for small shears (Wyant 1974) since the shear is proportional to wavelength λ (one beam experiences an angular deviation of λ/p where p is the grating period). The conditions to obtain nearly achromatic properties for these LSIs, are detailed by Wyant (1974) and Koliopoulos (1980). In general, broadband white light results in a contrast reduction of the fringe pattern. In practical applications for turbulence compensation, Eq. (5.4) requires a shear less than r_0 (for the definition of r_0, see Chapter 2). In the same way, the 2π ambiguity in Eq. (5.2) must be removed by the proper choice of s. For an extended source, the fringe contrast is reduced uniformly across the superimposed images of the pupil as stated by the Van Cittert–Zernike theorem (Wyant 1974; Koliopoulos 1980). This accordingly degrades the sensitivity of the LSI. In fact, an optimization of the shear distance s is required once again, depending on the spatial coherence factor of the object μ_{12}. Using the Van Cittert–Zernike theorem we have

$$\mu_{12}(s) = |\tilde{O}(s)|/|\tilde{O}(0)|, \qquad (5.5)$$

where $\tilde{O}$ is the Fourier transform of the brightness distribution in the object O. The quantity μ_{12} is the term reducing the fringe visibility and therefore the signal-to-noise ratio (SNR) of the phase measurement (see Section 5.5). The smaller the value of s, the larger the fringe visibility.

5.3.1.2 Implementation

In order to determine completely the WF, two interferograms having shear in orthogonal directions (x and y) are required. At the entrance of the sensor, the WF is usually beamsplit into two similar channels, one with a x-shear device and the other with a y-shear device. Each channel is equipped with a detector array to measure a map of the WF gradient. Each detector corresponds to an area in the telescope pupil called a subaperture. The two detector planes must be divided into contiguous subapertures for maximum light efficiency. Therefore, the detector array directly determines the spatial sampling of the WF. The area of one detector also provides a spatial filtering of the phase gradients. Finally, we can say that the measurement represents the average slope of the OPD in the shear direction, over each subaperture. To eliminate the need for detector calibration, a classical technique is to use heterodyne modulation (Koliopoulos 1980). For instance, the OPD can be modulated on one 'arm' of the interferometer at a higher frequency than the required control bandwidth. The interference pattern is then modulated and the phase lag of the signal of each detector directly represents the WF slope expressed by Eq. (5.4). A convenient system, developed by ITEK for AO, is the Ronchi grating LSI (Hardy *et al.* 1977; Koliopoulos 1980). It makes use of a rotating radial Ronchi (square-wave) grating placed in a focal plane and diffracting the incoming light. The multiple orders are interfered. The grating rotation generates the fringe modulation at a number of temporal frequencies depending on the order. The fundamental frequency corresponds to the interference between the ± 1 and 0 diffraction orders and is separated out by synchronous detection (Hardy *et al.* 1977). A good feature of this device is its capability to range the shear by changing the grating period p, from a small shear at the outer radius of the disk to a larger shear toward the inner radius. As previously pointed out, optimization of shear with observing conditions is very important in order to obtain the best performance of the LSI. Note that this device was used by the Lincoln Laboratory in the early developments of AO (Greenwood and Primmerman 1992). It has also equipped the compensating imaging system at Air Force Maui Optical Station since 1982 (Hardy, 1993).

Since two channels are required for x-slope and y-slope measurements, the LSI has at least two detectors per subaperture. But for maximum light efficiency, it requires four detectors as indicated by Hardy *et al.* (1977). Moreover, the efficiency is limited to about 70% owing to light losses in higher diffracted orders. The LSI also leads to relatively complex hardware and implementation difficulties. For these reasons the LSI is not used for astronomical applications but is replaced by the Shack–Hartmann WFS.

5.3.2 The Shack–Hartmann wave-front sensor

5.3.2.1 Principle and signal analysis

The principle of the Shack–Hartmann (SH) WFS is presented in Fig. 5.3. A lenslet array is placed in a conjugate pupil plane in order to sample the incoming WF. If the WF is plane, each lenslet forms an image of the source at its focus (Fig. 5.3(a)). If the WF is disturbed, to a first approximation each lenslet receives a tilted WF and forms an off-axis image in its focal plane (Fig. 5.3(b)). The measurement of the image position gives a direct estimate of the

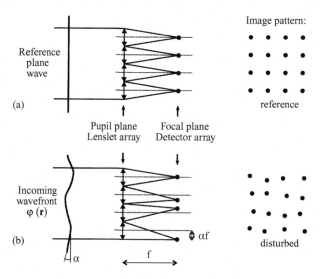

Fig. 5.3. Principle of the Shack–Hartmann wave-front sensor: (a) plane wave, (b) disturbed wave.

angle of arrival of the wave over each lenslet. As for the LSI, a map of WF slopes is obtained on an array of subapertures, here defined by the lenslets. Note that the SH WFS usually requires a reference plane wave generated from a reference source in the instrument, in order to calibrate precisely the focus positions of the lenslet array.

The positions of the SH images formed by the lenslet array can be measured by a number of methods. The simplest technique is to use a four quadrant detector (quad-cell) for each subaperture (Schmutz *et al.* 1979). Another solution is to use a charged-coupled device (CCD) to record all the images simultaneously. The good features of a CCD are that it determines pixel positions perfectly and has a 100% fill-factor. CCDs allow calculation of the center of gravity of the spot for the price of a larger number of pixels per subaperture than quad-cells, but remove the drawbacks of the latter (see below). It is even possible to use CCDs as an array of quad-cells if required.

A number of position-estimators have been studied in the literature for tracking systems (see for instance: Winick 1986; Gerson and Rue 1989). A simple estimation of the center of gravity position (c_x, c_y) is

$$c_x = \frac{\sum_{i,j} x_{i,j} I_{i,j}}{\sum_{i,j} I_{i,j}} \text{ and } c_y = \frac{\sum_{i,j} y_{i,j} I_{i,j}}{\sum_{i,j} I_{i,j}}, \tag{5.6}$$

where $I_{i,j}$ and $(x_{i,j}, y_{i,j})$ are the signal and the position coordinates of the CCD pixel (i, j). The sum is made on all the pixels devoted to a lenslet field. Because of the normalization by $\sum_{i,j} I_{i,j}$, the sensor is relatively insensitive to scintillation. It is possible to show that by replacing the discrete sum by a continuous integral and neglecting the scintillation, Eq. (5.6) exactly determines the average WF slope over the subaperture of area $\mathscr{A}_{sa}$, i.e. the angle of arrival α_x (on the sky)

$$\alpha_x = \frac{c_x}{f \mathscr{M}} = \frac{\lambda}{2\pi \mathscr{A}_{sa}} \int_{\text{subaperture}} \frac{\partial \varphi}{\partial x} \, dx \, dy \tag{5.7}$$

where f is the lenslet focal length and $\mathscr{M}$ the magnification between the lenslet plane and the telescope entrance plane. The same equation can be written for the y-axis. This interpretation is very useful for the design of a WFS because the angle of arrival variance due to turbulence is well known (see Chapter 2). For a circular subaperture of diameter d, the variance is given by Tatarskii (1971)

$$\langle \alpha_x^2 \rangle = 0.98 \frac{6.88}{4\pi^2} \lambda^2 \, d^{-1/3} r_0^{-5/3}. \tag{5.8}$$

Note that $\langle \alpha_x^2 \rangle$ does not depend on the wavelength because r_0 is proportional to

$\lambda^{6/5}$. The dynamics or field of view (FOV) of the subapertures required to measure the turbulence fluctuations can be derived from Eq. (5.8). For a given sampling of the pupil, it determines the lenslet focal length.

For the case of quad-cell detectors assuming small image displacement and image size smaller than quadrant size, it can be shown that the measured angle of arrival α_x is expressed by:

$$\alpha_x = \frac{\theta_b}{2} \frac{I_1 + I_2 - I_3 - I_4}{I_1 + I_2 + I_3 + I_4}, \tag{5.9}$$

where θ_b is the spot size (angular size on the sky) and I_1, I_2, I_3, and I_4 the intensities detected by the four quadrants. Let us underline that to convert position measurement made by the quad-cell into an angle, the image size must be known. It is the case for point sources if the images formed by the lenslet array are diffraction-limited: $\theta_b = \lambda/d$. But when the images are seeing-limited, $\theta_b \simeq \lambda/r_0$, the spot size depends on the seeing conditions and is unknown. For extended sources, the spot size may also be unknown. Hence the quad-cell response is not calibrated which results in an uncertainty in the loop gain of the AO system, for instance. Therefore, the quad-cell response (or the loop gain) has to be calibrated on the source images themselves during observation. Another way is to defocus the star images on the quad-cells in order to keep a constant known spot size but this reduces the SNR (see Section 5.5). In addition to the spot-size dependent response, the main drawbacks of quad-cell detectors are generally a limited dynamic range and a non-linear response (Ma et al. 1989; Gerson and Rue 1989).

An important consequence of Eqs. (5.6) to (5.8) is that the SH sensor is achromatic because the OPD in the turbulence is achromatic: it works perfectly well with broadband white light. This is an important feature of the sensor. A second consequence is that they work with extended sources. Since this sensor is in fact a multiple imaging system, it can operate with an extended source if the FOV is adapted to the source size (Fontanella 1985; Rousset et al. 1987). Very extended sources, like the solar surface, have already been used for SH WF measurement and turbulence compensation (Title et al. 1987; Acton and Smithson 1992) (see also Chapter 10). It requires the use of a field stop and a correlation-based position estimator (Rousset et al. 1987; Title et al. 1987; Michau et al. 1992). A minimum contrast in the scene is needed to allow the measurement (Michau et al. 1992). As for phase diversity, since the SH CCD works in the image plane (lenslet focal plane), all the parameters of this plane can be obtained (Fontanella 1985; Rousset et al. 1987). In particular, field dependent WF estimates can be made in principle in the case of anisoplanatic imaging sytems. The field of the subapertures can be divided in subfields of the isoplanatic patch size and one WF can be determined within each subfield using all the subapertures (Rousset *et al.* 1987).

Recently, Roddier has discussed a general approach to the Hartmann sensor, which has led to consideration of this sensor as being an achromatic shearing interferometer (Roddier 1990a,b). He has proposed the use of the Fourier transform to process the array of images extending the spatial resolution and the dynamics of the method. Based on these principles, new WFS concepts are under investigation (Roddier and Roddier 1991a; Primot 1993; Cannon 1995).

5.3.2.2 *Implementation*

A good feature of the SH sensor is the simultaneous determination of the *x*- and *y*-slopes by the measurement of the image position *x*- and *y*-coordinates (cf. Eq. (5.6)). Only one channel per subaperture is required *a priori*. The drawbacks of the SH are possible misalignment problems and the calibration precision. Since there is no modulation, it could also be drift sensitive. These drawbacks are usually overcome by the sensor itself being very compact and also by the use of a sufficiently accurate plane wave reference calibration. The lenslet array can be made small enough to fit exactly the CCD size with no relay optics. The optical elements in front of the CCD can be as compact as a classical camera objective, e.g. as in the design proposed by Fontanella (1985). A field stop at the telescope focus, matching the CCD non-overlapping area allotted to each lenslet, is required for sky background reduction and for extended source imaging. This design is used in the COME ON systems (Rousset *et al.* 1990) and works well on extended astronomical sources, like Eta Carinae for instance (Rigaut *et al.* 1991). Different types of design can be found in the literature (Noethe *et al.* 1984; Allen *et al.* 1987; Acton and Smithson 1992; Barclay *et al.* 1992).

How many pixels are needed in a lenslet FOV and in the formed image? It depends on the observing conditions! A very few pixels are often enough in the image size, and a few more for the dynamics (FOV). Let us recall that the number of pixels and the lenslet focal length are parameters linked to subaperture diameter and wavelength. The key issue is often the SNR to achieve (see Section 5.5). Finally, centroiding algorithms usually use threshold-ing and/or windowing to process only the useful pixels. A drawback of the SH sensor could be the computing burden for centroiding when using a large number of pixels per subaperture. It is now easily overcome with the fast processors available on the market. In addition, for systems with a large number of degrees of freedom, this problem is less constraining than that of the command computer. Indeed, the computing power increases as N for the centroiding, where N is the number of subapertures (or degrees of freedom), while for the command computer, it increases as N^2.

To sum up, since the SH sensor directly measures the angles of arrival, it works very well with incoherent white light extended sources. The required number of detectors is a minimum of four per subaperture. In principle, it can work on anisoplanatic fields of view, as needed in the multi laser guide star scheme for instance (see Part 4). SH sensors have already been used in AO systems that have a large number of degrees of freedom (Fugate *et al.* 1991; Primmerman *et al.* 1991; Rousset *et al.* 1994).

5.3.3 The curvature sensor

5.3.3.1 Principle and signal analysis

The curvature sensor (CS) has been proposed and developed by Roddier (1988) to make WF curvature measurements instead of WF slope measurements. The Laplacian of the WF, together with WF radial tilts at the aperture edge, are measured, providing data to reconstruct the WF by solving the Poisson equation with the Neumann boundary conditions. An interesting feature of this approach is that a membrane or a bimorph mirror can be used directly to solve the differential equation, because of their mechanical behaviour, *a priori* removing any matrix multiplication in the feedback loop (Roddier 1988). The principle of this sensor is presented in Fig. 5.4. The telescope of focal length f images the source in its focal plane. The CS consists of two detector arrays placed out of focus. The first detector array records the irradiance distribution in plane P_1 at a distance l before the focal plane. The second records the irradiance distribution in plane P_2 at the same distance l behind the focus. A local WF curvature in the pupil produces an excess of illumination in one plane, for instance and a lack of illumination in the other (Fig. 5.4). A field lens is used for symmetry in order to reimage the pupil. The planes P_1 and P_2 can also be

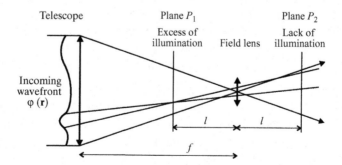

Fig. 5.4. Principle of the curvature sensor (adapted from Roddier (1988)).

seen as two defocused pupil planes. It can be shown that in the geometrical optics approximation, the difference between the two plane irradiance distributions is a measurement of the local WF curvature inside the beam and of the WF radial first derivative at the edge of the beam (Roddier 1987). The measured signal is the normalized difference between the illuminations $I_1(\mathbf{r})$ and $I_2(-\mathbf{r})$ in planes P_1 and P_2, and is related to the WF phase φ in the pupil plane by

$$\frac{I_1(\mathbf{r}) - I_2(-\mathbf{r})}{I_1(\mathbf{r}) + I_2(-\mathbf{r})} = \frac{\lambda f(f-l)}{2\pi l} \left[\frac{\partial \varphi}{\partial n} \left(\frac{f\mathbf{r}}{l} \right) \delta_c - \nabla^2 \varphi \left(\frac{f\mathbf{r}}{l} \right) \right], \tag{5.10}$$

where the quantity $\partial \varphi / \partial n$ is the radial first derivative of the WF at the edge, δ_c a linear impulse distribution around the pupil edge, and ∇^2 the Laplacian operator. In fact, Eq. (5.10) is the irradiance transport equation valid for paraxial beam propagation (Teague 1983; Streibl 1984). This equation provides a general description of the incoherent WF sensing methods (Roddier 1990b). Note that phase retrieval has been demonstrated using the irradiance transport equation (Teague 1983; Streibl 1984; Ichikawa *et al.* 1988). Let us remark that in Eq. (5.10) the normalization by $I_1(\mathbf{r}) + I_2(-\mathbf{r})$ yields to a sensor relatively insensitive to scintillation. In principle, the sensor is achromatic and, as we shall see, works with extended sources.

In order to understand the different conditions of use of the CS, we now discuss the choice of l. The distance l must be such that the validity of Eq. (5.10), i.e. the validity of the geometrical optics approximation, is ensured. The condition to be verified requires that the blur produced at the position of the defocused pupil image (in P_1 and P_2) must be small compared to the size of the WF fluctuations we want to measure, to avoid any smearing of the intensity variations. The size of the WF fluctuations is the subaperture size. For turbulence, the subaperture size may be of the order of r_0. Here, as for LSI, the subaperture array is directly defined by the detector array. Let d be the subaperture size. The detector size in P_1 and P_2 is then equal to ld/f (see Fig. 5.4). Denoting the blur angle θ_b, the geometrical optics approximation can be expressed by the following condition:

$$(f-l)\theta_b \leqslant ld/f. \tag{5.11}$$

The distance l is then given by

$$l \geqslant f[1 + d/(f\theta_b)]^{-1}. \tag{5.12}$$

For a very large blur angle, l tends to f and the measurement must be made in the pupil plane. However, the blur angle is generally small when compared to the inverse of the f-ratio of a subaperture outgoing beam d/f and l is given by

$$l \geq \theta_b \frac{f^2}{d}.$$

(5.13)

Let us now consider some typical cases.

- Point source, subaperture size $d > r_0$:
  Here, only the low order aberrations are measured. The blur angle is given by the turbulence: λ/r_0. Using Eq. (5.13), the condition is

$$l \geq \frac{\lambda f^2}{r_0 d}.$$

(5.14)

- Point source, subaperture size $d < r_0$:
  High order aberrations of spatial scale d must be measured. These aberrations diffract light over an angle λ/d. The blur angle to consider is no longer equal to λ/r_0. The condition is

$$l \geq \frac{\lambda f^2}{d^2}.$$

(5.15)

Let us note that this condition is very similar to the near-field approximation condition. For broadband white light, such condition defines the domain of validity of geometrical optics: $L \leq d^2/\lambda$ (in parallel beam), where L is the so-called Fresnel distance along the optical axis from the diffracting plane and λ a mean wavelength. Equation (5.11) is very similar but for a converging beam. Here the distance L is equal to $(f - l)f/l$, the distance of the pupil plane to the conjugate measurement planes on the incoming beam. We conclude that Eq. (5.11) is a near-field approximation condition and the diffraction effects can be neglected.

- Extended source of angular size $\theta > \lambda/r_0$:
  Here, the blur is given by the size of the source. We obtain

$$l \geq \theta \frac{f^2}{d}.$$

(5.16)

What can be derived from Eqs. (5.14) to (5.16)? First for high order aberration measurements, the distance l to focus must be larger than that for only low order aberration measurements. For extended sources, l must also be larger than for point sources. An increase of l means a decrease of the sensitivity (but an increase of the dynamics) of the CS as expressed by Eq. (5.10). The CS signal is proportional to l^{-1}. The distance l of the CS is very similar to the lenslet focal length of the SH sensor. Let us notice that when the distance l is decreased to the minimum, the CS is only able to measure tilts and can be reduced to a quad-cell. In this limiting case, the CS provides four edge measurements and no curvature.

Let us discuss the operation of CS in AO. Consider the case where $d > r_0$, as is most commonly encountered in astronomy. Before closing the loop, the distance l is imposed by condition (5.14) because of the large WF disturbances. But once the loop is closed, all low order aberrations are corrected. Therefore, the blur angle is then smaller and l can be reduced (Equation (5.15)), increasing the sensitivity of the sensor (Roddier 1995). To summarize, increasing the distance l increases spatial resolution on the WF measurement but decreases sensitivity. On the contrary, a smaller distance yields a higher sensitivity to low order aberrations. Note also that a smaller distance reduces the aliasing of the high order aberrations on the low order ones because of the diffraction effect. The use of CS for very extended sources, like the sun, is questionable when looking at Eq. (5.16). However, new data analysis is under investigation for such applications (Kupke *et al.* 1994). Then, the technique becomes similar to the phase retrieval.

5.3.3.2 Implementation

The setup proposed by Roddier *et al.* (1991) uses a variable curvature mirror placed at the focus of the telescope as a field lens. The inside and outside focus blurred pupil images can be reimaged on the same detector array by its concave or convex deformation. This produces a modulation of the illumination on the detector array. The signals are recovered by synchronous detection. The pixels inside the beam measure the local curvatures, the pixels on the edge of the beam the local WF slopes. The modulation frequency corresponds to the temporal sampling frequency of the WF and the deformation amplitude of the variable curvature mirror directly determines the distance l. A good feature of this device is its capability to modify the sensitivity of the sensor easily by changing the amplitude of the mirror vibration (i.e. l). This can be done in closed loop. Other set-ups have been proposed by Mertz (1990) and Forbes and Roddier (1991). Because of the low number of subapertures (and detectors) in their sensor, Roddier and his coworkers use photon-counting avalanche photodiodes (APD) as detectors, taking advantage of their high quantum efficiency and negligible electronic noise. For other applications such as testing of a ground-based optical telescope, CCD cameras are used to provide enough spatial resolution in the WF (Roddier and Roddier 1993).

To sum up, the CS works very well with incoherent white light. In principle, for curvature sensing only two detectors, one per measurement plane, are required per subaperture. Even if only one detector is used in practice, the light is split by the temporal modulation between the two measurement planes. For the SNR conditions, this is equivalent to splitting the light between two detectors. Finally, sensitivity and dynamics are easily adjusted by the distance l.

5.3.4 Comparison of slope and curvature sensors

In summary, two classes of WFS are available for AO: the slope sensors and the CS. As shown in Sections 5.3.1.1 and 5.3.2.1, both LSI and SH measure the WF slopes. Table 5.1 summarizes the characteristics of these two classes. A $k^{-11/3}$ spatial power spectrum of the phase in the turbulence is assumed. Note that all measurements are averaged over the subaperture area, this additional filtering is not taken into account in Table 5.1. A first important difference is related to the spatial spectrum of the measured quantities. The slope measurements with a power spectrum in $k^{-5/3}$ have a relatively large correlation length over the pupil. The slope sensors are more sensitive to low spatial frequencies. The power law also induces low aliasing. On the contrary, the curvature measurements with a power spectrum in $k^{1/3}$ are mainly decorrelated at two different points in the pupil (Roddier 1988). The CS has an equal sensitivity to all spatial frequencies. The power law induces aliasing of the high spatial frequencies but partly attenuated by the diffraction blur. As already underlined, the distance l must be adjusted to select the spatial resolution of interest. But also it may limit the aliasing effect. A decrease of l increases the blur by diffraction and reduces the aliasing of the high spatial frequencies. The variance of the slope is slowly varying with the subaperture diameter ($d^{-1/3}$) and is attenuated by the outer-scale of the turbulence L_0 (Fante 1975). On the contrary, the variance of the curvature strongly depends on the subaperture diameter ($d^{-7/3}$) (Roddier *et al.* 1990) and the effect of outer-scale is negligible. Because of the sensitivity of the CS to high spatial frequencies, the variance of the curvature could be affected by the inner-scale of the turbulence for small subapertures.

The temporal behaviour of the measured quantities is also very different. Once again, because of the importance of low temporal frequencies on the slope (spectrum in $f^{-2/3}$), its measurement presents a large correlation time and a low aliasing (spectrum in $f^{-11/3}$ at high frequency (Hogge and Butts 1976)). For curvature, the correlation time is much smaller and the aliasing is slightly higher: power spectrum in f^0 at low frequency and in $f^{-5/3}$ at high frequency (Conan *et al.* 1995). To sum up, it is clear that slope measurements are well adapted to the measurement of low spatial frequencies, with large correlation and low aliasing effects. On the contrary, curvature measurements are less redundant (small correlation) but present

Table 5.1. *Characteristics of slope and curvature sensor measurements: over-line denotes spatially average quantity, k spatial frequency, f temporal frequency*

Wave-front sensor		LSI and SH	CS
Measurements		$\partial\varphi/\partial x, \partial\varphi/\partial y$	$\overline{\nabla^2\varphi}, \overline{\partial\varphi/\partial n}$
Detectors per subaperture		4 or more	2 or less
Spatial behaviour	Turbulence spectrum	$k^{-5/3}$	$k^{1/3}$
	Correlation	relatively large	decorrelated
	Aliasing	low	relatively high
	Sensitivity	low spatial frequencies	all frequencies
	Variance	$d^{-1/3}(1-(d/L_0)^{1/3})$	$d^{-7/3}$
Temporal behaviour	Low frequency	$f^{-2/3}$	f^0
	High frequency	$f^{-11/3}$	$f^{-5/3}$
	Correlation	large	small
	Aliasing	low	slightly higher

higher aliasing. However, the aliasing effects can be reduced using a small distance *l* which attenuates high spatial frequencies by blurring. Computer simulations are necessary to find the proper specifications of a CS (Rigaut 1992; Rigaut *et al.* 1997). The comparison of these sensors in terms of SNR will be discussed in Section 5.5.

5.4 Wave-front reconstruction

For the three WFSs presented in the previous section, a reconstruction of the WF from the measurements is required. The general problem is the determination of the WF phase from a map of its gradient or Laplacian. It consists of the calculation of a surface by an integration-like algorithm. Even if in AO systems the WF phase itself is not explicitly sought, the same kind of problem must be solved for the derivation of the commands to be applied to the deformable mirror. The reconstruction problem can be expressed in a matrix-algebra framework. The unknowns, a vector ϕ of N commands or of N phase values over a grid, must be calculated from the data, a measurement vector $\mathbf{S}$ of M elements of slopes in two directions or Laplacians and edge slopes. The following general linear relation must be obtained:

$$\phi = \mathbf{BS}, \tag{5.17}$$

where $\mathbf{B}$ is the so-called reconstruction matrix (or command matrix). The subject of this section is how to derive the matrix $\mathbf{B}$. In AO systems, $\mathbf{S}$ is the error signal and ϕ is an increment of commands which slightly modifies the previous actuator state: this is the closed-loop operation (see Chapter 6). The determination of the phase is also of interest in open-loop, as for turbulence characterization or post-processing techniques (Primot *et al.* 1990). ϕ is then the total phase. A number of techniques are available to derive **B**. Two classes are well identified in the literature: the zonal methods and the modal methods (Southwell 1980). The matrix **B** is also dependent on the chosen minimization criterion. Usually, there are more measurements than unknowns, i.e. $M > N$, and a least-square fit is performed. More generally, the WF reconstruction problem is an inverse problem: how to estimate the WF phase (or commands) from the set of measured data? Firstly, linear relations can be written between data and unknown phases: the sensor model. Secondly, statistical properties of phase and measured data can be theoretically known or experimentally assessed. However, although turbulence can be considered as stationary in the short term (a few minutes), it is not true in the longer term: this brings additional difficulties.

5.4.1 Zonal and modal approaches

5.4.1.1 The zonal methods

In the zonal methods, the phase is determined on a discrete set of points (which can be the actuators themselves) distributed over the telescope aperture. A linear model of the WFS allows the linking of the measurements **S** to the incoming phase. The matrix equation between **S** and ϕ reads as

$$\mathbf{S} = \mathbf{A}\boldsymbol{\phi}. \tag{5.18}$$

What is the matrix **A** of $N \times M$ elements? For a given sensor discretization, the finite difference form can be used to express the slopes (the gradients) or Laplacians in terms of discrete phase values on a grid. Considering the SH with square subapertures for instance, Fried (1977) has proposed this model:

$$\mathbf{S}_{i,j}^{x} = [(\phi_{i+1,j+1} + \phi_{i+1,j}) - (\phi_{i,j} + \phi_{i,j+1})]/2d \tag{5.19}$$

$$\mathbf{S}_{i,j}^{y} = [(\phi_{i+1,j+1} + \phi_{i,j+1}) - (\phi_{i,j} + \phi_{i+1,j})]/2d, \tag{5.20}$$

where the $\phi_{i,j}$'s are the phase values at the four corners of the subaperture and $S_{i,j}^{x}$ and $S_{i,j}^{y}$ the measured slopes. The elements of **A** are $\pm(2d)^{-1}$ or 0. A similar model has been proposed by Hudgin (1977) for the LSI. In the case of a deformable mirror, the matrix **A** can be simply measured. The columns of **A** are the measurement vectors associated with each actuator when applying an unitary voltage to it and keeping all the others to zero. **A** is called the interaction matrix (see Chapter 6).

5.4.1.2 The least-square solution

This technique consists of the minimization of the measurement error ϵ_s and it is well adapted for a closed-loop operation. The measurement error is given by:

$$\epsilon_s = \|\mathbf{S} - \mathbf{A}\boldsymbol{\phi}\|^2, \tag{5.21}$$

where $\|\ \|$ is the norm of a vector. The WF phase $\boldsymbol{\phi}$ is estimated so that it minimizes ϵ_s. The least-square solution verifies

$$(\mathbf{A}^t\mathbf{A})\boldsymbol{\phi} = \mathbf{A}^t\mathbf{S}, \tag{5.22}$$

where $\mathbf{A}^t$ is the transpose of $\mathbf{A}$. Often, the standard solution cannot be used because $\mathbf{A}^t\mathbf{A}$ is singular. This is a consequence of the fact that the phase is determined only up to a constant by its derivatives. The WFS is unsensitive to a WF constant over the aperture, the so-called piston mode.

- Iterative methods have been proposed to solve Eq. (5.22). The solution is obtained to within an additive constant (Fried 1977; Hudgin 1977; Southwell 1980).
- Herrmann (1980) has shown that the best solution is the one with minimum norm, which has zero mean. The zero mean condition can be applied in different ways and yields a non-singular matrix (Ben-Israel and Greville 1980; Boyer *et al.* 1990) (see also Chapter 6).

It has been shown by Noll (1978) and Herrmann (1980) that Eq. (5.22) is in fact a discrete expression of the Poisson equation with the Neumann boundary conditions, even for the slope sensors. Therefore, all numerical techniques available for solving the Poisson equation can be used. Note that in the zonal approach, the WF phase itself is never derived, only discrete values are calculated. For instance, as the influence functions of the deformable mirror actuators are not explicitly used in the derivation of the command matrix $\mathbf{B}$, they may be unknown. But they are fully taken into account by the measured interaction matrix $\mathbf{A}$. The deformable mirror itself actually synthesizes the correction WF phase.

5.4.1.3 The modal methods

In the modal methods, the phase is represented by the coefficients of expansion in a set of basis functions Z_i, called modes. The reconstruction first calculates a vector of coefficients $\boldsymbol{\phi} = \{\phi_i\}$ using a relation similar to Eq. (5.17). Then the phase can be computed anywhere in the aperture by

$$\varphi(\mathbf{r}) = \sum_i \phi_i Z_i(\mathbf{r}), \tag{5.23}$$

where the sum is made from $i = 1$ to N where N is the number of modes in the expansion. Different sets of basis functions can be selected according to the need. For example, the deformable mirror influence functions or the system modes are commonly-used sets (see Chapter 6). In the modal approach, the interaction matrix $\mathbf{A}$ is usually calculated using the analytic expression of the modes $Z_i(\mathbf{r})$. For example, the two elements of $\mathbf{A}$ for the subaperture j and the mode i for a SH WFS are given by

$$A_{ij}^x = \frac{1}{\mathscr{A}_{sa}} \int_{\text{subaperture } j} \frac{\partial Z_i(\mathbf{r})}{\partial x} \, d\mathbf{r} \text{ and } A_{ij}^y = \frac{1}{\mathscr{A}_{sa}} \int_{\text{subaperture } j} \frac{\partial Z_i(\mathbf{r})}{\partial y} \, d\mathbf{r}. \quad (5.24)$$

As for the zonal approach Eq. (5.17) can be derived using the least-square technique. It is generally possible to avoid the singularity of $\mathbf{A}^t\mathbf{A}$ in Eq. (5.22). Indeed, the piston mode is always part of the basis ($i = 1$). It can be simply discarded in the expansion, only considering a subspace of the solution space. Therefore in matrix $\mathbf{A}$, the index i only varies from 2 to N. A number of sets of basis functions have been proposed in the literature. The Zernike polynomials are a well-known basis (Noll 1976), but the Karhunen–Loève modes are theoretically the optimum set of functions considering the turbulence statistics (Wang and Markey 1978) (see Chapter 2). From a practical point of view, it is not usually possible to generate these modes (Zernike or Karhunen–Loève) with the deformable mirror. Therefore, the mechanical mirror modes are of interest and must be orthogonalized through phase variance minimization within the aperture (Gaffard and Ledanois 1991). Diagonalizing their turbulence covariance matrix allows us to work with modified Karhunen–Loève-type mirror modes (Gendron 1993). In case of a square sampling of the telescope aperture, it may be useful to work with the complex exponentials as a set of basis functions (Freishlad and Koliopoulos 1986) and a Fast Fourier transform can be applied (Roddier and Roddier 1991b; Marais *et al.*, 1991).

How many modes can we determine with a given WFS? First note that the array of M_s subapertures defines a spatial sampling of the WF phase and by Fourier transform, any spectrum is expanded in a set of M_s complex exponentials. For slope WFS: M measurements are obtained from $M_s = M/2$ subapertures. The phase spectrum must be estimated from the two spectra of slopes in x- and y-directions which are redundant in fact. Indeed, the spectrum of the slopes in x- (or y-) direction is only k_x (or k_y) times the phase spectrum (spatial frequency $\mathbf{k} = (k_x, k_y)$). In slope WFS with twice as many measurements as subapertures, there is *a priori* just redundant information. In CS, the number of measurements M is equal to the number of subapertures M_s. Therefore whatever the WFS, the maximum number of modes $N_{\max}$ which can be determined from the measurements is of the order of the number of subapertures M_s, i.e. the number of degrees of freedom of the AO system.

5.4.2 Wave-front reconstruction as an inverse problem

5.4.2.1 The inverse problem

Such an approach has already been considered by several authors (Wallner 1983; Sasiela and Mooney 1985; Cho and Petersen 1989; Downie and Goodman 1989; Fried 1993; Bakut *et al.* 1994). A synthetic view of the problem is proposed here. The basic relation between the measured data $\mathbf{S}$ and the unknowns $\boldsymbol{\phi}$ is given by

$$\mathbf{S} = \mathbf{A}\boldsymbol{\phi} + \mathbf{n}, \tag{5.25}$$

where $\mathbf{n}$ is an additive noise vector of zero mean. The problem can be stated as follows: given the set of data $\mathbf{S}$, estimate the unknowns $\boldsymbol{\phi}$, but here considering *a priori* knowledges about $\mathbf{S}$ and $\boldsymbol{\phi}$. Note that $\boldsymbol{\phi}$ may be the vector of actuator commands, or the zonal values of the phase, or the modal expansion coefficients.

- Firstly, the noise can be assumed to be uncorrelated with the unknowns and to have Gaussian statistics. In fact, this statistic results from the complex combination of a number of random variables representing the detection noise which is Gaussian in case of electronic noise and/or Poissonian in case of signal and background photon noise. The noise statistics can be quantified from theoretical derivations (see Section 5.5), or better, for practical systems directly from the data set.
- Secondly, the statistics of the vector $\boldsymbol{\phi}$ is Gaussian and has a zero mean. It can also be quantified theoretically from well-known turbulence properties (see Chapter 2) or even experimentally when necessary.

To derive the 'best' estimation of the phase, the theory of the inverse problems provides us with two points of view: the maximum likelihood technique and the maximum *a posteriori* technique (Katsaggelos 1991).

The essence of the maximum likelihood technique is to determine the set of unknown parameters that maximizes the probability of producing the measurements. The likelihood function is directly given by the probability function $P(\mathbf{S}|\boldsymbol{\phi})$ governing data $\mathbf{S}$ given the phase $\boldsymbol{\phi}$:

$$P(\mathbf{S}|\boldsymbol{\phi}) = \frac{1}{\sqrt{(2\pi)^M |\mathbf{C}_n|}} \exp(-(\mathbf{S} - \mathbf{A}\boldsymbol{\phi})^{\mathrm{t}} \mathbf{C}_n^{-1} (\mathbf{S} - \mathbf{A}\boldsymbol{\phi})/2), \tag{5.26}$$

where $\mathbf{C}_n$ is the covariance matrix of the noise $\mathbf{n}$ and $|\mathbf{C}_n|$ its determinant. Maximizing the likelihood function is the same as maximizing its logarithm:

$$\frac{\partial}{\partial \boldsymbol{\phi}} \ln(P(\mathbf{S}|\boldsymbol{\phi})) = 0. \tag{5.27}$$

It yields to the generalized least-square solution first given by Sasiela and Mooney (1985):

$$\phi = (\mathbf{A}^t \mathbf{C}_n^{-1} \mathbf{A})^{-1} \mathbf{A}^t \mathbf{C}_n^{-1} \mathbf{S} \tag{5.28}$$

where the inversibility of $\mathbf{A}^t \mathbf{C}_n^{-1} \mathbf{A}$ must be checked in any case. The covariance matrix $\mathbf{C}_n$ can be diagonal if there is no correlation between the subaperture measurements. In addition if we assume that the noise variance is uniform on all the subapertures, Eq. (5.28) reduces to Eq. (5.22) already discussed. In practical systems, the correlated noise must be analyzed carefully when using photodiode arrays or CCDs because it may significantly degrade the performance. The uniform variance of the noise is not always observed since the pupil boundary subapertures are often truncated and therefore collect less photons. For a large system, the number of truncated subapertures is usually small.

Let us now consider the second point of view. The maximum *a posteriori* technique consists in the determination of the set of unknowns that maximizes the *a posteriori* probability of the unknowns $P(\phi|\mathbf{S})$ given both the measured data and *a priori* background knowledge about ϕ. From the Bayes' theorem we can write

$$P(\phi|\mathbf{S}) = \frac{P(\phi)P(\mathbf{S}|\phi)}{P(\mathbf{S})}, \tag{5.29}$$

where the *a priori* knowledge is the statistics of ϕ. Note that since the data are known $P(\mathbf{S}) = 1$. Therefore, we have

$$P(\phi|\mathbf{S}) = \frac{1}{\sqrt{(2\pi)^{M+N}|\mathbf{C}_n||\mathbf{C}_\phi|}} \exp(-(\phi^t \mathbf{C}_\phi^{-1} \phi + (\mathbf{S} - \mathbf{A}\phi)^t \mathbf{C}_n^{-1}(\mathbf{S} - \mathbf{A}\phi))/2),$$

$$\tag{5.30}$$

where $\mathbf{C}_\phi$ is the covariance matrix of the vector ϕ given by the statistics of the turbulence. Minimizing the logarithm of the *a posteriori* probability $P(\phi|\mathbf{S})$, it yields to (Sasiela and Mooney 1985; Fried 1993)

$$\phi = (\mathbf{A}^t \mathbf{C}_n^{-1} \mathbf{A} + \mathbf{C}_\phi^{-1})^{-1} \mathbf{A}^t \mathbf{C}_n^{-1} \mathbf{S}, \tag{5.31}$$

which is a Wiener-type solution (Rousset 1993). Let us note that if the unknowns are the Karhunen–Loève coefficients, the covariance matrix $\mathbf{C}_\phi$ is diagonal. Moreover, the coefficient variances (diagonal elements) decrease with the order of the modes following the turbulence power spectrum in $k^{-11/3}$. For the actuator commands, the variances are generally uniform but the covariance matrix is no longer diagonal because of the spatial correlation of the WF. If $\mathbf{C}_n$ is assumed to be diagonal and equal to $\sigma_s^2 \mathbf{I}$ where $\mathbf{I}$ is the matrix identity and σ_s^2 the noise variance in $\mathbf{S}$ (see Section 5.5), Eq. (5.31) can be rewritten

$$\phi = (\mathbf{A}^t \mathbf{A} + \sigma_s^2 \mathbf{C}_\phi^{-1})^{-1} \mathbf{A}^t \mathbf{S} \tag{5.32}$$

Observe that Eq. (5.32) reduces to Eq. (5.22) if σ_s^2 is very small compared to the turbulent variances of ϕ, i.e. in the case of high SNR. Let us note that with such techniques, it is possible to consider larger vector ϕ, i.e. $N > M$, by computing from *a priori* knowledge the matrices $\mathbf{A}$ and $\mathbf{C}_\phi$. The limit value for N is given by the non-invertibility of $\mathbf{A}^t\mathbf{C}_n^{-1}\mathbf{A} + \mathbf{C}_\phi^{-1}$. In closed-loop mode, note that the matrices $\mathbf{C}_n$ and $\mathbf{C}_\phi$ can not be calculated theoretically with the conventional statistical properties of the noise and the turbulence valid only in open loop. Finally in practical systems, the main problem encountered is the difficulty in obtaining reliable knowledge of the statistics of both noise and turbulence, in particular the non-stationarity of the turbulence.

5.4.2.2 Wave-front residual variance minimization

In optics, there is a well-known optimization criterion: the WF error spatial variance minimization. Indeed, minimizing the WF error is maximizing the Strehl ratio in the images (Born and Wolf 1980). Noting $\hat{\varphi}$ the WF correction produced by the deformable mirror (or the estimated WF), we want to minimize in a statistical average the residual WF error variance ϵ_φ over the telescope aperture (of area $\mathscr{A}_{ap}$) (Wallner 1983):

$$\epsilon_\varphi = \frac{1}{\mathscr{A}_{ap}} \int_{aperture} \langle (\varphi(\mathbf{r}) - \hat{\varphi}(\mathbf{r}))^2 \rangle \, d\mathbf{r}, \tag{5.33}$$

where $\langle \rangle$ denotes the ensemble average. Consider the influence functions $f_i(\mathbf{r})$ of the actuators as the basis functions (any other set of modes could be considered). The estimated phase is given by the linear combination

$$\hat{\varphi}(\mathbf{r}) = \sum_i \phi_i f_i(\mathbf{r}), \tag{5.34}$$

where the vector $\phi = \{\phi_i\}$ is given by Eq. (5.17). The minimization, first done by Wallner (1983), consists of the determination of the matrix coefficients B_{ij} which minimize ϵ_φ, taking into account the statistics of noise and turbulence. Using the already defined matrices, it leads to

$$\mathbf{B} = \mathbf{C}_f^{-1}\mathbf{C}_\phi\mathbf{A}^t(\mathbf{A}\mathbf{C}_\phi\mathbf{A}^t + \mathbf{C}_n)^{-1}, \tag{5.35}$$

where $\mathbf{C}_f$ is the $N \times N$ matrix of the scalar products of the actuator influence functions over the aperture. We would like here to underline the similitude between the result of Eq. (5.35) and the result given by Eq. (5.31). It was first done for the Zernike polynomials by Law and Lane (1996). For an orthonormal function basis, the matrix $\mathbf{C}_f$ reduces to $\mathbf{I}$. In addition it can be shown Eq. (5.35) is identical to Eq. (5.31). That demonstrates the similarity of the two points of view given by the inverse problem theory and the WF variance criterion. This criterion has also been used to determine the optimal influence functions of the mirror, or more generally the optimal set of modes to reconstruct the phase (Cho and Petersen 1989). It was also applied to a segmented mirror control in closed-loop, using Zernike polynomials as a set of basis functions (Downie and Goodman 1989).

5.5 Wave-front errors

5.5.1 Measurement noise

Two kinds of fundamental noise can be considered: the signal photon noise and the sky background photon noise. In addition, the detector may bring some supplementary noise due to the dark current and the read-out electronics, called hereafter the electronic noise. From the literature (Wyant 1975; Fontanella 1985; Roddier *et al.* 1988), one can see that all WFSs have similar behaviour in terms of signal photon noise (Rousset 1993). In this chapter for any WFS type, the measurement noise variance is denoted σ_s^2 and expressed in WF error units (radians of phase). The general form of σ_s^2 due to signal photon noise is (see Eq. (3.60)):

$$\sigma_s^2 \propto \frac{1}{n_{\mathrm{ph}}} \left(\frac{\theta_b d}{\lambda} \right)^2 \text{(radian}^2\text{)}, \tag{5.36}$$

where n_{ph} is the number of photoelectrons per subaperture and exposure time, θ_b the angular size of the source image, and d the subaperture diameter. The SNR in the subaperture is given by $n_{\mathrm{ph}}^{1/2}$. Let us underline that σ_s^2 is proportional to the inverse of the square of the SNR and to the ratio of two angles: the blur angle θ_b and the subaperture diffraction angle λ/d. θ_b characterizes the observing conditions of the WFS, it is equal to λ/d for the diffraction-limited case and λ/r_0 for the seeing-limited case for an unresolved source, or to θ the angular size of a resolved source.

5.5.1.1 Noise in LSI

Wyant (1975) derived this expression:

$$\sigma_s^2 \propto \frac{1}{n_{\mathrm{ph}}} \left(\frac{d}{s\mu_{12}} \right)^2 \text{(radian}^2\text{)}, \tag{5.37}$$

where μ_{12} is the fringe visibility determined by the coherence of the source (Eq. (5.5)). For an extended object, the visibility varies as $\lambda/(s\theta)$, which leads to $d/(s\mu_{12}) = \theta d/\lambda$. For a point source $\mu_{12} = 1$, the shear s is of the order of r_0, therefore $d/(s\mu_{12}) \simeq d/r_0$. Finally for high order measurement ($d < r_0$), s is of the order of d and $d/(s\mu_{12}) = 1$.

5.5.1.2 Noise in SH

The centroid variance is given from Eq. (5.6) by

$$\mathrm{Var}(c_x) = \frac{1}{n_{\mathrm{ph}}^2} \sum_{i,j} x_{i,j}^2 \, \mathrm{Var}(I_{i,j}), \tag{5.38}$$

where the pixel coordinates $(x_{i,j}, y_{i,j})$ are centered and normalized by the pixel pitch. For signal photon noise, the Poisson statistics leads to $\mathrm{Var}(I_{i,j}) = \langle I_{i,j} \rangle$. Therefore, the variance of the center of gravity is related to the size of the image in a subaperture determined by $\sum_{i,j} x_{i,j}^2 I_{i,j}$. The measurement error is given by (Rousset *et al.* 1987)

$$\sigma_{\mathrm{s}}^2 = \frac{\pi^2}{2} \frac{1}{n_{\mathrm{ph}}} \left(\frac{N_{\mathrm{T}}}{N_{\mathrm{D}}} \right)^2 (\mathrm{radian}^2), \tag{5.39}$$

where N_{T} is the image full width at half maximum (fwhm) and N_{D} is the fwhm of the diffraction pattern of a subaperture. In the following both N_{T} and N_{D} are expressed in terms of the number of pixels. This expression holds if thresholding or windowing is used to eliminate the background around the central core of the image. Note that the undersampling of the image introduces a limitation of the accuracy which is not taken into account in Eq. (5.39) (see Goad *et al.* 1986). For comparison with Eq. (5.36), let us point out that the ratio $N_{\mathrm{T}}/N_{\mathrm{D}}$ represents the term $\theta_{\mathrm{b}} d/\lambda$. For an extended source, $N_{\mathrm{T}}/N_{\mathrm{D}} = \theta d/\lambda$. For a point source, Table 5.2 summarizes the dependence of σ_{s} for the two limit cases where the subapertures are either diffraction-limited ($d < r_0$) or seeing-limited ($d > r_0$). In the seeing-limited case, the measurement error depends on r_0. The seeing enlargement of the image degrades the system performance (Rousset *et al.* 1987; Welsh and Gardner, 1989; Séchaud *et al.* 1991). For $d < r_0$, $n_{\mathrm{ph}} = 50$ is required to achieve a Strehl ratio (SR) of 90%.

For electronic noise (or detector noise), $\mathrm{Var}(I_{ij}) = \sigma_{\mathrm{e}}^2$, σ_{e} is the rms number of noise electrons per pixel and per frame. This noise is due to the read-out and the dark current of the detector. The measurement error is given by (Rousset *et al.* 1987)

$$\sigma_{\mathrm{s}}^2 = \frac{\pi^2}{3} \frac{\sigma_{\mathrm{e}}^2}{n_{\mathrm{ph}}^2} \frac{N_{\mathrm{S}}^4}{N_{\mathrm{D}}^2} (\mathrm{radian}^2), \tag{5.40}$$

where N_{S}^2 is the total number of pixels used in the center-of-gravity calculation. Here the SNR in a subaperture is given by $n_{\mathrm{ph}}/(\sigma_{\mathrm{e}} N_{\mathrm{S}})$. In the case of thresholding or windowing, N_{S} may be reduced down to the order of $2N_{\mathrm{T}}$. The two limit cases are summarized in Table 5.2. Let us underline that N_{D} is an instrumental parameter which has to be chosen by the designer of the sensor. N_{D} depends on the lenslet focal length. In the case $d > r_0$, the seeing enlargement of the image

Table 5.2. *Noise behaviour for the Shack–Hartmann*

	Photon noise		Electronic noise	
	Diffraction-limited	Seeing-limited	Diffraction-limited	Seeing-limited
σ_s	$n_{ph}^{-1/2}$	$n_{ph}^{-1/2}(d/r_0)$	$\sigma_e N_D n_{ph}^{-1}$	$\sigma_e N_D n_{ph}^{-1}(d/r_0)^2$

degrades significantly the performance of the SH (Rousset *et al.* 1987; Séchaud *et al.* 1991). For an intensified CCD, σ_e must be replaced by σ_e/G in Eq. (5.40) where G is the effective gain in CCD electron/photoelectron (usually $G \sim 1000$). For sky background photon noise, the expression is similar to Eq. (5.40) where σ_e^2 is equal to the average number of photoelectrons per pixel from the background (Rigaut 1993). Let n_{bg} be the total number of photoelectrons from the sky background distributed over N_S^2 pixels, we have $n_{bg} = \sigma_e^2 N_S^2$. If we assume $N_S^2 \simeq 4N_T^2$, we finally obtain

$$\sigma_s^2 = \frac{4\pi^2}{3} \frac{n_{bg}}{n_{ph}^2} \left(\frac{N_T}{N_D}\right)^2 \text{(radian}^2\text{)}. \tag{5.41}$$

Here the SNR is given by $n_{ph}/\sqrt{n_{bg}}$. Equation (5.41) is very similar to Eq. (5.39) and shows a higher sensitivity to sky background noise than to signal photon noise. To reduce sky background and/or electronic noise limitations, it is very important in SH to numerically window the FOV of each subaperture around the image current position in order to minimize the number of pixels N_S^2 in Eq. (5.40). Indeed, the total FOV is only necessary in open loop.

For the case of quad-cell detectors assuming signal photon noise and using Eq. (5.9), the measurement error is given by:

$$\sigma_s^2 = \frac{\pi^2}{n_{ph}} \left(\frac{\theta_b d}{\lambda}\right)^2 \text{(radian}^2\text{)}, \tag{5.42}$$

where θ_b is the spot size. Now assuming detector noise, the measurement error is:

$$\sigma_s^2 = 4\pi^2 \frac{\sigma_e^2}{n_{ph}^2} \left(\frac{\theta_b d}{\lambda}\right)^2 \text{(radian}^2\text{)}, \tag{5.43}$$

where σ_e is the number of noise electrons per quadrant. For sky background noise, $4\sigma_e^2$ is replaced by n_{bg} in Eq. (5.43). Equations (5.42) and (5.43) are in good agreement with other expressions in the literature (Tyler and Fried 1982; Parenti and Sasiela 1994). Slightly different numerical coefficients may be found depending on the expression of the spot size derived from the irradiance distribution of the image in a subaperture. Let us notice also that these equations are relatively similar to the ones obtained from the centroid variance (Eqs. (5.39) and (5.41)). However, the quad-cell presents a larger error due to photon noise than the center of gravity approach, a similar sensitivity to the sky background noise and generally a smaller error due to detector noise.

5.5.1.3 Noise in CS

The variance of the phase Laplacian is evaluated in terms of the number of photoelectrons. From Eq. (5.10), it is possible to derive (Roddier *et al.* 1988)

$$\text{Var}(\nabla^2\varphi) = \frac{4\pi^2 l^2}{(f-l)^2 f^2 \lambda^2} \frac{1}{n_{\text{ph}}}. \tag{5.44}$$

Because f is large in comparison to l, $\text{Var}(\nabla^2\varphi)$ is directly proportional to l^2. Therefore, the reduction of the distance l reduces the measurement noise. Expressing Eq. (5.44) in terms of phase error and using Eq. (5.11), we obtain

$$\sigma_{\text{s}}^2 = \pi^2 \frac{1}{n_{\text{ph}}} \left(\frac{\theta_b d}{\lambda}\right)^2 \text{(radian}^2), \tag{5.45}$$

where $\theta_b = \lambda/d$ for $d < r_0$, $\theta_b = \lambda/r_0$ for $d > r_0$, and $\theta_b = \theta$ for an extended source. This demonstrates the equivalence of slope and curvature sensors for the photon noise behaviour. Let us now consider the case of sky background photon noise. In Eq. (5.45), the subaperture SNR $\sqrt{n_{\text{ph}}}$ is replaced by $n_{\text{ph}}/\sqrt{n_{\text{bg}}}$ where n_{bg} is the number of sky photons collected in the FOV of the CS. We obtain

$$\sigma_{\text{s}}^2 = \pi^2 \frac{n_{\text{bg}}}{n_{\text{ph}}^2} \left(\frac{\theta_b d}{\lambda}\right)^2 \text{(radian}^2). \tag{5.46}$$

This result also is very similar to that of the SH sensor (Eq. (5.41)).

Roddier (1995) has pointed out the possibility of reducing the distance l, once the loop is closed on a point source. Doing so further decreases the amount of noise in the loop (and the aliasing effect) but produces a reduced range of linearity to the disturbances. This capability is not available in a SH since the focal length of a lenslet array cannot be changed during closed loop operation.

5.5.1.4 WFS noise in AO for IR imaging on large telescopes

There is a specific application which needs to be underlined: AO for IR imaging on large telescopes. Generally, the WF sensing is made in the visible taking advantage of the high detectivity of the available detectors, the low sky background level and the achromatic optical path difference induced by turbulence. Because imaging is made at the wavelength λ_{IM}, the phase measurement error σ^2_{sIM} must be expressed as

$$\sigma^2_{sIM} = \sigma^2_{sWFS}(\lambda_{WFS}/\lambda_{IM})^2 (\text{radian}^2) \qquad (5.47)$$

where σ^2_{sWFS} is given by Eqs. (5.39) to (5.41) and (5.45) to (5.46), and λ_{WFS} is the WFS working wavelength. Such systems are scaled for IR imaging with only a reduced number of subapertures. This leads to the use of the WFS in the regime $d > r_0(\lambda_{WFS})$. Looking at Table 5.2, we see that σ_s depends on the ratio d/r_0, for both photon and electronic noise. For electronic noise it depends on d/r_0 because we assume that the number of pixels used for the calculation of the center-of-gravity is directly related to the size of the image by thresholding or windowing. Using Table 5.2, the limiting magnitude m can be given for a specified WF error σ_{sIM} by

$$\text{photon noise } m = -2.5 \log_{10}[\cdots \lambda^2_{WFS}/(\lambda^2_{IM}\sigma^2_{sIM}r^2_0(\lambda_{WFS})\eta\tau\Delta\lambda)]$$

$$\qquad (5.48)$$

$$\text{electronic noise } m = -2.5 \log_{10}[\cdots \sigma_e N_D \lambda_{WFS}/(\lambda_{IM}\sigma_{sIM}r^2_0(\lambda_{WFS})\eta\tau\Delta\lambda)],$$

where η is the product of optical throughput and quantum efficiency, τ is the WFS exposure time, and $\Delta\lambda$ the spectral bandwidth. Note that these limiting magnitudes do not depend on either the telescope or subaperture diameter but only on $r_0(\lambda_{WFS})$. Note also that increasing the imaging wavelength λ_{IM} allows higher magnitudes to be reached. For photon noise, m depends only slightly on λ_{WFS}: $\lambda^{-2/5}_{WFS}$. For electronic noise, the magnitude depends on N_D, which is the only parameter scaling the design of the sensor.

5.5.2 Noise propagation in the reconstruction process

Noise is propagated from the measurements to the commands in the reconstruction process. From Eq. (5.17) considering the noise covariance matrix $\mathbf{C}_n$, the mean WF error σ^2_n over the aperture after reconstruction is

$$\sigma^2_n = \frac{1}{N}\sum_i \text{Var}(\phi_i) = \frac{1}{N}\text{trace}(\mathbf{B}\mathbf{C}_n\mathbf{B}^t), \qquad (5.49)$$

where $\mathbf{B}\mathbf{C}_n\mathbf{B}^t$ is the noise covariance matrix of $\boldsymbol{\phi}$. Assuming the ideal case $\mathbf{C}_n = \sigma^2_s\mathbf{I}$, we obtain the classical result (Southwell 1980)

$$\sigma^2_n = \frac{1}{N}\text{trace}(\mathbf{B}\mathbf{B}^t)\sigma^2_s. \qquad (5.50)$$

As emphasized by Fried (1977), Hudgin (1977), and Southwell (1980) for slope sensors, the error propagator coefficient $\text{trace}(\mathbf{B}\mathbf{B}^t)/N$ is usually of the order of or lower than 1. This demonstrates the efficiency of the reconstruction techniques, a result mainly due to the uncorrelated noise assumption. Note that for modes which are already normalized over the aperture, σ_n^2 is the sum of the mode variances (no division by N) and the error propagator coefficient is only given by trace($\mathbf{BB}^t$). For slope sensors, it has been pointed out by Herrmann (1980) that the noise can be decomposed into an irrotational and solenoidal part. Only the irrotational part is propagated on the command vector whereas the solenoidal part is eliminated by Eq. (5.22), the Poisson equation. The curl operator is a way to estimate the noise present in the slope measurements (Herrmann 1980). A comparison of the noise propagation for CS and the slope sensor can be found in Roddier *et al.* (1988). For CS, the error propagation coefficient increases as N while for the slope sensor it is well known that it increases as $\ln(N)$ (Fried 1977; Hudgin 1977; Noll 1978). Until recently, the error propagation as N was considered to be the main drawback of the CS, limiting its use to relatively low degree-of-correction systems (Roddier *et al.* 1991). However, Roddier (1995) has shown that this limitation is more than compensated by the possible reduction of the distance l once the loop is closed. Compared to SH systems, no decrease with N is observed in the performance of CS systems, as determined by computer simulations (Rigaut 1992; Rigaut *et al.* 1997).

For a slope sensor, if white noise is measured on the slopes, the reconstruction multiplies the white noise spectrum by k^{-2} to find a phase spectrum in k^{-2}. The phase variance being the two-dimensional integral of the phase spectrum, a $\ln(N)$ dependence is obtained (Noll 1978). For CS, if white noise is measured on the Laplacians, the reconstruction multiplies the white noise spectrum by k^{-4} to find a phase spectrum in k^{-4}. Therefore, the noise is essentially propagated on the low order aberrations. After integration, an N dependence is obtained for the phase variance. These $\ln(N)$ and N behaviors are very general and do not depend on the reconstruction algorithm. It has been shown that for Zernike polynomials the noise propagation depends on $(n + 1)^{-2}$ for SH (Rigaut and Gendron 1992), n being the radial degree of the polynomial. A $(n + 1)^{-4}$ dependence has also been observed for CS (Rigaut 1992). Noting that n is a kind of characteristic spatial frequency for the Zernike polynomials, these dependences are in perfect agreement with the results of the above spectral analysis.

Note that in the first approximation, the level of the noise spectrum in a set of modes does not depend on the number of subapertures, i.e. the spatial sampling frequency. For example, it intrinsically varies as k^{-2} for SH. As a consequence in a given AO system for IR imaging, there is no advantage in the first approximation, in reducing the number of subapertures to adapt the system for low light level conditions (see Eq. (5.49)). In principle, the best way to reduce the amount of noise on the commands is to reduce the number of modes, or better, to use an optimized modal control (Gendron and Léna 1994) (see Chapter 6). But for practical reasons, at very low light level it is sometimes advantageous to increase the subaperture area in addition to the exposure time.

5.5.3 The other wave-front errors

The other sources of errors on the WF are of different kinds: spatial, temporal, and angular. In the spatial errors the dominant term is the fitting error (see Chapter 2), the other smaller term is related to the aliasing effect. Because the measurements are evaluated on a finite grid, in all the above-mentioned reconstruction techniques aliasing of high order aberrations occurs (sampling theorem). This aliasing happens during the measurement by the WFS. It will therefore influence the choice of the subaperture configuration. Let us note that the spatial integration by subaperture area contributes to the reduction of the aliasing (Goad *et al.* 1986). For CS aliasing effects are important, so when the distance l can not be shortened, it could be very helpful to average neighbour subapertures in order to increase the spatial filtering but keeping the same sampling frequency. In addition the measurement vector does not estimate perfectly the incoming WF because of temporal delays and filtering in the closed-loop operation (see Chapter 6). Finally, because the WF to be corrected may come off-axis from the guide star direction, the estimated WF correction does not properly fit that WF due to the limited angular correlation of the turbulent WF (see Chapter 2).

5.6 Detectors for wave-front sensing

This section is dedicated to a brief presentation of the detectors which may be implemented in the WFSs previously described. The WFS performance strongly depends on the characteristics of the detector, it is the key element. The detector parameters to consider can be deduced from the WF measurement error analysis stressed in Section 5.5, that is:

- the spectral bandwidth,
- the quantum efficiency which determines, with the spectral bandwidth, the number of photons detected for WF sensing,
- the detector noise including dark current, read-out, and amplifier noises,
- the time lag due to the read-out of the detector,
- the array size and the spatial resolution.

5.6.1 Photoelectric detectors

In the photoelectric effect, the photon energy is used to extract an electron by an electric field from a photocathode in vacuum. The measurement of the current produces the signal needed for the photon detection. In the case of low light level devices, the electron is multiplied so that the photon noise remains predominant compared to the other sources of signal fluctuations. Nevertheless, the gain variations of the electron multiplier amplify the photon noise (Lemonier *et al.* 1988). The ultimate sensitivity of the photoelectric effect detectors is fixed by the available photocathode properties: a poor quantum efficiency compared to solid-state detectors and a spectral bandwidth limited to the UV and visible spectra.

5.6.1.1 Photomultiplier tubes

Photomultiplier tubes (PMT) were developed for low light level detection. A photocathode is coupled to a high gain electron multiplier. The detector noise remains negligible compared to the photon noise. For that reason, PMTs were used in the early developments of adaptive optics system (Hardy *et al.* 1977, Hardy 1993). However, these detectors are no longer attractive for low light level WF sensing as they combine poor quantum efficiency and low spatial resolution although small integrated arrays of PMTs are now available.

5.6.1.2 MAMA Camera

The multi anode microchannel array (MAMA) consists of a photocathode, a microchannel plate (MCP) for electron multiplication, and an anode array for event detection (Timothy 1993). Basically, this camera is dedicated to photon counting with a limitation of 10^6 counts per second for the total array. The available anode array may reach 1024×1024 pixels.

5.6.1.3 ICCD

The intensified CCD camera (ICCD) consists of an image intensifier coupled to a CCD camera (Lemonier *et al.* 1988). Due to the spatial properties of CCDs (geometry, number of pixels) and to their ability to work in analog mode rather than in photon counting mode, this device has been extensively used for low light level WF sensing, especially for SH sensors. However, the charge transfer in the CCD and the intensifier phosphor persistence may be a limitation for high frame rate systems.

Derived from ICCD, the electron bombarded CCD (EBCCD) consists of a first generation image intensifier where the phosphor screen is replaced by a thinned backside bombarded CCD. Compared to ICCD, this device presents some advantages for low light level WF sensing: the photon noise amplification due to the electron multiplier gain dispersion and the temporal problems related to phosphor persistence become negligible because of the electron bombardment approach (Cuby *et al.* 1990), for an example see Chapter 8.

5.6.2 Solid-state detectors

In recent years, rapid advances have been made in solid-state detector technology. These devices present high quantum efficiency. They offer different spectral bandwidths. Moreover, a read-out noise of a few electrons may be obtained with the arrays.

5.6.2.1 Avalanche photodiodes

In the avalanche photo-diode (APD), the classical photodiode high quantum efficiency is combined with an internal gain by operating the photodiode in a Geiger mode. Single element photon counting devices are now available using various semiconductors operating in the visible or in the near infrared spectrum (Zappa *et al.* 1996). Single element APD assemblies are used in the existing curvature sensors because of the required small number of detectors (see Chapter 9). The implementation of large assemblies are complex and APD arrays are under development.

5.6.2.2 Back-illuminated bare CCD

The back illumination of thinned CCD combined with backside coatings has dramatically improved the CCD quantum efficiency and the spectral bandwidth. The CCD technology is well suited to large array integration and low noise. CCDs with low detector noise, high frame rate, and a typical size of 64×64 pixels have recently been developed specifically for adaptive optics (Twichell *et al.* 1990). At 500 kpixel/s, a detector noise of the order of $3\bar{e}$ rms have been measured with such devices (Beletic 1996). They are multi-output. Therefore, the pixel rate per output is reduced and consequently the read-out noise. Moreover, the CCD read-out time lag may be also reduced.

5.6.2.3 Infrared photodiode arrays

Although they exhibit a high quantum efficiency, the use of IR photodiode arrays with faint sources was questionable for WF sensing due to their read-out noise (Rigaut *et al.* 1992). Nevertheless, significant improvements have been made today and different arrays are available with less than $10\bar{e}$ noise but at low read-out frequency. Basically, the available photodiodes are made with InSb which has a spectral bandwidth of 1 to 5 μm and a quantum efficiency of 90% or with HgCdTe which exhibits a spectral bandwidth of 1 to 2.5 μm and a quantum efficiency of 50%. Beyond the spectral bandwidths which are different from CCDs, the read-out mode of these arrays using CMOS chips is very powerful compared to CCD. In fact, only the photodiodes useful for the measurements may be read, reducing the read-out time lag.

5.6.3 Comparison between bare CCD and intensified CCD

Using Eq. (5.49), the choice between low noise back-illuminated bare CCD and intensified CCD can be studied (Séchaud *et al.* 1991). Considering a given system and a required accuracy $\sigma_{\text{sIM}} = 2\pi/X$, it is possible to evaluate the maximum electronic noise admitted on a bare CCD (being electronic-noise limited) to be equivalent to an intensified CCD (being photon-noise limited): i.e. reaching the same limiting magnitude. We have

$$\sigma_{\text{e}} \leqslant \frac{\sqrt{3}}{16} \frac{\eta_{\text{CCD}} \Delta\lambda_{\text{CCD}}}{\eta_{\text{PK}} \Delta\lambda_{\text{PK}}} \frac{X}{N_{\text{D}}} \frac{\lambda_{\text{WFS}}}{\lambda_{\text{IM}}} \qquad (5.51)$$

where the 'CCD' subscript is for the bare CCD and 'PK' for the intensified CCD photocathode. For $\lambda_{\text{WFS}} = \lambda_{\text{IM}}$, $N_{\text{D}} = 2$, $\eta_{\text{CCD}}\Delta\lambda_{\text{CCD}}/\eta_{\text{PK}}\Delta\lambda_{\text{PK}} = 20$ and $X = 20$ (a very good accuracy), the required maximum electronic noise is $\sigma_{\text{e}} \leqslant 20\overline{e}$. But for an IR system dedicated to astronomy, we may have $\lambda_{\text{WFS}} = \lambda_{\text{IM}}/4$, $N_{\text{D}} = 1$ and $X = 4$ (partial correction regime) keeping the other parameters unchanged, then $\sigma_{\text{e}} \leqslant 2\overline{e}$. This specification is very difficult to achieve and requires specific developments to be fulfilled today (see Section 5.6.2). To sum up, for AO systems requiring high WF accuracy, the high efficiency bare CCD is well suited. But for partial correction (i.e. low accuracy at high magnitude), photon counting detectors may still provide good performance even with low quantum efficiency.

5.6.4 Discussion

First note that all of the presented detectors must generally be cooled to limit their dark current to the order of a few hundreds of electrons per pixel per second. Hence reducing the contribution of the dark current to the detector noise.

In order to achieve the ultimate performance of an AO system in terms of limiting magnitude, the WFS detector must be quasi-noiseless, i.e. only limited by the signal photon noise. Therefore today, the APDs are the best detectors. Because of their high quantum efficiency, they are better than PMTs and intensified CCDs. A dedicated optics is required to couple the incoming photons to the sensitive area of each APD. In addition, each APD requires its own electronics and connection to the real-time computer. Such detector configuration is tractable for relatively low order AO systems where small numbers of detectors are needed, such as the systems developed by the University of Hawaii and the Canada–France–Hawaii Telescope (Chapter 9). With these systems stable closed-loop operation has been demonstrated on $m_V \sim 18$ guide stars, although with little gain in Strehl.

The read-out noise is the main drawback of the back-illuminated bare CCDs, still limiting their application to brighter objects $m_V < 16$. However, they can have very high quantum efficiency and very large spectral bandwidth, in fact a better efficiency than APDs. In addition for high order AO systems, CCD provides the required multiplexing of the pixel read-out, significantly reducing the cost of the WFS. On-chip binning of pixels is also very efficient to adapt the number of pixels to read, the read-out noise being incurred only once per binned group of pixels. This is a useful tool to accomodate the WFS camera to the various observing conditions. The drawback of the CCD read-out mode in AO is the time lag. Indeed, the time lag limits the bandwidth of the servo-loop when observing bright objects (Chapter 6). This is especially true for frame transfer CCDs. After exposure, the image is rapidly transferred into the on-chip memory area which is generally read during the next exposure. The time lag is then equal to the time needed to read-out the memory. On the contrary for discrete detectors (as APDs), there is no time lag problem *a priori*. The time lag for self-scanned solid-state detector arrays can be limited when using line transfer or random access read-out (cf. IR photodiode arrays). But such read-out modes are not available on the high performance CCDs. Another way is the use of multioutput CCDs. Note that to achieve 1 or $2\bar{e}$ noise, the CCD must be read at low pixel rate (a few tens of kHz), also increasing the required number of outputs to keep a sufficiently high frame rate for WF sensing. Therefore for bright objects using higher pixel rates, the time lag of such types of CCDs may be significantly reduced.

To sum up, the low noise back-illuminated bare CCD is the best choice for high order AO systems but with limitations toward the highest magnitude and in servo-loop bandwidth. An associated low order WFS equipped with APDs can be helpful to complement the CCD performance for the access to the highest magnitude range. For low order AO systems, APDs are the best detectors.

5.7 Acknowledgements

I would like to thank M. Séchaud and V. Michau for very helpful discussions and for their contribution to the section 'Detectors for wave-front sensing'. I am grateful to P-Y. Madec, J-M. Conan, C. Dessenne, and L. Mugnier for their careful reading of the manuscript and their comments. Particular thanks also go to F. Roddier.

References

Acton, D. S. and Smithson, R. C. (1992) Solar imaging with a segmented adaptive mirror. *Appl. Opt.* **31**, 3161–9.

Allen, J. G., Jankevics, A., Wormell, D. and Schmutz, L. (1987) Digital wave-front sensor for astronomical image compensation. Proc. SPIE 739, pp. 124–8.

Angel, R. (1994) Ground-based imaging of extrasolar planets using adaptive optics. *Nature* **368**, 203–7.

Armitage, J. D. and Lohmann, A. (1965) Rotary shearing interferometry. *Optica Acta* **12**, 185–92.

Bakut, P. A., Kirakosyants, V. E., Loginov, V. A., Solomon, C. J. and Dainty, J. C. (1994) Optimal wavefront reconstruction from a Shack-Hartmann sensor by use of a Bayesian algorithm. *Optics Comm.* **109**, 10–15.

Barclay, H. T., Malyak, P. H., McGonagle, W. H., Reich, R. K., Rowe, G. S. and Twichell, J. C. (1992) The SWAT wave-front sensor. *Lincoln Lab. J.* **5**, 115–29.

Beletic, J. W. (1996) Review of CCDs and controllers: status and prospects. In: *Adaptive Optics*, OSA Technical Digest Series 13, p. 216. Optical Society of America, Washington DC.

Ben-Israel, A. and Greville, T. N. (1980) *Generalized Inverses: Theory and Applications*. R. E. Krieger Publ. Co., Huntington.

Born, M. and Wolf, E. (1980) '*Principles of Optics*'. 6th edn. Pergamon Press, Oxford.

Boyer, C., Michau, V. and Rousset, G. (1990) Adaptive optics: Interaction matrix measurements and real-time control algorithms for the COME-ON project. Proc. SPIE 1237, pp. 406–21.

Cannon, R. C. (1995) Global wave-front reconstruction using Shack-Hartmann sensors. *J. Opt. Soc. Am. A* **12**, 2031–9.

Cho, K. H. and Petersen, D. P. (1989) Optimal observation apertures for a turbulence-distorted wave front. *J. Opt. Soc. Am. A* **6**, 1767–75.

Conan, J. M., Rousset, G. and Madec, P. Y. (1995) Wave-front temporal spectra in high-resolution imaging through turbulence. *J. Opt. Soc. Am. A* **12**, 1559–70.

Cuby, J-G., Richard, J-C. and Lemonier, M. (1990) Electron Bombarded CCD: first results with a prototype tube. Proc. SPIE 1235, pp. 294–304.

Downie, J. D. and Goodman, J. W. (1989) Optimal wave-front control for adaptive segmented mirrors. *Appl. Opt.* **28**, 5326–32.

Fante, R. L. (1975) Electromagnetic beam propagation in turbulent media. *Proc. IEEE* **63**, 1669–92.

Fienup, J. R. (1982) Phase retrieval algorithms: a comparison. *Appl. Opt.* **21**, 2758–69.

Fontanella, J-C. (1985) Analyse de surface d'onde, déconvolution et optique active. *J. Optics* (Paris) **16**, 257–68.

Forbes, F. and Roddier, N. (1991) Adaptive optics using curvature sensing. Proc. SPIE 1542, pp. 140–7.

Freischlad, K. R. and Koliopoulos, C. L. (1986) Modal estimation of a wave front from difference measurements using the discrete Fourier transform. *J. Opt. Soc. Am.* **3**, 1852–61.

Fried, D. L. (1977) Least-square fitting a wave-front distortion estimate to an array of phase-difference measurements. *J. Opt. Soc. Am. A* **67**, 370–75.

Fried, D. L. (1993) Atmospheric turbulence optical effects: understanding the adaptive optics implications. In: *Adaptive Optics for Astronomy*, eds D. M. Alloin, J.-M. Mariotti, NATO ASI Series C 423, pp. 25–57. Kluwer Academic Publishers; Dordrecht.

Fugate, R. Q., Fried, D. L., Ameer, G. A., Boeke, B. R., Browne, S. L., Roberts, P. H., *et al.* (1991) Measurement of atmospheric wave-front distortion using scattered light from a laser guide star. *Nature* **353**, 144–6.

Gaffard, J. P. and Ledanois, G. (1991) Adaptive optical transfer function modeling. Proc. SPIE 1542, pp. 34–45.

Gendron, E. (1993) Modal control optimization in an adaptive optics system. In: *Active and Adaptive Optics*, ed. F. Merkle, ICO 16 Satellite Conf., ESO Conf. Proc. 48, pp. 187–92. European Southern Observatory, Garching.

Gendron, E. and Léna, P. (1994) Astronomical adaptive optics I, modal control optimization. *Astron. Astrophys.* **291**, 337–47.

Gerson, G. and Rue, A. K. (1989) Tracking system. In: *The Infrared Handbook*, eds W. L. Wolfe, G. J. Zissis, Environmental Research Institute of Michigan, Ann Arbor. Chap. 22, pp. 22.1–22.108.

Goad, L., Roddier, F., Beckers, J. and Eisenhardt, P. (1986) National Optical Astronomy Observatories (NOAO) IR Adaptive Optics Program III: criteria for the wave-front sensor selection. Proc. SPIE 628, pp. 305–13.

Greenwood, D. P. and Primmerman, C. A. (1992) Adaptive optics research at Lincoln Laboratory. *Lincoln Lab. J.* **5**, 3–23.

Gonsalves, R. A. (1976) Phase retrieval from modulus data. *J. Opt. Soc. Am.* **66**, 961–4.

Hardy, J. W. (1993) Twenty years of active and adaptive optics. In: *Active and Adaptive Optics*, ed. F. Merkle, ICO 16 Satellite Conf., ESO Conf. Proc. 48, pp. 29–34. European Southern Observatory, Garching.

Hardy J. W., Lefebvre J. E. and Koliopoulos C. L. (1977) Real-time atmospheric compensation. *J. Opt. Soc. Am.* **67**, 360–9.

Herrmann, J. (1980) Least-square wave front errors of minimum norm. *J. Opt. Soc. Am.* **70**, 28–35.

Hogge, C. B. and Butts, R. R. (1976) Frequency spectra for the geometric representation of wave-front distortions due to atmospheric turbulence. *IEEE Trans. Antennas. Propag.* **AP-24**, 144–54.

Hudgin, R. H. (1977) Wave-front reconstruction for compensated imaging. *J. Opt. Soc. Am.* **67**, 375–8.

Ichikawa, K., Lohmann, A. W. and Takeda, M. (1988) Phase retrieval based on the irradiance transport equation and the Fourier transform: experiments. *Appl. Opt.* **27**, 3433–6.

Katsaggelos, A. K. (1991) Introduction. In: *Digital Image Restoration*, ed. A. K. Katsaggelos, pp. 1–20. Springer Verlag, Berlin.

Kendrick, R. L., Acton, D. S. and Duncan, A. L. (1994) Phase diversity wave-front sensor for imaging systems. *Appl. Opt.* **33**, 6533–46.

Koliopoulos, C. L. (1980) Radial grating lateral shear heterodyne interferometer. *Appl. Opt.* **19**, 1523–8.

Kupke, R., Roddier, F. and Mickey, D. L. (1994) Curvature-based wave-front sensor for use on extended patterns. Proc. SPIE 2201, pp. 519–27.

Law, N. F. and Lane, R. G. (1996) Wave-front estimation at low light levels. *Optics Comm.* **126**, 19–24.

Lemonier, M., Richard, J-C., Riou, D. and Fouassier, M. (1988) Low light level TV imaging by intensified CCDs. In: *Underwater Imaging*, Proc. SPIE 980, pp. 27–35.

Lloyd-Hart, M., Wizinowich, P., McLoed, B., Wittman, D., Colucci, D., Dekany, R., *et al.* (1992) First results of an on-line adaptive optics system with atmospheric wave-front sensing by an artificial neural network. *Astrophys. J.* **390**, L41–L44.

Marais, T., Michau, V., Fertin, G., Primot, J. and Fontanella, J. C. (1991) Deconvolution from wave-front sensing on a 4-m telescope. In: *High Resolution Imaging by Interferometry II*, ed. J. M. Beckers, F. Merkle, ESO Conf. Proc. 39, pp. 589–97. European Southern Observatory, Garching.

Maréchal, A. and Françon, M. (1970) *Diffraction, Structure des Images.* Masson, Paris.

Ma, J., Sun, H., Wang, S. and Yan, D. (1989) Effects of atmospheric turbulence on photodetector arrays. *Appl. Opt.* **28**, 2123–6.

Mertz, L. (1990) Prism configuration for wave-front sensing. *Appl. Opt.* **29**, 3573–4.

Michau, V., Rousset, G. and Fontanella, J. C. (1992) Wavefront sensing from extended sources. In: *Real Time and Post-Facto Solar Image Correction* ed. R. R Radick, NSO/SP Summer Workshop Series No. 13, pp. 124–8. National Solar Observatory, Sacramento Peak.

Noethe, L., Franza, F., Giordano, P. and Wilson, R. N. (1984) Optical wave-front analysis of thermally cycled 500 mm metallic mirrors. *IAU Colloqium* 79, pp. 67–74.

Noll, R. J. (1976) Zernike polynomials and atmospheric compensation. *J. Opt. Soc. Am.* **66**, 207–11.

Noll, R. J. (1978) Phase estimates from slope-type wave-front sensors. *J. Opt. Soc. Am.* **68**, 139–40.

O'Meara, T. R. (1977) The multidither principle in adaptive optics. *J. Opt. Soc. Am.* **67**, 306–15.

Parenti, R. R. and Sasiela, R. J. (1994) Laser-guide-star systems for astronomical applications. *J. Opt. Soc. Am. A*, **11**, 288–309.

Paxman, R. G. and Fienup, J. R. (1988) Optical misalignment sensing and image reconstruction using phase diversity. *J. Opt. Soc. Am. A*, **5**, 914–23.

Paxman, R. G., Schulz, T. J. and Fienup, J. R. (1992) Joint estimation of object and aberrations by using phase diversity. *J. Opt. Soc. Am. A*, **9**, 1072–85.

Primmerman, C. A., Murphy, D. V., Page, D. A., Zollars, B. G. and Barclays, H. T. (1991) Compensation of atmospheric optical distortion using a synthetic beacon. *Nature* **353**, 141–3.

Primot, J., Rousset, G. and Fontanella, J-C. (1990) Deconvolution from wave-front sensing: a new technique for compensating turbulence-degraded images. *J. Opt. Soc. Am. A*, **7**, 1598–608.

Primot, J. (1993) Three-wave lateral shearing interferometer. *Appl. Opt.* **32**, 6242–9.

Restaino, S. R. (1992) Wave-front sensing and image deconvolution of solar data. *Appl. Opt.* **31**, 7442–9.

Rigaut, F. (1992) CFHT AO Bonnette system simulation. *CFHT Internal Report.* Waimea: Canada France Hawaii Telescope.

Rigaut, F. (1993) Astronomical reference sources. In: *Adaptive Optics for Astronomy*, ed. D. M. Alloin, J-M. Mariotti, NATO ASI Series C 423, pp. 163–74. Kluwer Academic Publishers, Dordrecht.

Rigaut, F., Rousset, G., Kern, P., Fontanella, J. C., Gaffard, J. P., Merkle, F., *et al.* (1991) Adaptive optics on a 3.6-m telescope: results and performance. *Astron. Astrosphys.* **250**, 280–90.

Rigaut, F. and Gendron, E. (1992) Laser guide star adaptive optics: the tilt determination problem. *Astron. Astrophys.* **261**, 677–84.

Rigaut, F., Cuby, J-G., Caes, M., Monin, J-L., Vittot, M., Richard, J-C., *et al.* (1992) Visible and infrared wave-front sensing for astronomical adaptive optics. *Astron. Astrophys.* **259**, L57–L60.

Rigaut, F., Ellerbroek, B. and Northcott, M. (1997) Comparison of curvature-based and Shack–Hartmann-based adaptive optics for the Gemini telescope. *Appl. Opt.* **105**, 2856–68.

Roddier, F. (1981) The effect of atmospheric turbulence in optical astronomy. In: *Progress in Optics*, ed. E. Wolf, **XIX**, pp. 281–376. North Holland, Amsterdam.

Roddier, F. (1987) Curvature Sensing: a Diffraction Theory, NOAO R&D Note No. 87-3.

Roddier, F. (1988) Curvature sensing and compensation: a new concept in adaptive optics. *Appl. Opt.* **27**, 1223–5.

Roddier, F. (1990a) Variations on a Hartmann theme. *Opt. Eng.* **29**, 1239–42.

Roddier, F. (1990b) Wave-front sensing and the irradiance transport equation. *Appl. Opt.* **29**, 1402–03.

Roddier, F. (1995) Error propagation in a closed-loop adaptive optics system: a comparison between Shack–Hartmann and curvature wave-front sensors. *Opt. Comm.* **113**, 357–9.

Roddier, C. and Roddier, F. (1991a) New optical testing methods developed at the University of Hawaii: results on ground-based telescopes and Hubble Space Telescope. Proc. SPIE 1531, pp. 37–43.

Roddier, F. and Roddier, C. (1991b) Wave-front reconstruction using iterative Fourier transforms. *Appl. Opt.* **30**, 1325–7.

Roddier, C. and Roddier, F. (1993) Wave-front reconstruction from defocused images and the testing of ground-based optical telescopes. *J. Opt. Soc. Am. A*, **10**, 2277–87.

Roddier, F., Roddier, C. and Roddier, N. (1988) Curvature sensing: a new wave-front sensing method. Proc. SPIE 976, pp. 203–209.

Roddier, F., Graves, J. E. and Limburg, E. (1990) Seeing monitor based on wave-front curvature sensing. Proc. SPIE 1236, pp. 474–9.

Roddier, F., Northcott, M. and Graves, J. E. (1991) A simple low-order adaptive optics system for near-infrared applications. *Publ. Astron. Soc. Pac.* **103**, 131–49.

Rousset, G., Primot, J. and Fontanella, J-C. (1987) Visible wavefront sensor development. In: *Workshop on Adaptive Optics in Solar Observations*, Lest Foundation Technical Report 28, pp. 17–34. University of Oslo, Oslo.

Rousset, G., Fontanella, J. C., Kern, P., Léna, P., Gigan, P., Rigaut, F., *et al.* (1990) Adaptive optics prototype system for infrared astronomy I: system description. Proc. SPIE 1237, pp. 336–44.

Rousset, G. (1993) Wave-front sensing. In: *Adaptive Optics for Astronomy*, . ed. D. M. Alloin, J.-M. Mariotti, NATO ASI Series C 423, pp. 115–37. Kluwer Academic Publishers, Dordrecht.

Rousset, G., Beuzit, J-L., Hubin, N., Gendron, E., Madec, P-Y., Boyer, C., *et al.* (1994) Performance and results of the Come on + adaptive optics system at the ESO 3.6 meter telescope. Proc. SPIE 2201, pp. 1088–98.

Sandler, D. G., Cuellar, L., Lefebvre, M., Barrett, R., Arnold, R., Johnson, P., *et al.* (1994) Shearing interferometry for laser-guide-star atmospheric correction at large D/r_0. *J. Opt. Soc. Am. A*, **11**, 858–73.

Sasiela, R. J. and Mooney, J. G. (1985) An optical phase reconstructor based on using a multiplier-accumulator approach. Proc. SPIE 551, p. 170–6.

Séchaud, M., Rousset, G., Michau, V., Fontanella, J. C., Cuby, J. G., Rigaut, F., *et al.* (1991) Wave-front sensing in imaging through the atmosphere: a detector strategy. Proc. SPIE 1543, pp. 479–90.

Shack, R. B. and Platt, B. C. (1971) Production and use of a lenticular Hartmann Screen (abstract). *J. Opt. Soc. Am.* **61**, 656.

Schmutz, L. E., Bowker, J. K., Feinleib, J. and Tubbs, S. (1979) Integrated imaging irradiance (I^3) sensor: a new method for real-time wavefront mensuration. Proc. SPIE 179, pp. 76–80.

Southwell, W. H. (1980) Wave-front estimation from slope measurements. *J. Opt. Soc. Am.* **70**, 998–1006.

Streibl, N. (1984) Phase imaging by the irradiance transport equation of intensity. *Opt. Commun.* **49**, 6–9.

Tatarskii, V. I. (1971) *The effects of the Turbulent Atmosphere on Wave Propagation.* National Technical Information Service, Springfield VA.

Teague, M. R. (1983) Deterministic phase retrieval: a Green's function solution. *J. Opt. Soc. Am.* **73**, 1434–41.

Timothy, J. G. (1993) Imaging photon-counting detector systems for ground-based and space applications. Proc. SPIE 1982, pp. 4–8.

Title, A. M., Peri, M. L., Smithson, R. C. and Edwards, C. G. (1987) High resolution techniques at Lockheed solar observatory. In: *Workshop on Adaptive Optics in Solar Observations*, Lest Foundation Technical Report 28, pp. 107–16. University of Oslo, Oslo.

Twichell, J. C., Burke, B. E., Reich, R. K., McGonagle, W. H., Huang C. M., Bautz, M. W., *et al.* (1990) Advanced CCD imager technology for use from 1 to 10000 Å. *Rev. Sci. Instrum.* **61**, 2744–6.

Tyler, G. A. and Fried, D. L. (1982) Image-position error associated with a quadrant detector. *J. Opt. Soc. Am.* **72**, 804–8.

Underwood, K., Wyant, J. C. and Koliopoulos, C. L. (1982) Self-referencing wavefront sensor. Proc. SPIE 351, pp. 108–14.

Von der Luhe, O. (1987) Photon noise analysis for a LEST multidither adaptive optical system. In: *Workshop on Adaptive Optics in Solar Observations*, Lest Foundation Technical Report 28, University of Oslo, Oslo. pp. 255–63.

Von der Luhe, O. (1988) Wave-front error measurement technique using extended, incoherent light sources. *Opt. Eng.* **27**, 1078–87.

Wallner, E. P. (1983) Optimal wave-front correction using slope measurements. *J. Opt. Soc. Am.* **73**, 1771–6.

Wang, J. Y. and Markey, J. K. (1978) Modal compensation of atmospheric turbulence distortion. *J. Opt. Soc. Am.* **68**, 78–87.

Welsh, B. M. and Gardner, C. S. (1989) Performance analysis of adaptive-optics systems using laser guide stars and slope sensors. *J. Opt. Soc. Am. A*, **6**, 1913–23.

Winick, K. A. (1986) Cramér-Rao lower bounds on the performance of charge-coupled-device optical position estimators. *J. Opt. Soc. Am. A*, **3**, 1809–15.

Wyant, J. C. (1974) White light extended source shearing interferometer. *Appl. Opt.* **13**, 200–2.

Wyant, J. C. (1975) Use of an AC heterodyne lateral shear interferometer with real-time wave-front correction systems. *Appl. Opt.* **14**, 2622–6.

Zappa, F., Lacaita, A. L., Cova, S. D. and Lovati, P. (1996) Solid-state single-photon detectors. *Opt. Eng.* **35**, 938–45.

6

Control techniques

PIERRE-YVES MADEC

Office National d'Études et de Recherches Aérospatiales (ONERA) Châtillon, France

An adaptive optics system (AOS) can be defined as a multi-variable servo-loop system. Like classical servos, it is made of a sensor, the wave-front sensor (WFS), a control device, the real time computer (RTC), and a compensating device, the deformable mirror (DM). The goal of this servo is to compensate for an incoming optical wavefront distorted by atmospheric turbulence. It is designed to minimize the residual phase variance in the imaging path, i.e. to improve the overall telescope point-spread function (PSF). The input and the output of this servo are respectively the wave front phase perturbations and the residual phase after correction.

The design and the optimization of an AOS is a complex problem. It involves many scientific and engineering topics, such as understanding of atmospheric turbulence, image formation through turbulence, optics, mechanics, electronics, real time computers, and control theory. The goal of this chapter is to provide the basis of spatial and temporal controls. The reader will not find a tutorial on control theory but its application to an AOS. So the reader is assumed to be familiar with classical control theory (see for instance Franklin *et al.* 1990).

In the following, no restriction is made about the WFS or DM used. The principles are kept unchanged in the case of Shack–Hartmann, curvature sensor, or other WFS, and stacked array, bimorph, segmented DMs, or other wave front compensation devices.

In a first part, the control matrix determination and the modal control analysis of an AOS are described. Then the AOS temporal behavior is described by means of a transfer function representation. The AOS closed-loop optimization is discussed for the cases of a bright star and a faint star reference source. Finally, general considerations on the RTC are given.

6.1 Adaptive optics system: a servo-loop

Figure 6.1 gives a schematic view of an AOS, using classical block diagram representation, with the control computer (CC), the digital to analog converters (DAC), and the high voltage amplifiers (HVA). As shown, the input of this servo is the uncompensated wave front, whose value at a point (x, y) and at a time t is denoted by $\varphi_{tur}(x, y, t)$. The correction introduced by the DM is denoted by $\varphi_{corr}(x, y, t)$. $\varphi_{res}(x, y, t)$ is the value of the residual phase after correction, so that

$$\varphi_{res}(x, y, t) = \varphi_{tur}(x, y, t) - \varphi_{corr}(x, y, t). \qquad (6.1)$$

In closed-loop operation, the AOS minimizes $\varphi_{res}(x, y, t)$ with the finite spatial and temporal resolution given by the local measurements of the wave front. To achieve this minimization, the WFS must be placed in the optical path after the DM in such a way it measures $\varphi_{res}(x, y, t)$. The WFS spatial resolution is given by its number of useful subapertures. Its temporal resolution is determined by the sampling frequency of its detector, which is generally set by the exposure time T.

In the following, bracket notations refer to vectors. $| \rangle$ represents a column vector and $\langle |$ a row vector. Matrices are represented in italic capital fonts. With these notations, the WFS measurements φ_{res} is represented by $|P\rangle$ which is called the measurement vector. This is an m-dimensional vector where m is the number of WFS measurements. The control voltages to be applied to the DM are deduced from $|P\rangle$. They can be described by a vector called the control vector, and denoted by $|V\rangle$. This is an n-dimensional vector, where n is the number of DM actuators. When applied to the DM, these voltages control the shape of its reflecting surface, described by the corrected phase φ_{corr}. One should notice that the control of the DM from the WFS measurements is a physical way to reconstruct a wave front.

In an AOS, the control voltages are determined from the WFS measurements through a control law which can be split into two parts: a static part, which mainly deals with the reconstruction of the WFS measurements on the basis of

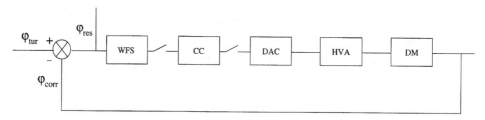

Fig. 6.1. Block-diagram representation of an AOS.

the DM actuators, and a dynamic part which ensures the stability and the accuracy of the closed loop. The first part, the static part, estimates the control vector $|v\rangle$ containing the control voltages that, if they were sent to the DM, would give the best fit of the WFS measurements $|P\rangle$. This estimation is performed through the multiplication of $|P\rangle$ by a control matrix denoted by D^*. This matrix is deduced from the calibration of the AOS. This is the topic of Section 6.2. Since in closed loop $|P\rangle$ is a measurement of the residual phase error φ_{res}, $|v\rangle$ is a correction increment and cannot compensate for the turbulent phase φ_{tur}. A second part, the dynamic part, is therefore required to determine the effective voltages $|V\rangle$.

The dynamic part evaluates the control vector $|V\rangle$ which will correct for φ_{tur}. It is deduced from the time sequence of the correction increments $|v\rangle$ and of the effective voltages $|V\rangle$ through a control algorithm which may be defined in a general manner by:

$$|V\rangle_k = -\sum_{i=1}^{l} a_i |V\rangle_{k-i} + \sum_{j=0}^{p} b_j |v\rangle_{k-j} \qquad (6.2)$$

The values of a_i and b_j determine the closed-loop performance, i.e. the trade-off between its stability and its accuracy. This is the topic of Section 6.5.

6.2 Control matrix determination

The control matrix D^* is used in closed-loop to compute the control vector $|v\rangle$ from the measurement vector $|P\rangle$. The determination of this control matrix assumes that the AOS is a linear system, at least for small values of $|P\rangle$ in closed-loop. Its behavior is fully described by the knowledge of the optical interaction matrix (Boyer 1990). This matrix, denoted by D, defines the sensitivity of the WFS to the DM deformations. This is a $m \times n$ matrix, where n is the number of actuators and m is the number of WFS measurements. The nth column is the measurement vector corresponding to a unit voltage applied to the nth DM actuator, that is to the nth actuator influence function (see Section 4.4). Let $|V\rangle$ be any control vector (static or dynamic) and $|P\rangle$ the corresponding measurement vector. We have

$$|P\rangle = D|V\rangle. \qquad (6.3)$$

The problem of the control matrix determination is the inversion of Eq. (6.3). Since the matrix D is rectangular (in the general case, there are less actuators than measurements), Eq. (6.3) cannot be inverted. This means that a criterion must be defined to compute the pseudo-inverse control matrix D^*. Let us define a norm in the vector space of the WFS measurements as

$$\||P\rangle\|^2 = \langle P|P\rangle = \sum_{i=1}^{m} P_i^2 \tag{6.4}$$

where P_i is the ith component of $|P\rangle$.

A common choice for this criterion is the following: for a given measurement vector $|P\rangle$, D^* must lead to a control vector $|V\rangle$ minimizing the norm of the residual measurement vector. The best estimate of $|V\rangle$ is therefore

$$|V_{\text{best}}\rangle = D^*|P\rangle, \tag{6.5}$$

and the residual measurement vector is given by

$$|\epsilon\rangle = |V\rangle - |V_{\text{best}}\rangle = (I - DD^*)|P\rangle, \tag{6.6}$$

where I is the identity matrix.

The minimization of $\||\epsilon\rangle\|^2$ with respect to the coefficients of the control matrix gives a set of linear equations (known as the normal equation), from which the classical least square solution is obtained

$$D^* = (D^t D)^{-1} D^t, \tag{6.7}$$

where ^{t} is the transposition operator. Eq. (6.7) shows that the control matrix D^* exists only if the square symetric matrix $D^t D$ is invertible. $(D^t D)$ is a covariance matrix characterizing the coupling between the DM actuators through the WFS measurements. The diagonalization of this matrix provides a set of independent modes in the WFS space. They define an orthonormalized basis for the vector space of the DM actuators. If E is the matrix of the n eigenmodes of $(D^t D)$ (each eigenmode is a column vector), and Δ is the diagonal matrix of the corresponding eigenvalues, we have

$$D^t D = E\Delta E^t. \tag{6.8}$$

Let us give some physical properties of these modes. Each of them can be represented as a particular shape of the DM reflecting plate. The norm of the WFS measurement vector $|P_q\rangle$ associated with the qth eigenmode $|E_q\rangle$ can be deduced from Eq. (6.3) and Eq. (6.4).

$$\||P_q\rangle\|^2 = \langle E_q|D^t D|E_q\rangle. \tag{6.9}$$

which yields, using the orthonormalized properties of the eigenmodes

$$\||P_q\rangle\|^2 = \lambda_q, \tag{6.10}$$

where λ_q is the eigenvalue associated with the eigenmode $|E_q\rangle$.

Each eigenvalue thus represents the WFS sensitivity to the corresponding eigenmode. The assumption that $D^t D$ is invertible is equivalent to the absence of zeros in its set of eigenvalues. Unfortunately, there is generally at least one null eigenvalue. This value corresponds to the so-called piston mode. This mode is a translation of the DM reflecting surface. It is optically of no interest, except in the case of interferometry, and cannot be detected by the WFS.

The singularity of the $D^t D$ matrix means that some modes of the DM are not detected by the WFS: the inversion of Eq. (6.3) leads to an infinity of solutions minimizing the norm of the residual measurement vector. Considering $|V\rangle$ a control vector and $|E_{\text{piston}}\rangle$ the piston eigenmode, Eq. (6.3) can be written as

$$|P\rangle = D(|V\rangle + \alpha|E_{\text{piston}}\rangle), \qquad (6.11)$$

where $|P\rangle$ is independent of the value of α. There is an infinity of control vectors giving rise to the same WFS measurements. To determine a control matrix uniquely, a choice has to be made between all the possible solutions. A reasonable choice is to cancel the unseen DM modes which is equivalent to taking the control vector of minimal norm. In operator theory terms, this choice corresponds to the generalized inverse of D. As a consequence, D^* is still given by Eq. (6.7) with the following definition for the pseudo-inverse of $D^t D$

$$(D^t D)^{-1} = E\Delta'^{-1} E^t \qquad (6.12)$$

where Δ'^{-1} is a diagonal matrix whose elements are equal to $1/\lambda_q$ if λ_q is not null, and 0 otherwise. This is an example of modal filtering, since it removes the eigenmodes with a zero eigenvalue from the control of the AOS. Finally, the control matrix D^* is given by

$$D^* = E\Delta'^{-1} E^t D^t, \qquad (6.13)$$

which can be rewritten as

$$D^* = E\Delta'^{-1}(DE)^t, \qquad (6.14)$$

It is worth noting that the measurement of the optical interaction matrix D, i.e. the calibration of the system, must be carefully performed but cannot be perfect. The effect of the error in its determination is reduced in closed-loop operation. The corresponding error in the correction at step k is taken into account in further steps forward. As a matter of fact, it is experimentally found to be small. But the precise analysis of the effect of the calibration error still remains an open issue which is, to our knowledge, not yet addressed in the open literature.

6.3 Modal control analysis

Modal analysis usually refers to the control of specific overall mirror deformations, the modes, whereas the so-called zonal analysis refers to the control of local deformations at specific locations, e.g. at the actuator locations. In fact, zonal control may be considered as a particular case of modal control. The advantage of modal analysis is shown in the following sections.

6.3.1 *Filtering of low WFS sensitivity modes*

A first advantage of modal analysis is to offer the possibility of filtering modes sensitive to the measurement noise. Let $|n\rangle$ be a measurement noise vector, and $|b\rangle$ the associated control coefficients. We have

$$|b\rangle = D^*|n\rangle. \qquad (6.15)$$

If the measurement noise level is assumed to be the same in each subaperture and if the noise level is uncorrelated from one subaperture to the other, the covariance matrix of $|b\rangle$ due to the measurement noise denoted by C_b can be easily derived

$$C_b = \sigma^2 D^* D^{*t}, \qquad (6.16)$$

where σ^2 is the variance of the measurement noise.

The ratio of the noise variance propagated on the control modes (diagonal elements of C_b) over the measurement noise variance σ^2 are the so-called noise propagation coefficients. From Eq. (6.16) these coefficients can be shown to be equal to the sum of the squares of the corresponding D^* row of coefficients.

If E is chosen as the control mode basis, it can be shown from Eqs.(6.8), (6.13) and (6.16) that the noise propagation coefficient on a control mode is proportional to the inverse of the corresponding eigenvalue: the lower this eigenvalue, the greater the control noise. A very efficient use of modal analysis consists in filtering modes presenting a low WFS sensitivity, i.e. a high sensitivity to the noise. In the previous section, it was shown that filtering of the piston mode, insensitive to the WFS, is required to determine the control matrix. It represents the extreme case of low WFS sensitivity.

6.3.2 *Control of several wave front correction devices*

Another advantage of modal analysis is that it allows control of several wave front correction devices. Generally, their related control vector spaces are not independent: these phase correctors are coupled. In order to control them with good stability, it is important to use a control mode basis where some modes fully define the coupling between the various devices. It is then possible to filter these coupled modes from all the wave front correction device controls but one, or to share the control of the coupled modes between two (or more) phase correctors by means of a temporal frequency filtering.

As an example, consider the case of tip/tilt correction. The major part of the turbulent wave front fluctuations are tip and tilt. To reduce the mechanical stroke required for the DM actuators, a steering mirror is generally used. If the bandwidth of the steering mirror is sufficiently high (see Section 4.5), there is no need for the DM to correct for tip and tilt. Otherwise it is useful to compensate for high stroke slowly evolving tip and tilt with the steering mirror, and for small stroke quickly evolving tip and tilt with the DM. In both cases, one must determine a set of control modes containing at least tip and tilt to filter them from the DM control (first case), or to control them at high frequency (second case).

6.3.3 Multi-input/multi-output servo decoupling

An AOS is a multi-input/multi-output servo-system. Since the inputs are the DM control voltages, and the outputs the WFS measurements, it appears that both inputs and outputs are spatially coupled. The study of the temporal behavior of the AOS is then quite complex. A classical way to make this analysis easier is to find a new set of inputs and outputs which are independent (O'Meara 1977; Winocur 1982; Gaffard and Ledanois 1991). The control modes studied in Section 6.2 define such a set of independent inputs. It is also easy to show that the WFS measurements corresponding to these modes realize a set of independent outputs. So, the AOS control loop can be split into n independent single-input/single-output servos working in parallel. The temporal behavior of each channel can be analyzed independently. This topic will be addressed in Section 6.4.

6.4 Adaptive optics temporal behavior

Figure 6.1 gives a classical block-diagram representation of an AOS control-loop. The WFS gives measurements of the residual optical phase. Whatever the principle of the WFS, the associated detector integrates the photons coming from the guide star during a time T, then delivers an intensity measurement. The WFS measurements, derived from this analog signal, are then available only at sampling times whose period is T. Consequently an AOS is a servo using both continuous and sampled data.

The wave-front computer (WFC) derives the WFS measurements from the analog signal of the detector. The CC calculates the DM control voltages from the WFS measurements and DACs are used to drive the DM HVAs.

In the following, the transfer function of each element is discussed (Gaffard and Boyer 1990; Boyer and Gaffard 1991; Demerlé *et al.* 1993). Then a general expression of the overall open-loop transfer function is established.

Finally, all the transfer functions defining the temporal behavior of this servo are studied.

6.4.1 Transfer function of adaptive optics system elements

6.4.1.1 Wave-front sensor transfer function

From a temporal standpoint, the main characteristic of the WFS is the integration time T of the detector. The output of the sensor, denoted by $WFS(t)$ is the average of $|P\rangle$ from t to $t + T$,

$$WFS(t) = \frac{1}{T} \int_{t-T}^{t} |P\rangle(t)\,dt. \tag{6.17}$$

Eq. (6.17) can also be written as

$$WFS(t) = \frac{1}{T} \int_{t-T}^{\infty} |P\rangle(t)\,dt - \frac{1}{T} \int_{t}^{\infty} |P\rangle(t)\,dt, \tag{6.18}$$

showing that the temporal behavior of the detector is the difference between an infinite integral of $|P\rangle(t)$ and the same integral but with a pure time delay T. Since the Laplace transform of an integrator is $1/s$, and the Laplace transform of a pure time delay τ is $\exp(-\tau s)$, with $s = j\omega$ ($j^2 = -1$ and $\omega = 2\pi f$, where f is the temporal frequency), $WFS(s)$, the transfer function of the WFS, is then

$$WFS(s) = \frac{1 - e^{-Ts}}{Ts}. \tag{6.19}$$

Note that both the input and the output signals of the WFS are continuous.

6.4.1.2 Wave-front computer transfer function

The WFC is a real time computer which reads and digitizes the detector signals, and applies specific algorithms to derive the wave-front measurements. The main temporal characteristic of this device is a pure time delay due to the read-out of the detector and the computation. Let τ be this delay, and $WFC(s)$ the transfer function of the WFC. We have

$$WFC(s) = e^{-\tau s}. \tag{6.20}$$

In the case of a Shack–Hartmann WFS, τ is dictated by the read-out time of the CCD sensor. Generally $\tau = T$ (in case of a frame transfer read-out), but it can be reduced by using parallel and fast read-out electronics and powerful computers. In the case of a curvature WFS, τ depends on the computing time.

In the following, it will be shown that this time delay is a major limitation for the AOS performance.

One important feature of the WFC is that its input signals are continuous, and its output signals are sampled, at a frequency defined by the WFS detector integration time T.

6.4.1.3 Control computer transfer function

The main task of the CC is to perform a matrix multiplication so that the DM control voltages can be derived from the WFS measurements delivered by the WFC. The CC also applies the temporal controller to optimize the AOS closed-loop response. To reduce the control voltage computing time, the matrix multiplication can be performed during the read-out time of the detector, as soon as any WFS measurement is available. In this case, there is no significant additional time delay, and the main temporal characteristics of this computer are determined by the implemented temporal controller, whose effect will be studied in the next sections. Following Eq. (6.2) this controller is defined by the recurrent formula

$$O(nT) = \sum_{j=0}^{p} b_j I((n-j)T) - \sum_{i=1}^{l} a_i O((n-i)T), \qquad (6.21)$$

where $O(kT)$ and $I(kT)$ are respectively the output and the input of the controller at the kth control step, and a_i and b_j the coefficients of the controller.

The temporal characteristics of the controller depend on the values of a_i and b_j. Since the controller works in the sampled time domain, its transfer function is defined by its Z-transform.

Let us recall that

$$Z(O((n-k)T)) = z^{-k} Z(O(nT)), \qquad (6.22)$$

where $Z(O(nT))$ is the Z-transform of O.

From Eqs. (6.21) and (6.22) the transfer function of the CC can be written as

$$CC(z) = \frac{\displaystyle\sum_{j=0}^{p} b_j z^{-j}}{1 + \displaystyle\sum_{i=1}^{l} a_i z^{-i}}. \qquad (6.23)$$

6.4.1.4 Digital analog converter transfer function

DACs are synchronized by the integration time of the WFS detector. They hold the control voltages of the DM constant during T, until the next voltages are available from the CC. The transfer function of this device called a zero-order holder, denoted by $DAC(s)$ is

$$DAC(s) = \frac{1 - e^{-Ts}}{Ts}.$$ (6.24)

It is physically different from the integration process of the WFS, though the transfer function is the same. Input signals are sampled, but output signals are continuous.

6.4.1.5 High voltage amplifier transfer function

The HVAs amplify the low voltage outputs of the DACs to drive the actuators of the DM. They are characterized by their temporal bandwidth, and can be considered as first or second order analog filters. Generally, the bandwidth of the HVAs is adjusted so that it is greater than the frequency domain of interest of the AOS. Within this domain, their transfer function, denoted by $HVA(s)$, is close to unity. Both input and output signals are continuous.

6.4.1.6 Deformable mirror transfer function

The DM temporal behavior is given by its mechanical response, characterized by resonance frequencies and damping factors. Usually, it is a second order filter. Typically, the first resonance frequency of DMs is greater than a few kHz, and is damped by the HVA. In the following, its transfer function denoted by $DM(s)$ will be considered equal to one over the frequency domain of interest. This assumption does not reduce the generality of the results, and simplifies the following analysis. Both input and output signals are continuous.

6.4.2 Overall adaptive optics system transfer function

6.4.2.1 Open-loop transfer function

Assuming the linearity of the system, its open-loop transfer function can be written as the overall product of the previously defined transfer functions (Demerlé et al. 1993). But one major problem is the mixing of continuous and sampled elements. From a physical standpoint, the input and the output of the AOS servo are respectively the turbulent wave front fluctuations and the residual phase errors after correction by the DM. Both these signals are continuous: so in the following this servo will be studied by means of the so-called Laplace variable s.

In the expression of the AOS open-loop transfer function, the Laplace transform and the Z-transform are both needed. Nevertheless it can be assumed that, since the Z-transform is the expression of the transfer function of a process taking place in the sampled time domain, it can be studied in the continuous time domain using the classical transformation $z = e^{Ts}$. This assumption is valid within the low frequency domain, which is of interest for an AOS. As shown in Section 6.5.1.2, this domain is restricted to one-quarter of the sampling frequency. Finally, the overall open-loop transfer function of an AOS, denoted by $G(s)$, can generally be written as

$$G(s) = \frac{e^{-\tau s}(1 - e^{-Ts})^2}{T^2 s^2} CC(z = e^{Ts}), \tag{6.25}$$

which is a good approximation for many systems. This analysis has given a good estimation of many AOS transfer function, such as the COME-ON (Rousset *et al.* 1990; Rigaut *et al.* 1991) and the COME-ON PLUS systems (Rousset *et al.* 1993), the IfA AOS (Roddier *et al.* 1991), but it can also be applied to the fringe tracker of the ASSI experiment (Robbe *et al.* 1997) and the granulation tracker of the THEMIS telescope (Molodij *et al.* 1996).

The expression of $G(s)$ consists of two parts:

- $CC(z)$ which defines the controller. It will be optimized by the AOS servo engineer to reach the best possible correction efficiency within the limit of the overall system stability. This optimization is discussed in section 6.5.
- $e^{-\tau s}(1 - e^{-Ts})^2/(T^2 s^2)$ which represents the transfer functions of the basic components required to realize an AOS. In opposition to $CC(z)$, this transfer function cannot be optimized, except during the preliminary design phase of the system.

Considering this second part, it is straightforward to show that the gain of $(1 - e^{-Ts})/(Ts)$ is equal to $\sin(\omega T/2)/(\omega T/2)$ and its phase to $-\omega T/2$. In the low frequency domain $\sin(\omega T/2)/(\omega T/2) \approx 1$ and

$$\frac{1 - e^{-Ts}}{Ts} \approx e^{-Ts/2}, \tag{6.26}$$

which represents a pure time delay of $T/2$.

From this result, it can be seen that all the components of an AOS but the controller exhibit an overall behavior which can be represented as a pure time delay $T + \tau$. The one frame time delay T is due to the exposure time of the detector, and to the DAC: it cannot be reduced. In the following, τ will be referred to as the AOS time delay. It will be shown that the correction efficiency of an AOS is highly dependent on the value of τ.

Consider the example of the COME-ON system (see Chapter 8). The control frequency is 100 Hz and the time delay is about 10 ms, mainly due to the CCD read-out. Figure 6.2 shows the gain and the phase of its theoretical open-loop transfer function considering a simple pure integrator. In this case the general expression of the transfer function $CC(z)$ reduces to $K_i/(1 - z^{-1})$, and the theoretical expression of $G(s)$ becomes

$$G(s) = K_i \frac{e^{-\tau s}(1 - e^{-Ts})}{T^2 s^2},$$

(6.27)

where K_i is the integrator gain.

6.4.2.2 Closed-loop transfer function

Let $H(s) = \varphi_{\text{corr}}(s)/\varphi_{\text{tur}}(s)$ be the closed-loop output transfer function. We have

$$H(s) = \frac{G(s)}{1 + G(s)}$$

(6.28)

Figure 6.3 represents the gain of the COME-ON theoretical closed-loop transfer function. In this example, it exhibits an overshoot which is a general feature of closed-loop responses. To ensure good stability, one usually limits this overshoot to a maximum value of 2.3 dB.

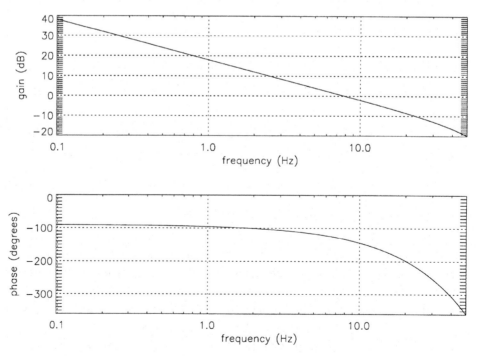

Fig. 6.2. Bode diagram of the open-loop transfer function.

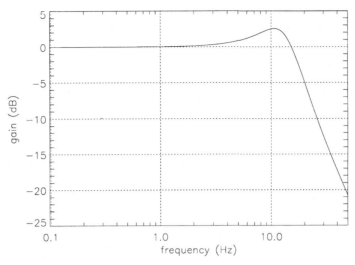

Fig. 6.3. Bode diagram of the closed-loop transfer function.

6.4.2.3 Closed-loop error transfer function

This function, denoted by $\epsilon(s)$, is defined as the transfer function between the residual phase and the turbulent wave front fluctuations. It represents the ability of the AOS to compensate for phase perturbations as a function of frequency. It is the most important feature of such a servo since it characterizes its correction efficiency. It is related to the open-loop transfer function $G(s)$ by

$$\epsilon(s) = \frac{1}{1 + G(s)}. \tag{6.29}$$

Figure 6.4 represents the gain of the COME-ON theoretical closed-loop error transfer function.

6.4.2.4 Noise propagation transfer function

A fundamental limitation of an AOS is the WFS measurement noise, whatever its physical origin, i.e. photon noise, sky background, or electronic read-out detector noise. From a servo standpoint, it can be represented by an additive white spectrum signal introduced after the WFS (see Fig. 6.5).

To characterize its effect on the residual phase after correction, the noise transfer function is defined as the ratio between the residual phase due to noise propagation and the measurement noise. Let $N(s)$ be this transfer function. Figure 6.5 shows that it can be written as

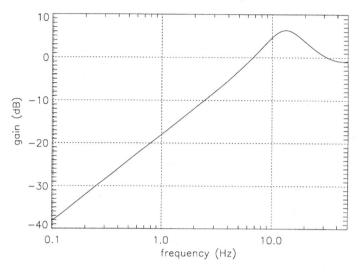

Fig. 6.4. Gain of the closed-loop error transfer function.

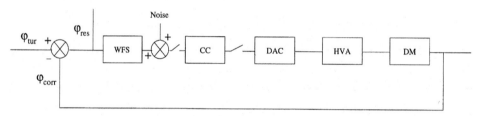

Fig. 6.5. Block-diagram of an AOS including noise representation.

$$N(s) = \frac{G(s)}{\mathrm{WFS}(s)(1 + G(s))}. \qquad (6.30)$$

It can be seen from Eqs. (6.19) and (6.26) that

$$N(s) \approx H(s)\mathrm{e}^{-Ts/2}. \qquad (6.31)$$

As a consequence, the gain of $N(s)$ is equal to the gain of the closed-loop transfer function at low frequencies.

6.4.3 *Adaptive optics system bandwidth definition*

Bandwidth values can be found in many papers describing AOSs and their performance, but only a few papers give a precise definition of the bandwidth being used.

Three different AOS bandwidths can be considered:

- The first one is the classical −3 dB closed-loop cut-off frequency. This definition is well suited to a classical servo where the output is the main parameter. In the case of an AOS, such a bandwidth is of little interest, except from a commercial standpoint; this is the highest bandwidth which can be defined.
- The second one is the 0 dB open-loop cut-off frequency. For frequencies lower than this bandwidth, the AOS is able to apply a gain in the loop, i.e. to compensate for perturbations. For higher frequencies, the AOS attenuates the signals in the loop: no more correction can be obtained. This bandwidth gives a first idea of the frequency domain where the AOS is efficient: it is useful to compare the intrinsic capability of different AOSs. A more precise estimation of the AOS efficiency is given by the following bandwidth definition.
- The third one is the 0 dB closed-loop error cut-off frequency. This definition is of great interest since it is related to the residual optical phase after correction. For frequencies lower than this bandwidth, the AOS attenuates the turbulent perturbations. For higher frequencies, the AOS first amplifies the turbulent wave front perturbations, and then has simply no effect. The knowledge of this bandwidth gives the frequency domain where the AOS is efficient.

Considering the COME-ON system example, the −3 dB closed-loop bandwidth is found to be equal to 18 Hz, the 0 dB open-loop bandwidth is equal to 7.7 Hz, and the 0 dB closed-loop error bandwidth is equal to 6.7 Hz. There is almost a factor of three between the bandwidth defining the real performances of the system, and the highest bandwidth definition. It is worth noting that the 0 dB open-loop and closed-loop error bandwidths have roughly the same value, and the same meaning. When using a pure integrator as a corrector, these two bandwidths can be considered equivalent.

6.5 Optimization of the adaptive optics closed-loop performance

In the previous section, the temporal behavior of an open loop AOS was described. This knowledge can now be used to optimize the closed-loop response of the servo. In the following, the control optimization in the case of a bright or a faint reference source is considered. The difference is given by the measurement noise level.

6.5.1 Control optimization with a bright reference source

Over recent years, many authors have studied this topic (Greenwood and Fried 1976; Greenwood 1977; Tyler 1994). In this case, the measurement noise is neglected. At first, an expression for the considered optimization criterion is given. Then the effect of a pure time delay on the AOS performance is studied, and finally some examples of the optimization of $CC(z)$ are given.

6.5.1.1 Optimization criterion

As previously discussed, the output of interest of an AOS is the residual phase after correction. It defines the quality of the AOS PSF, which is characterized by its full width half maximum (fwhm) or its Strehl ratio (SR). The smaller the fwhm; the higher the SR; the better the AOS response. These parameters are both related to the variance of the residual phase after correction, denoted by $\sigma^2_{\varphi\text{res}}$.

Because an AOS is designed to attenuate the turbulent phase variance and to minimize the residual phase variance, the criterion chosen for an AOS optimization is the relative residual phase variance, i.e. the ratio between these two variances. The value of this criterion depends both on the AOS correction efficiency and on the speed of the turbulence.

From the knowledge of the temporal behavior of wave front fluctuations (Conan *et al.* 1995), and of the AOS transfer functions, the residual phase variance, denoted by $\sigma^2_{\varphi\text{resrel}}$ can be deduced. Parseval's theorem yields

$$\sigma^2_{\varphi\text{resrel}} = \frac{\displaystyle\int_{-\infty}^{+\infty} |W(f)|^2 |\epsilon(f)|^2 \, df}{\displaystyle\int_{-\infty}^{+\infty} |W(f)|^2 \, df}, \tag{6.32}$$

where $|W(f)|^2$ is the power spectral density (PSD) of the wave front phase fluctuations. The temporal behavior of Zernike polynomials can be split into two frequency domains. For frequencies lower than a given cut-off frequency, $|W(f)|^2$ is characterized by a $f^{-2/3}$ law in case of tip or tilt, and a f^0 law in other cases. For frequencies higher than the cut-off frequency, $|W(f)|^2$ is characterized by a $f^{-17/3}$ law. The cut-off frequency is related to the wind velocity $\bar{v}$ averaged along the optical path

$$f_{\text{cut-off}}(n) \approx 0.3(n+1)\frac{\bar{v}}{D_{\text{tel}}}, \tag{6.33}$$

where D_{tel} is the telescope pupil diameter and n is the radial degree of the Zernike polynomial. Equations 6.29 and 6.27 allow the optimization of the digital controller transfer function $CC(z)$ to minimize $\sigma^2_{\varphi\text{res}}$.

6.5.1.2 Time delay effect on the adaptive optics system correction efficiency

It is a well-known result that time delay dramatically reduces AOS performance (Fried 1990; Roddier *et al.* 1993). The time delay is defined here by Eq. (6.20).

To illustrate its effect on the AOS bandwidth, the case of a pure integrator is considered. The bandwidth is here defined as the 0 dB open-loop cut-off frequency. The gain is adjusted in such a way that the closed-loop overshoot is lower than 2.3 dB. Figure 6.6 gives the relative bandwidth, defined as the ratio between the AOS bandwidth and the control sampling frequency, as a function of the relative time delay, defined as the ratio between the AOS time delay τ, and the control period T. This figure shows that a reduction in time delay brings a significant gain in bandwidth. The highest obtainable bandwidth is equal to one-quarter of the control frequency. If there is no time delay in the AOS, for example with a curvature WFS or with a Shack–Hartmann WFS using a very short read-out time CCD, the bandwidth is very sensitive to the computation time needed to compute the DM control voltages from the WFS measurements.

The reduction of the time delay in an AOS is a very efficient way to increase its bandwidth without changing the control frequency, i.e. the number of photons detected by the WFS. This problem must be considered during the preliminary design phase of the system, since it is related to the WFS detector read-out mode and to the real time control design system.

With the same assumptions, Fig. 6.7 represents the relative residual phase variance, as defined by Eq. (6.32), versus the atmospheric temporal cut-off frequency as defined by Eq. (6.33). The PSD of the wave-front phase fluctua-

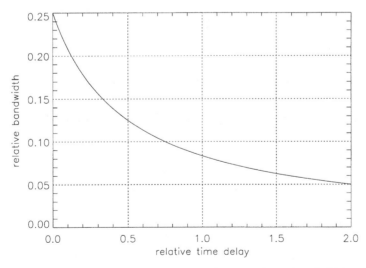

Fig. 6.6. Effect of time delay on the AOS 0 dB open-loop bandwidth; the relative bandwidth is the ratio of the bandwidth over the control sampling frequency; the relative time delay is the ratio of the AOS time delay over the control period.

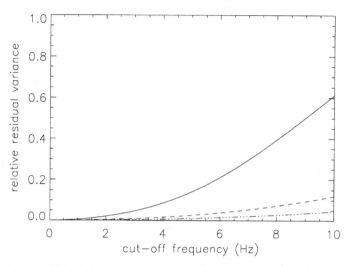

Fig. 6.7. Relative residual variance versus atmospheric temporal cut-off frequency for various AOS time delays: one frame (solid line), one-quarter of a frame (dashed line), and no delay (dash–dot line).

tions considered in this figure is the tip/tilt PSD. The solid line is plotted for a one frame time delay. The dashed line corresponds to one-quarter of a frame time delay. The dash–dot for no time delay. In all cases, it can be verified that the higher the cut-off frequency, i.e. the speed of the turbulence, the poorer the AOS correction quality. For a given cut-off frequency, the correction quality is better in the case of a smaller time delay, i.e. in the case of a higher AOS bandwidth. There is a residual phase variance reduction of more than a factor 10 when reducing the time delay from one frame down to zero, and this reduction is still of a factor 6 for a quarter of one frame time delay.

Figures 6.6 and 6.7 clearly illustrate the advantages of reducing the time delay in an AOS. As previously pointed out, the reduction of this parameter by careful design of the controller is crucial. The use of a predictive controller is a promising way to solve this problem. It has been demonstrated (Aitken and McGaughey 1995) that WFS measurements are predictable. The idea is to use theoretical and experimental knowledge of the temporal evolution of the wave-front phase fluctuations to compensate for the time delay (Paschall and Anderson 1993; Wild 1996). Open-loop demonstrations of the efficiency of such methods have been performed (Lloyd-Hart and McGuire 1995; Jorgenson and Aitken 1992, 1993).

Recently, Dessenne has demonstrated theoretically and experimentally closed-loop operation of predictive controllers for AOS control (Dessenne 1997).

6.5.1.3 Temporal controller optimization

In this section, the optimization of $CC(z)$ is studied, considering a one frame time delay in the loop. In all cases considered, the controller parameters are adjusted in such a way that the closed-loop overshoot is less than 2.3 dB.

In the previous section the time delay is shown to be the main limitation in an AOS in the case of a bright reference source. This is due to the induced phase lag whose effect is to reduce the frequency domain where the AOS is efficient. The goal of the $CC(z)$ optimization is to reduce this phase lag. In the following, two examples of classical controllers which increase the AOS performance are given.

A PID (proportional-integrator-derivative) controller is classically used to reduce the phase lag. In fact, it can be shown that, since the phase lag is proportional to the frequency, the derivative part of a PID is of no interest. So, the first example is a proportional integrator (PI) whose Z-transform is

$$CC(z) = \frac{K_i}{1 - z^{-1}} + K_p, \tag{6.34}$$

where K_i is the integrator gain, and K_p is the proportional gain.

On Fig. 6.8, the relative residual phase variance is plotted as a function of the atmospheric cut-off frequency in the case of a pure integrator (solid line) with $K_i = 0.5$ and of a PI controller (dashed line) with $K_i = 0.5$ and $K_p = 0.3$.

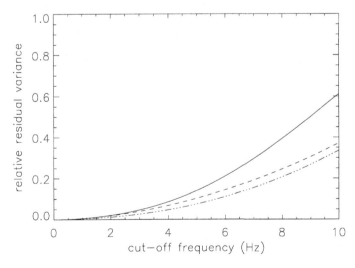

Fig. 6.8. Relative residual phase variance versus atmospheric cut-off frequency for a one frame AOS time delay and for various temporal controllers; pure integrator with $K_i = 0.5$ (solid line); PI controller with $K_i = 0.5$ and $K_p = 0.3$ (dashed line); Smith predictor with $K_i = 1.1$ and $K_{Sp} = 0.7$ (dash–dot line).

Compared to a pure integrator, a PI can bring a 40% gain in AOS efficiency for a 10 Hz cut-off frequency.

The second example is a Smith predictor, which is a controller especially dedicated to servos exhibiting a time delay. It is designed to reduce the effect of the time delay. One possible expression for its Z-transform is

$$CC(z) = \frac{K_i}{(1 - z^{-1})(1 + K_{Sp}z^{-1})},$$ (6.35)

where K_i is the integrator gain, and K_{Sp} is the Smith predictor gain. One should note that for a K_{Sp} value greater than 1, this controller is unstable.

On Fig. 6.8 the relative residual phase variance is plotted as a function of the atmospheric cut-off frequency in the case of a Smith predictor (dash–dot line) with $K_i = 1.1$ and $K_{Sp} = 0.7$. Compared to a pure integrator, the use of a Smith predictor can bring a 44% gain in AOS efficiency for a 10 Hz cut-off frequency.

6.5.2 Control optimization with a faint reference source

The WFS measurement noise depends on the reference source magnitude. The higher the magnitude, the higher the measurement noise. This noise is propagated through the AOS control all the way to the residual wave front fluctuations. In Section 6.4.1 this propagation was shown to be characterized by a transfer function whose gain is almost equal to the closed-loop one. This gain can be defined as 1 for frequencies lower than the closed-loop cut-off frequency, and 0 in the high frequency domain. The closed-loop cut-off frequency is roughly proportional to the AOS bandwidth, defined as the 0 dB open-loop cut-off frequency.

A new optimization criterion in minimizing the residual phase variance is defined, taking into account the noise propagation through the AOS control loop. Considering a white spectrum measurement noise with a PSD value denoted by PSD_{noise}, the total residual phase variance can be written as

$$\sigma^2_{\varphi res} = \int_{-\infty}^{+\infty} |W(f)|^2 |\epsilon(f)|^2 \, df + \int_{-\infty}^{+\infty} PSD_{noise} |N(f)|^2 \, df,$$ (6.36)

where the first integral corresponds to the residual turbulent phase variance, and the second one to the propagated noise variance.

It can be easily shown that the higher the AOS bandwidth, the lower the residual turbulent phase variance but the higher the propagated noise variance. It is then possible to find an optimal AOS bandwidth minimizing $\sigma^2_{\varphi res}$, i.e. roughly equalizing the two terms of Eq. (6.36). This optimal bandwidth depends on both the WFS SNR and the temporal behavior of turbulence. It should be noticed that, since neither the SNR nor the temporal behavior of the turbulence are constant (due to r_0 and wind speed fluctuations), this optimal bandwidth evolves with time. From a practical standpoint, the optimal bandwidth of the AOS needs to be regularly updated.

Considering the phase expansion on the Zernike polynomials (Noll 1976), the modal SNR and the temporal behavior of turbulence depend on the radial degree of each polynomial (Conan et al. 1995). For any other set of modes defining an orthonormalized basis in the telescope pupil, it could be shown similarly that both SNR and temporal behavior depend on the mode considered. It means that the optimal bandwidth is different for each of these modes. This is the so-called *modal control optimization* (Ellerbroek et al. 1994, Gendron and Léna 1994, 1995).

6.6 Real time computers

A wide variety of real time computers (RTCs) have been realized. It must be remembered that, in the early developments of AOSs, analog RTCs were used to provide the short computing time required. This was the case in the first operational system, the compensated imaging system installed at the Air Force AMOS station on top of Mount Haleakala crater on Maui, Hawaii. The second generation of computers were based on hybrid hard-wired/digital computers. Digital RTCs have now replaced them, taking benefit from the exponential increase in their computing power and allowing a large flexibility in the software developments. In particular, digital RTCs allow one to update and adapt the control laws in real time, according to the changes in the conditions of observation.

The computing power CP is, at least, determined by the operation rate required to compute the control vector from the WFS measurements. The number of operations (multiplication + addition) per unit computing time is of the order of n^2, with n the number of DM actuators. The required computing time is related to the relative time delay (see Section 6.5.1.2 and Fig. 6.6). For a WFS with a negligible time delay, the computing time should be $\approx 0.1\ T$ to limit the decrease of the relative bandwidth to about 20 % and this yields

$$CP \approx 10\,\frac{n^2}{T}. \qquad (6.37)$$

In case of a WFS with a one frame time delay, due to the read-out of the detector for example, the computer can take advantage of this delay to begin the matrix multiplication, and then

$$CP \approx \frac{n^2}{T}.$$ (6.38)

Taking typical values $n = 200$ and $T = 2$ ms yields $CP = 200$ M Flops in the worst case. This computing power corresponds to the present state of the art of processors.

If n and T are determined by r_0, i.e. are proportional to $(D/r_0)^2$ and $r_0/\overline{v}$ respectively, it is worth noting that CP varies as r_0^{-5}.

But n and T may also be determined by the magnitude of the reference source. The brighter the reference source, the higher n and $1/T$, therefore the higher CP. Besides, for very faint reference sources, a simple integrator may be a satisfactory corrector. But for bright reference sources, the optimized modal control and the use of complex correctors, and particularly predictive correctors, should be efficient, increasing the computing power required. A parallel architecture computer is needed to perform the matrix operations, with a capability to provide scalable solutions able to evolve towards higher performance. High speed multi-link processors are therefore well suited and digital signal processors are currently used.

6.7 Acknowledgements

I would like to express my gratitude to Marc Séchaud for his encouragement to write this chapter and his sensible advice, to Gérard Rousset for numerous fruitful discussions about the best way to control an AOS. I am indebted to Caroline Dessenne, Laurent Mugnier, and Jean-Marc Conan for their accurate reading of this chapter. I also want to thank my other colleagues from ONERA, especially Vincent Michau and Didier Rabaud, for their help and support.

References

Aitken, G. J. M. and McGaughey, D. (1994) Predictability of atmospherically-distorted stellar wavefronts. Proc. ESO 54, pp. 89–94.
Boyer, C. and Gaffard, J.-P. (1991) Adaptive optics, transfer loops modeling. Proc. SPIE 1542, pp. 46–61.
Boyer, C., Michau, V. and Rousset, G. (1990) Adaptive optics: interaction matrix measurement and real-time control algorithms for the Come-On project. Proc. SPIE 1271, pp. 63–81.
Conan, J.-M., Rousset, G. and Madec, P.-Y. (1995) Wave-front temporal spectra in high-resolution imaging through turbulence. *J. Opt. Soc. Am.* **12**(7), 1559–69.
Demerlé, M., Madec, P.-Y. and Rousset, G. (1993) Servo-loop analysis for

adaptive optics. In: *Adaptive Optics for Astronomy*, NATO ASI Series C, 423, pp. 73–88.

Dessenne, C., Madec, P.-Y. and Rousset, G. (1997) Modal prediction for closed-loop adaptive optics. *Opt. Lett.* **20**, 1535–7.

Ellerbroek, B. L., Van Loan, C., Pitsianis, N. P. and Plemmons, R. J. (1994) Optimizing closed-loop adaptive-optics performance with use of multiple control bandwidths. *J. Opt. Soc. Am.* **11**(11), 2871–86.

Franklin, G. F., Powell, J. D. and Workman, M. L. (1990) *Digital Control of Dynamic Systems*, 2nd edn., Addison Wesley Publ. Reading, MA.

Fried, D. L. (1990) Time-delay-induced mean-square error in adaptive optics. *J. Opt. Soc. Am.* **7**(7), 1224–5.

Gaffard, J-P and Boyer, C. (1990) Adaptive optics: effect of sampling rate and time lags on the closed-loop bandwidth. Proc. SPIE 1271, pp. 33–50.

Gaffard, J.-P. and Ledanois, G. (1991) Adaptive optical transfer function modeling. Proc. SPIE 1542, pp. 34–45.

Gendron, E. and Léna, P. (1994) Astronomical adaptive optics: I. Modal control optimization. *Astron. Astrophys.* **291**, 337–47.

Gendron, E. and Léna, P. (1995) Astronomical adaptive optics: II. Experimental results of an optimized modal control. *Astron. Astrophys. Suppl. Ser.* **111**, 153–67.

Greenwood, D. P. and Fried, D. L. (1976) Power spectra requirements for wave front–compensative systems. *J. Opt. Soc. Am.* **66**(3), 193–206.

Greenwood, D. P. (1977) Bandwidth specification for adaptive optics systems. *J. Opt. Soc. Am.*, **67**(3), 390–3.

Jorgenson, M. B. and Aitken, G. J. M. (1992) Prediction of atmospherically-induced wavefront degradation. OSA, Tech. Digest Series 13, pp. 161–3.

Jorgenson, M. B. and Aitken, G. J. M. (1993) Wavefront prediction for adaptive optics. In: *Active and Adaptive optics*, ed. F. Merkle, Proc. ESO 48, p. 143.

Lloyd-Hart, M. and McGuire, P. (1995) Spatio-temporal prediction for adaptive optics wavefront reconstructors. Proc. ESO 54, pp. 95–101.

Molodij, G., Rayrole, J., Madec, P.-Y. and Colson, F. (1996) Performance analysis for T.H.E.M.I.S. image stabilizer optical system. I. *Astron. Astrophys., Suppl. Ser.* **118**, 169–79.

Noll, R.-J. (1976) Zernike polynomials and atmospheric turbulence. *J. Opt. Soc. Am.* **66**(3), 207–11.

O'Meara, T. R. (1977) Stability of an N-loop ensemble-reference phase control system. *J. Opt. Soc. Am.* **67**(3), 315–18.

Paschall, R. N. and Anderson, D. J. (1993) Linear quadratic Gaussian control of a deformable mirror adaptive optics system with time-delayed measurements. *App. Opt.* **32**(31), 6347–58.

Rigaut, F. *et al.* (1991) Adaptive optics on a 3.6 m telescope: results and performance. *Astron. Astrophys.* **250**, 280–90.

Robbe, S. *et al.* (1997) Performance of the angle of arrival correction system of the I2T– ASSI stellar interferometer. *Astron. Astrophys., Suppl. Ser.* **125**, 1–21.

Roddier, F., Northcott, M. J. and Graves, J. E. (1991) A simple low–order adaptive optics system for near-infrared applications. *Pub. Astron. Soc. Pacific*, **103**, 131–149.

Roddier, F., Northcott, M. J., Graves, J. E. and McKenna, D. L. (1993) One-dimensional spectra of turbulence-induced Zernike aberrations: time-delay and isoplanicity error in partial adaptive optics compensation. *J. Opt. Soc. Am.* **10**(5), 957–65.

Rousset, G. *et al.* (1990) First diffraction-limited astronomical images with adaptive optics. *Astron. Astrophys.* **230**, L29–L32.

Rousset, G. *et al.* (1993) The Come-On Plus adaptive optics system: results and performance. In: *Active and Adaptive Optics*, ed. F. Merkle, Proc. ESO 48, pp. 65–70.

Tyler, G.A. (1994) Bandwidth considerations for tracking through turbulence. *J. Opt. Soc. Am.* **11**(1), 358–67.

Wild, J. W. (1996) Predictive optimal estimators for adaptive-optics systems. *Opt. Lett.* **21**(18), 1433–5.

Winocur, J. (1982) Modal compensation of atmospheric turbulence induced wave front aberrations. *App. Opt.* **21**(3), 433–8.

7

Performance estimation and system modeling

MALCOLM J. NORTHCOTT

Institute for Astronomy, University of Hawaii, USA

7.1 Introduction

Adaptive optics systems are often expensive and complex instruments, which need to work under a wide range of operating conditions. It therefore behooves the designer of an adaptive optics system to develop a model of the systems in order to best allocate the available resources to the project. The existence of an accurate computer model of an AO system is probably crucial for commissioning a system. Such a model helps one diagnose problems and artifacts in the system, and explore possible solutions. This chapter deals with methods for simulating the optical performance of the adaptive optical components of the system. Simulation of the mechanical aspects of the system design are not covered in this chapter.

We start the chapter with a discussion of approximate methods that can be used in the initial design phases of a project to constrain the parameters of the AO system. We then move to slightly more complex techniques, which may be used to verify results to greater accuracy. Next we discuss complete optical simulations which may be used to model the interaction of various system components, taking optical diffraction into account. Finally we discuss the problem of comparing measured results with simulation results. Measuring the performance of an AO system is a difficult undertaking, and is often poorly done.

7.2 Linear error budget analysis

A simple error budget analysis serves to elucidate the first order performance of an AO system on a given site and at a given telescope. In Table 7.1 we list errors which affect an AO system. These errors and their magnitudes are discussed in the indicated sections of this book. As one can see, there are a

Table 7.1. *Sources of wave-front error in an AO system*

Error source	Time scale	Sections
Mirror fitting	Fast	2.2–3, 3.2–3, 4.4
Mirror hysteresis	Slow	4.2.1.4
Mirror saturation	Slow	4.7 (Table 4.1)
Time delay	Fast	2.2–3, 3.4.2, 6.4.3, 6.5.1.2
WFS noise	Fast	3.5, 5.5.1, 6.5.2
WF reconstruction	Fast	5.5.2, 6.4.2.4, 6.5.1
Non-isoplanicity	Fast	2.4, 3.4.3
Scintillation	Fast	2.4
Chromaticity (†)	Fast/Slow	2.4
Cone effect (*)	Fast	12.3
Mean focus (*)	Slow	12.3
Atm. dispersion (‡)	Slow	Can be compensated
Telescope optics	Slow/Fixed	To be measured or modeled
Instrument optics	Fixed	"
AO system optics	Fixed	"
Mechanical flexures	Slow	"
Thermal expansion	Slow	"
Calibration errors	Slow	To be estimated

(†) if wave-front sensing and imaging are done at a different wavelength.
(‡) for wide band imaging at short wavelength.
(*) for laser guide source systems only.

large number of sources of error in an AO system, and it is important not to neglect any of them. Although the rms wave-front errors usually add as random errors, the loss in Strehl due to the errors is multiplicative. Since the Strehl ratio is usually the desired performance metric, even small sources of error may be deemed important.

The time scales of the errors are important, since they affect the symmetry of the PSF. Errors which occur on a fast time scale will average over a typical astronomical exposure to give a smooth halo to the AO PSF. Slow or fixed aberrations will not average, so will produce speckles in the PSF. For slowly varying aberrations, the aberrations may vary quickly enough to change the PSF over a series of exposures, which makes deconvolution difficult. If one is not concerned with non-linearities in the optical and mechanical systems, a straightforward linear systems model can be used to combine the above errors. Such a model can be computed easily using a variety of shrink-wrapped mathematical and engineering analysis tools, provided that expressions can be computed for the most important sources of error (see Table 7.1. In this approach, the feedback system is most easily modeled in frequency space, using standard transfer function techniques (see Chapter 6). As mentioned later, the results from more complete AO simulations can be used to improve the accuracy of a simple error analysis.

If a reasonably accurate model of the control system is included in this analysis, the analysis can be used to compute the optimal feedback gains for the system under a given set of observing conditions. This approach is currently used by several groups with working AO systems.

The most immediate shortcoming of the simple error analysis, is that it does not take into account the statistics of the wave-front errors. A more complete linear analysis can be carried out by combining the theoretical expressions for the wave-front covariance due to various system components. Given the wave-front covariance, an accurate PSF can be computed for the AO system, rather than just obtaining the Strehl ratio. The difficulty of carrying out this analysis is quite considerable, since one has to compute the values of non-trivial multi-dimensional integrals.

If non-linear effects are important, it is probably more straightforward to build a complete non-linear system model. In these models it is most convenient to simulate feedback in the time domain, since this simplifies the task of incorporating the physics of the various non-linear effects into the model.

7.3 Non-linear computer modeling

In order to properly understand the functioning of an AO system, a more thorough performance analysis than an error budget analysis is required. The most straightforward way to do this is to carry out a Monte Carlo simulation of the system, using a simulated atmosphere as input to the model. Fortunately modern workstations are fast enough to permit a very thorough analysis of an AO system, including diffraction effects. The forgoing error budget analysis was based upon the assumption that the various error sources are independent. In practice there are interdependencies between the error measurements. For instance telescope aberrations can reduce the SNR for atmospheric wave-front measurements. Non-linearities, such as actuator saturation and hysteresis may easily be modeled in the computer, but are otherwise difficult to quantify.

The main difficulty in performing computer modeling is to simplify the problem sufficiently to make it tractable, while incorporating all of the important physics. A common choice amongst system modelers, has been to break their system up into four components. The atmospheric simulation, the deformable mirror simulation, the wave-front sensor simulation, and optical propagation code to tie the previous three components together. This splits the problem into manageable sub-problems, and as a benefit allows one to interchange various model components relatively easily. A typical flow chart is shown in Fig. 7.1.

As in all numerical work it is a good idea to check the impact of various approximations on the final result. This is particularly important in cases where it is difficult to compute analytically the order of magnitude of the error introduced by a particular approximation. For approximations which involve coarseness of sampling, the impact should always be evaluated, by checking that changing the resolution does not greatly alter the simulation result.

For instance if a simulation is carried out with 128 square arrays using a 30-pixel radius pupil, at least one run of the simulation, using a 256 square array and a 60-pixel pupil should be carried out. The results of the two simulations should then be compared, to estimate the size of the error introduced by sampling.

Due to the slow speed at which these types of simulations run, it is often a good idea to use the results from a complete simulation to augment a simpler wave-front error analysis. A good example of this would be to use the simulation to compute an on-axis PSF, and then obtain isoplanatic behavior by degrading the Strehl ratio using the theoretical expressions for anisoplanatic error.

Throughout the following discussions we will pay attention to the level of approximation used, and the steps that could be taken to compute their impact.

7.4 Optical propagation

The central requirement of all of these simulations is optical propagation code. This code is normally the most CPU intensive part of the simulation. The main choice which must be made at this stage is to decide if it is important to model diffraction effects. Modeling diffraction effects is absolutely essential for correct modeling of the curvature wave-front sensor. For the shearing inter-

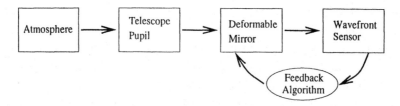

Fig. 7.1 Typical flow of control in an adaptive optics simulation.

ferometer (SI) and the Shack–Hartmann (SH) type of wave-front sensors, diffraction effects are less important.

7.4.1 Ray tracing

For SI and SH type wave-front sensors first order calculations can be carried out using ray tracing techniques. For this a commercial ray tracing package could be used. The main requirement is that an element can be inserted in the telescope pupil plane to represent an atmospheric phase screen. Most modern ray tracing packages have a facility for doing this.

An advantage to this technique is that every optical component in the AO system can be included in the simulation. The primary disadvantage is a poor simulation of diffraction effects in the PSF. Estimates of Strehl ratios from ray tracing become increasingly poor, when the residual wave-front errors exceed 1 radian rms. A further advantage of ray tracing is that non-point sources can also be simulated with ease.

7.4.2 Fresnel propagation

If diffraction effects must be taken into account, one needs in general to apply a Fresnel propagation algorithm. For some simple systems one may be able to make do with using a simple Fourier transform to propagate between the image and pupil planes. For the more general case there are several techniques for efficiently computing Fresnel propagation. The most straightforward is the Fourier technique described below. Another popular technique which might be used is propagation code based upon Gaussian beams, which offers greater numerical stability than the Fourier technique. However in the case of AO systems, we are usually dealing with relatively small Fresnel numbers, and a small number of propagation steps, a regime in which the simple Fourier technique works well. It is problematical to include every optical element in a full diffraction calculation. Modeling the internal optics would dramatically slow the simulation, and considerably increase the numerical error. If there are significant internal aberrations in the optical systems, these should be determined using a ray tracing technique, and aberrations added to the propagated wave front at an appropriate point. For instance, telescope aberrations can be added to the wave front at the entrance pupil of the telescope. Diffraction effects can be modeled only at a single wavelength. For many simulations the monochromatic approximation is sufficient. A wide bandwidth model may be approximated by combining the results from several monochromatic propagation computations.

For the simplest SH systems, all that is required is to propagate from the pupil plane to the image plane. This is easily accomplished using a Fourier transform. For curvature AO systems, and for accurate multi-layer atmospheric models, slightly more complex Fresnel propagation will be needed. The Fresnel–Kirchoff diffraction formula assuming a system illuminated with a plane wave is written as follows:

$$A(\mathbf{x}, z) = \int_{-\infty}^{\infty} \frac{(1 + \cos(\chi))}{2s} W(\mathbf{u})\exp(iks) \, d\mathbf{u}, \tag{7.1}$$

where $W(\mathbf{u})$ is the complex pupil function and $A(\mathbf{x}, z)$ is the complex amplitude in a plane at distance z, k is the modulus of the optical wave-vector, and the distance $s = [|\mathbf{x} - \mathbf{u}|^2 + z^2]^{1/2}$. For our purposes the angles of propagation are normally small enough that the obliquity factor and path length dependence of the amplitude can be ignored, namely set $(1 + \cos(\chi))/2s$ to a constant. Assuming $z \gg |\mathbf{x} - \mathbf{u}|$, the distance s can be written $s \simeq z + |\mathbf{x} - \mathbf{u}|^2/2z$, ignoring terms in the expansion of order $O[(\mathbf{x} - \mathbf{u})^4/z^3]$. Substituting this expression for s in Eq. (7.1) gives

$$A(\mathbf{x}, z) \propto \exp(ikz) \int_{-\infty}^{\infty} W(\mathbf{u})\exp(ik(\mathbf{x} - \mathbf{u})^2/2z) \, d\mathbf{u}. \tag{7.2}$$

There are two ways to solve Eq. (7.2), the most obvious method is to expand the square giving:

$$A(\mathbf{x}, z) \propto \exp(ikz)\exp(ikx^2/2z) \int_{-\infty}^{\infty} W(\mathbf{u})\exp(iku^2/2z)\exp(-ik(\mathbf{x} - \mathbf{u})/z) \, d\mathbf{u}, \tag{7.3}$$

which is simply a Fourier transform with a spherical phase factor (defocus) added to the pupil phase. Since the Fourier transform of the complex pupil function $W(\mathbf{u})$ gives the complex amplitude in the image plane, one can think of the above result as a defocused image. The second method is to realize that Eq. (7.2) has the form of a Fourier convolution integral:

$$A(\mathbf{x}, z) \propto W(\mathbf{x}) \star \exp(ik|\mathbf{x}|^2/2z), \tag{7.4}$$

and can thus be solved using the convolution theorem which states that the Fourier transform of $A(\mathbf{x}, z)$ is the product of the Fourier transform of $W(\mathbf{x})$ with the Fourier transform $\kappa(\mathbf{u})$ of $\exp(ik|\mathbf{x}|^2/2z)$. By definition

$$\kappa(\mathbf{u}) = \int_{-\infty}^{\infty} \exp((ikx^2/2z) - i\mathbf{x} \cdot \mathbf{u}) \, dx. \tag{7.5}$$

Completing the square gives:

$$\kappa(\mathbf{u}) = \exp(-i\mathbf{u}^2 z/2k) \int_{-\infty}^{\infty} \exp(ik(\mathbf{x} - \mathbf{u}z/k)^2/2z) \, dx. \tag{7.6}$$

The integral in Eq. (7.6) is the famous Fresnel, or Cornu spiral integral, which asymptotes to the value 1 over an infinite range. Hence

$$\kappa(\mathbf{u}) = \exp(-i\mathbf{u}^2 z/2k), \tag{7.7}$$

is also a spherical phase factor. Physically one can think of the convolution approach, as first transforming to the image plane, then transforming back to a defocussed pupil plane.

When implemented numerically these two methods have complementary properties. Both techniques, due to the physical approximations inherent in their derivation, fail for short propagation distances. The main numerical limitation on both techniques is due to aliasing. Enough pixels must be placed across the pupil to allow adequate sampling of the pupil wave front. The spherical phase factor necessary for Fresnel propagation must also be adequately sampled. Generally this means that we can allow a phase step of no more than 1 radian between adjacent pixels, that contain significant amplitude. The Fourier technique excels at modeling propagation very close to the image plane. The convolution technique fills in the rest of the space between the pupil to near the image plane. An advantage of the convolution technique, which is particularly useful for modeling curvature AO systems or propagation through a multi-layer atmosphere, is that size of the geometric pupil remains unchanged by the convolution.

The only practical way to simulate wide band propagation is to compute the propagation at several different wavelengths and combine them. This makes wide optical bandwidth simulations of an AO system quite time consuming. Simulating the effects of extended (incoherent) sources is problematical. If all that is required is the illumination in the focal plane, then convolving a PSF with the object brightness distribution is the correct approach. However for Fresnel propagation the situation is more complex. If the Fresnel image is close to the image plane, then convolution with the image brightness distribution will produce a reasonable approximation.

7.4.3 Aliasing considerations

As with all Fourier techniques aliasing errors are a major concern with the above method. Aliasing can occur in both the image plane or the pupil plane. Unfortunately the atmospheric phase variations are not spatially band limited, which makes it formally impossible to avoid introducing aliasing when sampling the atmospheric phase. However the aliasing can be reduced to any level desired by increasing the sampling. A rule of thumb that is widely used and usually gives acceptable results, is that there should be no more than one r_o area per pixel at the sensor wavelength. This assumption should always be checked by comparing calculations using different wavefront sampling.

A second source of aliasing is due to the spherical phase factors which are needed in the Fresnel propagation algorithm. The same rule of thumb also applies to sampling the Fresnel phase factor, which limits the Fresnel number that may be attained with a given array size. In the case of the convolution Fresnel propagation, we may be able to relax the spherical phase factor sampling requirement, by noting that it is applied to the system point-spread function, which is spatially localized, even though it never becomes zero. Again any decision regarding sampling in this domain, should be checked by running simulations with different sampling.

It is normally the case that the image plane sampling in an AO system need be no better than critically sampled. A notable exception is the case of modeling a spectrograph, where fine sampling in the image plane may be required to properly calculate the light distribution across the spectrograph slit. In this case the Fourier technique is likely to require very large arrays in order to avoid aliasing and a Gaussian beam propagation code may be more appropriate.

7.5 Modeling the atmosphere

In this section we will discuss the problem of modeling the Kolmogorov turbulence spectrum. We will restrict the discussion to the modeling of a single turbulence layer moving at fixed velocity. Generalizing to a multi-layer model can be achieved by combining an appropriate number of single layer models. First order combination can be achieved simply by adding the path contributions from the various layers. A better approach is to use Fresnel propagation code to propagate the complex amplitude between atmospheric layers.

We will also make the assumption of infinite outer scale. This is generally not a concern for telescopes of 4-m or smaller. All of the methods discussed can be modified to approximate the effects of a finite outer scale. We will discuss three methods of modeling here, the spectral approach, a Karhunen–Loève polynomial model, and a fractal model.

7.5.1 Spectral method

The most straightforward way to model the atmosphere is to generate a phase screen with the correct power spectrum

$$\Phi(\mathbf{u}) = \frac{0.023}{r_o^{5/3}} |\mathbf{u}|^{-11/3}, \tag{7.8}$$

by direct filtering of an array of uncorrelated Gaussian random numbers (McGlamery 1976). When using this technique the piston term (zero frequency) is set to zero, thus avoiding the singularity in the spectrum in Eq. (7.8) at zero frequency.

There are several problems with this method. The most serious being that low order terms are underestimated, and due to the repetitive nature of the FFT, areas near opposite edges of the array will be correlated.

With modern fast computers one can simply use large arrays of data alleviating the cyclical nature of the FFT by only using the central portion of the resulting phase screens. The underestimation of low order terms can be corrected by subtracting these terms, and then adding them back with the correct statistical weight (Wampler *et al.* 1994). We can use the Karhunen–Loève model described next to carry out this correction.

With these two precautions taken data from this type of model can be very good. It is difficult to use this method to produce a long time series of data, since this would mandate a large Fourier transform.

The underlying computational efficiency of this scheme is dominated by the Fourier transform time, provided an efficient random number generator is used, and is thus of order $N \log(N)$. However if a continuous time series of data is required, the arrays may need to be of substantial size.

7.5.2 Karhunen–Loève model

The Karhunen–Loève modes of a process are a unique set of orthogonal functions, which have statistically independent weights. These modes are interesting for many reasons, one being that the mode weights provide the most compact description of the process. Fortuitously it turns out that the Karhunen–Loève modes for Kolmogorov turbulence over a circular aperture can be expressed analytically in terms of the well-understood Zernike polynomials (see Noll 1976; Roddier 1990; and Chapter 3 of this book). The first step in this computation is to construct a Zernike covariance matrix over the atmospheric turbulence. This matrix can then be diagonalized, the resulting eigenvectors being the Karhunen–Loève modes. An expression for the Zernike covariance matrix is given in Chapter 3 [Eqs. (3.13) and (3.14)]. The efficiency of this scheme is of the order of $N \times Z_n^2$, where Z_n is the maximum order of Zernikes used. The coefficients of the Zernike terms grow very rapidly with increasing order, making rounding error a significant concern, once more than about 50 orders are required. This technique is thus well suited to the generation of wave fronts for simulating small telescopes.

The disadvantages of this approach, are that it is not completely straightforward to model time evolution, and there is a truncation error. The theoretical time evolution of the various modes is known, so time evolution can be carried out by constructing a random sequence of weights for each mode, with the correct time spectrum. The truncation error is more difficult to deal with. Although this is quite negligible for an uncorrected wave front, it can have a large effect on the rms of the wave front after correction. A simple *ad hoc* approach is to add delta-correlated random noise to the wave front to add in the energy lost due to truncation error. A much better solution is to combine this technique with the spectral method, using the Karhunen–Loève method to generate the low frequency components, and the spectral method to generate the high frequency components. The combination of the two methods generates very high fidelity wave fronts, in a fairly computationally efficient manner.

7.5.3 Fractal method

The fractal method is a very elegant method for generating atmospheric phase screens. It relies upon the observation that the turbulence structure is fractal in nature. This leads to a very computationally efficient procedure, which typically involves of the order of $N \log(N)$ operations. However in this case no over-sizing of the array is required (Lane 1992; Schwartz *et al.* 1994; Jaenisch *et al.* 1994).

A nice advantage of this method is that it can be used to simulate turbulence over dispersed arrays, since fine sampling need only be carried out on and immediately around each aperture, coarser sampling sufficing to tie the apertures together. This method is probably the only practical method for simulating turbulence over a large array of telescopes.

It is possible to model the whole energy cascade of Kolmogorov turbulence using this method, by allowing the node weights at different spatial sampling levels to evolve with the appropriate statistics. At the time of writing the fractal algorithm has not been widely studied or used.

7.5.4 Multi-layer models

Fortunately for those of us simulating AO systems, it is generally the case that atmospheric turbulence occurs in discrete thin layers. This is fortuitous indeed, since it means that a number of the preceding two-dimensional models can be combined to produce a rather good model of the three-dimensional atmosphere.

Typical turbulent layer thicknesses that have been reported in the literature are on the order of 100 m or less. For a 100-m layer we do not need to worry about diffraction within a layer for structure bigger than about 1 cm. For some non-astronomical sites, boundary layer turbulence can be very deep, and thus require more complex modeling. We will not discuss poor site modeling here.

A first order atmospheric model, will simply add up the phase contribution from the various heights along the beam propagation direction. At this level we accurately model isoplanatic effects, but neglect the effect of scintillation. It is important to remember to include projection effects when looking off zenith. For the typical field of view of an astronomical telescope, all field angles can be considered to have the same projection angle. The correction factor therefore consists of multiplying the turbulence rms by $\cos(\phi)$, where ϕ is the zenith angle.

For high order AO systems, scintillation effects may well be important. To model these it is necessary to propagate the optical complex amplitude between turbulence layers and eventually to the telescope. This requires the use of Fresnel propagation code, as discussed earlier. This additional level of complexity will considerably slow the progress of simulations.

7.6 Wave-front sensor simulation

Although the wave-front sensor simulation is at the core of the AO simulation, its actual implementation is relatively simple. We describe here the procedures for simulating the most common wide bandwidth wave-front sensors. Other types of sensors should be equally easy to simulate.

7.6.1 The Shack–Hartmann wave-front sensor

The most straight forward way to simulate a SH wave-front sensor is to extract the piece of wave front feeding each SH lenslet, and independently propagate each to the image plane using a Fourier transform. One could also apply appropriate tilts to each of the subaperture regions and transform the complete pupil to the image plane in a single large Fourier transform step.

The latter approach has the advantage of correctly accounting for the diffraction interference between adjacent spots. Usually the spots in a SH sensor are well enough separated that the diffraction effects between individual spots are negligible. On some more recent SH systems, which use adjacent 2×2 CCD pixel groups for each spot, interference between spots may be a significant source of error.

7.6.2 The curvature wave-front sensor

The curvature wave-front simulation is slightly more complex, in that it requires a Fresnel diffraction calculation to compute the illumination in two defocused images. For this application the convolution Fresnel diffraction approach has the great benefit that it does not change the size of the pupil, which considerably eases the problem of dividing the pupil light into different wave-front sensor bins.

It is interesting to note that the mathematical operations involved in the convolution Fresnel calculation have a one to one correspondence to the optical components in the curvature wave-front sensor subsystem. The first lens in the system corresponds to the Fourier transform to the image plane. The membrane mirror multiplies the PSF by a spherical phase factor, and the final lens transforms back to the pupil plane.

7.6.3 The shearing interferometer

The shearing interferometer wave-front sensor can also be viewed as a variation of the knife-edge test. Viewed this way, the reason for its achromaticity is clear. This sensor is relatively easy to simulate, requiring a transform to the image plane, multiplication by a mask function, and a transformation back to the pupil plane.

7.7 The control loop

The whole of this discussion is predicated upon the assumption that the AO control loop is most easily simulated using a discrete time-step approach. This approach is generally more time consuming than a spectral approach, but it allows for relatively easy incorporation of non-linearities such as hysteresis.

As with many other aspects of the simulation, aliasing is a potential problem here. The simplest approach, is to compute the model at the natural feedback rate of the system, and apply corrections at the beginning of the next cycle. This approach approximates all of the phase lags in the servo system by a single time delay equal to the feedback cycle time. Phase delays in a typical system include the data processing delay, electrical and mechanical time constants. To simulate these phase delays more accurately requires computing multiple wave-front sensor signals over one integration period, while the signal works its way to the ultimate deformation of the active mirror. It may be important to simulate the phase delays more accurately for a system with a slow feedback rate. This is another aspect of the simulation where the expected error should be estimated from theory, or computed by changing the time resolution of the simulation.

There is no restriction to the complexity of the feedback algorithm that may be applied in an AO system. However most commonly a pure integrator is used, possibly with some small proportional term to roll off the d.c. gain. If sophisticated feedback algorithms are going to be simulated, great care must also be applied to the atmospheric simulation to ensure that it is delivering realistic wave-front time sequences.

7.8 The active mirror

Fortunately for all currently available active mirrors, the material movement is so small that we may accurately consider the influence function of individual actuators to be independent. This leads to a pseudo-linear description of the mirror surface generated by combining the influence functions for each actuator. The main cases of non-linearity in a mirror model are actuator saturation and hysteresis. Both of these sources of error have been reduced in modern mirrors, with PMN materials giving quite high stroke and better hysteresis behavior. However for some active mirrors the hysteresis is a strong function of temperature, and may be a significant factor if the mirror runs near 0 °C. Even with modern mirrors, stroke saturation can be seen in the presence of bad telescope aberrations, poor seeing, or incomplete tip/tilt correction.

Hysteresis is not generally an important effect unless the wave-front sensor measurement space does not completely span the deformable mirror space. In this case there will be modes of mirror deformation which are undetectable (in the sensor's null-space) by the wave-front sensor. In the presence of hysteresis, these modes can grow uncontrolled. Unmeasurable mirror modes will always occur if there are fewer detectors than actuators, which is a good reason for avoiding this situation. Unfortunately most Shack–Hartmann systems have an unmeasurable mirror mode known as the waffle or checkerboard mode in which adjacent actuators are actuated with opposite sign. Simulations of these must therefore include a mirror hysteresis simulation.

7.9 Including other effects

We have discussed only the most important aspects of an AO simulations. There are many other aspects which may be important in a particular case. In Table 7.2 we list the physical effects which may be important to consider in an AO simulation. We have included an estimate of the importance of each effect for a typical AO system, of ten to a few hundred actuators on a 4-m telescope.

Table 7.2. *Physical effects which may be important in the modeling of an AO system*

Physical effect	Relative importance
Sky background	high
Detector dark current	high
Detector read noise	high
Photon shot noise	high
Electrical output bandwidth	moderate
Computational time lag	high
Detector read-out time	high
Outer scale	moderate
Atmospheric dispersion	low
Scintillation	low
Active mirror hysteresis	moderate
Active mirror saturation	moderate
Active mirror inertia	low
Isoplanatic effects	moderate

For more specialized AO systems the relative importance of the items will change. For example if one is building a high order system specifically for extra-solar planet detection, scintillation will become a dominant error.

References

Jaenisch, H.-M., Handley, J.-W., Scoggins, J. and Carroll, M.-P. (1994) Atmospheric turbulence optical model (ATOM) based on fractal theory. Proc. SPIE 2120, pp. 106–29.

Lane, R.-G, Glindemann, A. and Dainty, J. C. (1992) Simulation of a Kolmogorov phase screen. *Waves in Random Media*, **2**(3), 209–24.

McGlamery, B. L. (1976) Computer simulation studies of compensation of turbulence degraded images. Proc. SPIE/OSA Conf. on Image Processing, 74, pp. 225–33.

Noll, R.-J. (1976) Zernike polynomials and atmospheric turbulence, *J. Opt. Soc. Am.* **66**(3), 207–11.

Roddier, N. (1990) Atmospheric wave-front simulation using Zernike polynomials. *Opt. Eng.* **29**(10), 1174–80.

Schwartz, C., Baum, G. and Ribak, E.-N. (1994) Turbulence-degraded wave fronts as fractal surfaces. *J. Opt. Soc. Am. A*, **11**, 444.

Wampler, S., Ellerbroek, B.-L. and Gillies, K.-K. (1994) A software package for adaptive optics performance analysis. Proc. SPIE 2201, pp. 239–45.

Part three

Adaptive optics with natural guide stars

8

The COME-ON/ADONIS systems

GÉRARD ROUSSET

Office National d'Études et de Recherches Aérospatiales (ONERA), France

JEAN-LUC BEUZIT

Observatoire de Grenoble, France

8.1 Introduction to the COME-ON program

In the mid 1980s several programs were undertaken in astronomy to implement adaptive optics (AO) for visible (Doel *et al.* 1990; Acton and Smithson 1992) and infrared (IR) (Merkle and Léna 1986; Beckers *et al.* 1986) imaging. Those were stimulated by the coming new generation of very large telescopes of diameter D around 8 m (Barr 1986) and by the availability of AO components developed by defense programs (see for instance: Hardy *et al.* 1977; Pearson 1979; Gaffard *et al.* 1984; Fontanella 1985; Parenti 1988). Initiated by P. Léna, F. Merkle, and J.-C. Fontanella on the basis of the existing competences in France and at the European Southern Observatory (ESO), the COME-ON project was started in 1986 with the aim of demonstrating the performance of AO for astronomy. The consortium in charge of the project was initially made of three French laboratories associated with ESO, COME-ON standing for: **C**GE, a French company now CILAS (formerly LASERDOT), **O**bservatoire de Paris-**M**eudon, **E**SO and **ON**ERA. The purpose of the project was initially to build an AO-prototype system based on the available technologies and test it at an astronomical site, in order to gather experience for the ESO Very Large Telescope (VLT) program, including multi-telescope interferometry with the VLT interferometer (VLTI). The main requirement was to achieve nearly diffraction-limited imaging at the focus of a 4-m class telescope at near IR wavelengths from 2 to 5 μm, depending on the seeing conditions. With the successful results obtained the project became a full development program (Fontanella *et al.* 1991) in order to turn the first prototype into a real astronomical instrument through successive phases which will be presented hereafter.

In October 1989, for the first time in ground-based astronomy, diffraction-limited images at a wavelength of $\lambda = 2.2$ μm (K band) were obtained in real time with the 19-actuator COME-ON prototype system on the 1.52-m telescope at the Observatoire de Haute-Provence (OHP) in France under average seeing conditions (about 1″) (Rousset *et al.* 1990a). In April 1990, resolutions down to 0.17″ and Strehl ratios up to 0.3 were achieved with this system at 2.2 μm on the ESO 3.6-m telescope at La Silla (Chile) under average seeing conditions (about 0.8″) (Rigaut *et al.* 1991a). During 1990 and 1991, a number of tests were carried out to improve the system in preparation for the next phase. In parallel, astronomical observations were performed demonstrating the great value of AO (Rigaut *et al.* 1992a; Malbet *et al.* 1993; Saint-Pé *et al.* 1993a,b).

The second generation of this system, so-called COME-ON-PLUS, was set up on the ESO 3.6-m telescope in December 1992 (Rousset *et al.* 1993). The first prototype was substantially modified with a 52-actuator deformable mirror, a high sensitivity wave-front sensor (WFS), a higher temporal bandwidth, and an optimized modal control (Rousset *et al.* 1994). The system has been offered on this telescope as a standard instrument to the European astronomical community since mid-1993. Finally, the last phase was the upgrade of the COME-ON-PLUS system into a common-user instrument, improving its scientific versatility and operational efficiency (Beuzit *et al.* 1995). This instrument is now called ADONIS (**AD**aptive **O**ptics **N**ear **I**nfrared **S**ystem). The implementation started in 1995 and was completed in 1996. Since 1993, a large number of astrophysical results obtained with this system have been published in the literature, see for instance the reviews made by Léna (1995a,b) and Chapter 15 of this volume.

8.2 System description

8.2.1 COME-ON: an AO-prototype system

The COME-ON prototype system was initially designed with the limited ambition to achieve diffraction-limited imaging at 3.8 μm (L band) on the ESO 3.6-m telescope (Rousset *et al.* 1990b). The number N of degrees of freedom of the system was derived from $N = (D/r_0(3.8\text{-}\mu\text{m}))^2$ in order to achieve a good correction of the turbulence, r_0 being the Fried diameter (Fried 1965). For a seeing of 1″, this leads to $N = 11$ at $\lambda = 3.8$ μm. Then 19 actuators were chosen for the deformable mirror and 20 subapertures for the WFS. The required temporal bandwidth was derived from $f_{bw} = \overline{v}/r_0$ (Greenwood 1977). For a mean wind speed $\overline{v} = 10$ m/s, $f_{bw} = 9$ Hz at 3.8 μm. A 100 Hz sampling frequency was chosen for the closed-loop operation taking into account a factor 10 between the sampling frequency and the resulting bandwidth for a digital control.

The system aims to correct the IR images by sensing the wave front (WF) from a reference source observed at visible wavelengths taking advantage of the achromatism of the WF distortions due to atmospheric turbulence. The WFS may therefore benefit from the high sensitivity of the detectors in the visible which can be photon-noise limited. But the subapertures are scaled by $r_0(\lambda_{IM})$ at the imaging IR wavelength λ_{IM} which is much larger than $r_0(\lambda_{WFS})$ at the WFS visible wavelength λ_{WFS}. Then, the subaperture diameter is much larger than $r_0(\lambda_{WFS})$. It results in a signal-to-noise ratio (SNR) of the WF sensing set by $r_0(\lambda_{WFS})$ and not by the subaperture diameter: the measurement is seeing-limited (see Chapter 5). Therefore, increasing the subaperture diameter does not increase the SNR.

The philosophy of the design of COME-ON was to construct a test bench, the main components of which could be exchangeable for future developments. The optical layout is shown in Fig. 8.1 (Kern *et al.* 1988). The instrument is set up at the Cassegrain $f/8$ focus of the ESO 3.6-m telescope. All the optical paths were initially in a single plane. Only the imaging path has been modified in a later stage (see Section 8.2.2).

The mirrors M2 and M5 are symmetric off-axis parabolas. M2 images the telescope pupil onto the deformable mirror (M3). A tip/tilt plane mirror (M4) compensates for the overall WF tilt fluctuations which are the largest disturbances generated by the turbulence (Fried 1965). To avoid extra mirrors in the optical set-up, the tip/tilt mirror is placed close to the deformable mirror but out of a pupil conjugate plane. It results in a slight IR background modulation

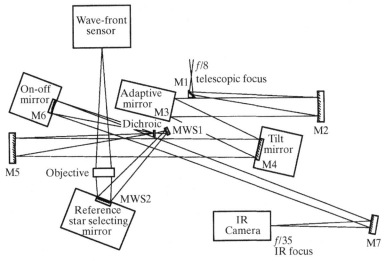

Fig. 8.1. Optical layout of the COME-ON system. Reproduced from Kern *et al.* (1988).

whose effect was found to be negligible on IR imaging detector arrays. A dichroic beam splitter reflects the IR part of the incoming light (wavelength > 0.95 μm) toward the imaging path while it transmits the visible counterpart toward the WFS channel. MWS1 and MWS2 are two folding flat mirrors. MWS2 is conjugate with the telescope pupil and is used as a WFS reference star selecting mirror. This field selecting mirror allows the use of an off-axis star in a 45″ field for the WFS when the observed on-axis object is to faint at visible wavelengths. The WFS is a Shack–Hartmann (SH) type, the lenslet array of which is conjugated with the deformable mirror, i.e. the telescope entrance aperture. On the imaging channel, the IR light is reflected by an off-axis ellipse (M6) producing a $f/35$ IR focus. M6 is conjugated with both the telescope pupil and the cold stop of the IR camera. It is used as a fast on–off chopping mirror allowing the recording of sky background between the source exposures without offsetting the telescope secondary mirror. Therefore, the AO loop is continuously closed even on the off position looking at the blank sky. All these schemes were defined during the first year of the system study and are still of interest for the design of such a system except for the dichroic plate, the best choice being today to transmit the IR light. A special set of mirrors (not shown in Fig. 8.1) was used to adapt the optical bench at the Coudé focus of the OHP 1.52-m telescope for the tests in 1989. The main characteristics of the COME-ON system are summarized in Table 8.1 (Rousset *et al.* 1990b).

The deformable mirror manufactured by CGE is a continuous facesheet mirror equipped with 19 piezoelectric actuators on an hexagonal pattern. Its first mechanical resonance frequency is around 3.5 kHz, well above the control loop sampling frequency. The tip/tilt mirror developed by Observatoire de Paris-Meudon has a two-axis gimbal mount equipped with four piezoelectric actuators working in push–pull (see Fig. 8.2). It has a first mechanical resonance around 200 Hz which was partly compensated by an internal servo-loop.

The SH WFS uses 20 subapertures on a 5×5 square grid. Figure 8.3 presents an example of an SH image pattern recorded when observing a binary star and the geometry of the pupil plane. Two lenslet arrays were provided by ONERA with two focal lengths to match the seeing conditions. They are directly mounted in front of the WFS camera without any additional optical element. The superimposition of the sky background imaged by neighbor lenslets on the camera is avoided by a field stop at the entrance of the WFS as proposed by Fontanella (1985). For example, the maximum field-of-view (FOV) alloted to each lenslet is 6″ for the smallest focal length. The WFS camera was initially a 100×100 pixel Reticon array equipped with two intensifier stages: a Proxitronic proximity focus as the first stage and a Philips Composants 1410 microchannel plate intensifier. The S25 type entrance photo-

Table 8.1. *Characteristics of the COME-ON AO-prototype system (1988–1992)*

Deformable mirror	Continuous facesheet
	19 actuators, hexagonal array
	65-mm pupil diameter
	±7.5 µm stroke for ±1500 V
	3.5 kHz first mechanical resonance
Tip/tilt mirror	Two-axis gimbal mount
	4 piezoelectric actuators
	$7''$ amplitude on the sky
	6 milli-arcsec resolution
	200 Hz first mechanical resonance
WF sensor	Shack–Hartmann
	20 subapertures, 5×5 square grid
	$6''$ maximum FOV per subaperture
	Two stage intensified Reticon array
	S25 photocathode
	100×100 pixels
	100 Hz frame rate
	Read-out noise limited
Real time computer	Centroid computation
WF computer	Two interconnected computers
	Dedicated hard wired
	8-bit digitization
	1.3 Mpixel/s maximum rate
Command computer	Command vector computation
	Motorola 68020 microprocessor
	VME bus
	12-bit digital-to-analog conversion
	100 Hz command rate
	9 Hz open loop bandwidth at 0 dB

cathode has a peak sensitivity at 0.5 µm, a bandwidth of 0.3 µm and a mean quantum efficiency of 7%. The frame rate of the camera which is the sampling frequency of the turbulence disturbances was 100 Hz. During the first year on the ESO 3.6-m telescope, this camera did not allow the loop to be closed on a star of magnitude $m_V > 9$. At low light level, this camera was read-out noise limited because of the detector noise level of the Reticon array. In addition, two other cameras for WF sensing were tested: an electron bombarded CCD (EBCCD) in collaboration with the Laboratoire d'Electronique Philips (LEP) and an IRCCD camera in collaboration with the Observatoire de Grenoble (Rigaut *et al.* 1992b). Characteristics and performance of the EBCCD will be given in Section 8.2.2. The IRCCD camera and the IR WFS path were far from

Fig. 8.2. The tip/tilt mirror of the COME-ON system. Reproduced from Rousset *et al.* (1990b).

being optimized but it was possible to obtain a stable closed-loop operation on a star of magnitude $m_K = 1.6$, hence demonstrating the feasibility of near-IR wave-front sensing.

Both adaptive mirrors were driven by a digital control loop. Two interconnected computers were used to process at a rate of 100 WF measurements per second. The first stage was the WF computer. It was in charge of the digitization and data reduction of the WFS camera signals in order to provide the WF slope vector to the second stage. It was a programmable dedicated hardwired computer developed by ONERA. The second stage was the command computer in charge of the calculation and the digital-to-analog conversion of the command vector to apply to the two adaptive mirrors. The calculation is the multiplication of the command matrix by the slope vector and the application of the temporal controller (see Chapter 6). This computer was based on the VME bus and the Motorola 68020 microprocessor and made of from-the-shelf boards. The command algorithm was a modal control where the system eigenmodes were used (see Chapter 6) (Boyer *et al.* 1990). The temporal bandwidth of the system was measured to be 9 Hz for the open-loop transfer function at 0 dB. The performance was limited by the important time lag in the

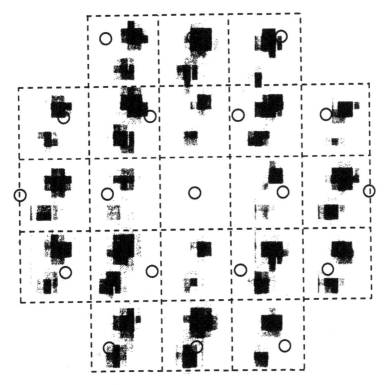

Fig. 8.3. SH image pattern (50 × 50 pixels) of the COME-ON system observing a binary star (separation 3″) with a field-of-view of 6″ per subaperture. The geometries of the subapertures (– – –) and of the deformable mirror actuators (○) are superimposed.

servo-loop due to the exposure and read-out times of the WFS camera (Boyer *et al.* 1990). This bandwidth was then extended up to 25 Hz for bright stars by the reduction of the time lag due to the pixel read-out (Rigaut *et al.* 1991b).

The COME-ON IR imaging camera was based on a 32 × 32 InSb charge injected device (CID) array developed by Société Anonyme de Télécommunication (SAT) as the detector of the 1−5 µm camera of the ISO satellite (Lacombe *et al.* 1989). The camera was read-out noise limited. Images were recorded using the standard photometric broad band filters at wavelengths of 1.2 µm (J), 1.68 µm (H), 2.23 µm (K), 3.87 µm (L′), and 4.75 µm (M). This camera was decommissioned at the beginning of 1993.

8.2.2 COME-ON-PLUS: an upgrade

The COME-ON-PLUS system is an upgraded version of the COME-ON prototype system (Rousset *et al.* 1992). Significant improvements were brought into the spatial and temporal correction capabilities, the sensitivity of the WFS, the throughput and the mechanical stability for long exposure on the imaging camera. The main features of the system are a 52-actuator deformable mirror, two selectable WFSs, one of which is dedicated to faint reference sources, and an optimized modal control to manage the low SNR cases in the servo-loop for such faint reference sources.

The optical layout was slightly modified (Hubin et al. 1992). A second WFS was set up. A translatable mirror allowed the selection between the two WFSs. The imaging channel was modified into a $f/45$ focus. The beam was folded down to cross the bench, thus providing under the bench a standard interface plate for visitor equipment. In April 1993, the SHARP II camera from the Max Planck Institut für Extraterrestrische Physik (MPIE) in Garching (Hofmann et al. 1995) was set up for the first time on COME-ON-PLUS using the new $f/45$ imaging channel (Rousset et al. 1993). A significant improvement in the optical efficiency was made between COME-ON and COME-ON-PLUS by the use of a new silver coating for all mirrors. The throughput was measured including all the optical elements (around 10 surfaces): for the WFS channel 40% over the 0.45−0.75 µm spectral range and for the imaging channel 80% between 1 and 5 µm. The main characteristics of the COME-ON-PLUS system are summarized in Table 8.2 (Rousset et al. 1994). The other components of COME-ON were not replaced.

The new continuous facesheet deformable mirror manufactured by LASER-DOT (now CILAS) is equipped with 52 piezoelectric stacked actuators on an 8×8 square grid (see Fig. 8.4) (Jagourel and Gaffard 1991). Its first mechanical resonance frequency is around 13 kHz.

The two SH WFSs work in the visible and use 32 subapertures on a 7×7 square grid. Figure 8.5 presents an example of SH image pattern recorded when observing a binary star and the geometry of the pupil plane. The WFS noise is still seeing-limited. The lenslet arrays were also provided by ONERA. The maximum FOV is limited to 6″. The maximum frame rate of the WFSs is 200 Hz. The WFSs are selected depending on the magnitude of the natural reference star. For relatively bright stars, the intensified Reticon array is used. The range of magnitude m_V is between 6 and 10. Because of the limitations of the Reticon based camera, a new WFS camera is provided to observe fainter sources. It is equipped with an EBCCD manufactured by LEP (Richard et al. 1990). The EBCCD is based on a first generation single stage triode intensifier tube. The standard output screen is here replaced by a thinned back-bombarded CCD. Thus, the accelerated photoelectrons are directly detected by the CCD. The photocathode is a S20R with a peak quantum efficiency of 10% at 0.55 µm. The electronic gain of the tube reaches 2000 for a 15 kV accelerating

Table 8.2. *Characteristics of the COME-ON-PLUS AO system*

Deformable mirror	Continuous facesheet
	52 actuators, square array
	65-mm pupil diameter
	± 5 µm stroke for ± 430 V
	13.5 kHz first mechanical resonance
Tip/tilt mirror	No change
WF sensor	Two Shack–Hartmann WFSs
	32 subapertures, 7×7 square grid
	$6''$ maximum FOV per subaperture
	56×56 pixels
Bright object camera	Two stage intensified Reticon array
	200 Hz frame rate
	$6 \leqslant m_V \leqslant 10$
Faint object camera	Electron bombarded CCD
	S20R photocathode
	25 to 200 Hz frame rate
	Photon-noise limited
	$10 \leqslant m_V \leqslant 15$
Real time computer	Two interconnected computers
WF computer	No change
Command computer	Motorola 68040 microprocessor
	+ Motorola 56000 DSP
	VME bus
	14-bit digital-to-analog conversion
	200 Hz command rate
	29 Hz maximum bandwidth
	Optimized modal control

voltage. The CCD read-out noise is around 100 electrons/pixel per frame. As a consequence, this camera is always photon-noise limited. Another key advantage of this camera is the very high stability of the gain leading to a high reliability on photo-event detection and count (Cuby *et al.* 1990). The read-out frame rate is selectable between 25 and 200 Hz depending on the incoming light level. In fact, 100 Hz is the maximum practical frame rate in order to achieve a proper long term lifetime of the EBCCD. The working range of magnitude m_V was found on the sky to be between 11 and 15 including all system losses.

The command computer was upgraded in order to achieve the calculation of the command for 52 actuators at a sampling frequency of 200 Hz. Note that the

Fig. 8.4. The 52-actuator deformable mirror of the COME-ON-PLUS system (since 1992). Reproduced from CILAS.

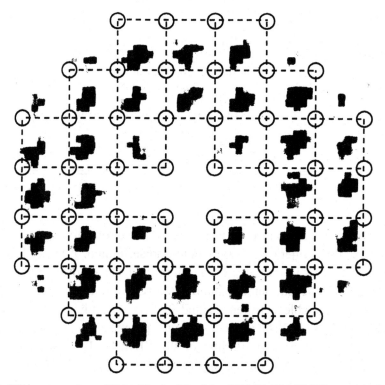

Fig. 8.5. SH image pattern (56 × 56 pixels) of the COME-ON-PLUS system observing a binary star (separation 1.3″) with a FOV of 6″ per subaperture. The geometries of the subapertures (– – –) and of the deformable mirror actuators (○) are superimposed.

WF computer was not changed during this phase. The computing power of the command computer was boosted by a dedicated Motorola 56000 DSP board developed by LASERDOT (now CILAS). For bright stars, the servo-loop performance is maximum. The temporal bandwidths of the system were measured to be for the open-loop transfer function at 0 dB: 29 Hz for the deformable mirror and 23 Hz and 27 Hz for the two axes tip/tilt mirror (Rousset *et al.* 1994). The bandwidths for the tip/tilt mirror are primarily limited by its mechanical response. For the deformable mirror, the limitations are due to both its high voltage amplifiers and the phosphor screen of the Reticon camera. With the EBCCD camera, the theoretical bandwidth is 11 Hz at a 100 Hz frame rate (i.e. sampling frequency).

For the COME-ON-PLUS system, a new optimized modal control algorithm has been developed to deal with low SNR in the WF measurements (Gendron and Léna 1994). The initial goal was to push the limiting magnitude of the system toward $m_V \approx 16$ for tip/tilt correction only (Gendron *et al.* 1991). The main feature of this algorithm is the possibility to adapt the number of corrected spatial modes and their gain in terms of turbulence and astronomical conditions such as: the seeing, the atmospheric correlation time, the reference source characteristics (brightness and size) and, in a future development, the angular separation between the reference star and the observed object. The set of modes must be chosen to deal with *decoupled* degrees of freedom. Possible modes are the mirror Karhunen–Loève modes (Gendron 1993). Each mode has a different SNR. These SNRs depend on the conditions listed above. The mode SNR is here the ratio of its turbulence variance to its noise variance. For COME-ON-PLUS, the SNR varies by a factor 100 on the set of modes. The lowest order, i.e. the two tilts, has the highest SNR while the highest order (i.e. the highest spatial frequency generated by the deformable mirror) has the lowest one. The command optimization consists in performing for each mode the adjustment of the loop gain in order to find the optimum bandwidth taking into account the SNR and the turbulence correlation time (Gendron and Léna 1994, 1995). Too high a bandwidth leads to an important propagation of the noise on the command while too low a bandwidth does not properly compensate for the turbulence. A new command matrix can be calculated for each new set of modal gains, which are derived as often as necessary from an open-loop measurement of the turbulence characteristics. This optimization is made off-line. Each new available command matrix is simply loaded in the real time computer (Rousset *et al.* 1994).

A new imaging IR camera, SHARP II has been provided by the MPIE in Garching (Hofmann *et al.* 1995) to replace the former 32×32 from April 1993 onwards. It has a 256×256 NICMOS3 HgCdTe detector array covering the J, H, and K bands with very low dark current ($\sim$ 1 electron per pixel and per second) and read-out noise ($\sim$ 40 electrons per read-out).

8.2.3 ADONIS: a user friendly system

The first observing results obtained with the COME-ON and COME-ON-PLUS systems clearly demonstrated the impressive potential of this technique in different fields of astrophysics (Rigaut *et al.* 1992a; Léna 1995a, and this volume). Nevertheless, a fairly large team of qualified personnel was still required to operate the system as several instrumental parameters had to be optimized depending on a number of astronomical requirements such as the magnitude, colour, and morphology of the reference source, the wavelength of observations, the angular separation between the object and the reference source, and depending also on atmospheric conditions, such as the atmospheric turbulence amplitude and coherence time, etc. With COME-ON-PLUS this optimization could not be carried out very easily and had given rise to a somewhat inefficient use of telescope time (typically < 25%). The experience gained with COME-ON and COME-ON-PLUS then led to the concept of ADONIS, which was intended to improve the performance, versatility and operational efficiency of the AO system, mainly by upgrading the real time computer and adding a so-called master computer running artificial intelligence software (AIS) to handle the overall instrument control and the interface with the user (Beuzit *et al.* 1994). In addition, two dedicated IR imaging cameras have been built to take full advantage of the high quality of the images generated by the AO system. A mechanical and optical interface allows visitor experiments like polarimeters, coronographs, spectrographs to benefit from the AO correction (see Section 8.2.4). Another objective of ADONIS was to develop operational procedures and test technical concepts that could be later applied to the AO system under study for the ESO Very Large Telescope.

The optomechanical layout of ADONIS has not been substantially changed from the one of COME-ON-PLUS. An infrared wave-front sensor channel has been added to offer the possibility of wave-front sensing in the infrared when neither a visible counterpart of the astrophyical target nor a reference star is available. The IR wave-front sensor itself was not part of the ADONIS instrument and is nowadays being developed by Observatoire de Paris. Its implementation is foreseen for end-1998. Many subsystems of the bench have to be set and modified according to the astronomical or atmospheric conditions. The master computer has a direct control over all of these functions via a 80C32-based micro-controller. Typical remote controlled functions include the WFS selection (low flux, high flux or, eventually, infrared), the WFS frame frequency, gain, neutral density filter settings and the dichroic mirror choice according to the reference star magnitude and spectral type. Other elements such as the infrared chopping mirror offset are also remotely controlled.

A new real time computer (RTC) was implemented on ADONIS. It has been developed specifically for AO applications by SHAKTI (France) and ONERA. It integrates in one system all functions previously achieved by the COME-ON-PLUS wave-front and command computers. Its modular architecture relies on a VME motherboard and dedicated DSP C40 modules. A remote rack enables the analog processing and digitization (12 bits/10 MHz converter) of the video signal directly at the WFS camera output, thus avoiding analog transmission noise. The RTC itself consists of: one interface board dedicated to the WFS data reduction (windowing, flat-fielding, dead pixel correction, thresholding); several C40 modules determining the WF x and y slopes and computing the mirror commands; the 12-bit analog conversion 64-output module needed to send the output voltages to both the deformable and tip/tilt mirrors amplifiers; and finally the graphical display of either the WFS images or the command vectors. The RTC carries out the complete processing for each 7×7 Shack–Hartmann subaperture of 8×8 pixels in less than 100 µs after the end of the WFS integration. An additional feature of this new RTC is the possibility of recording sets of WFS Shack–Hartmann images, x and y slopes, and mirror commands with the servo-loop either open or closed. This permits the correction to be optimized during observations by changes in the modal control, thus saving telescope time, and provides valuable real time assesment of the quality of the AO correction. It will finally allow the use of off-line PSF reconstruction algorithms (Véran *et al.* 1997, and Chapter 14 of this volume).

The overall control of the ADONIS instrument as well as the user's interface is managed by an Unix workstation (HP9000/720), called the master computer, which incorporates artificial intelligence software (AIS). Figure 8.6 illustrates the global architecture of the ADONIS control system and data flow (Demailly *et al.* 1994). The master computer is interfaced to each element of the adaptive optics system (optomechanical bench, micro-controller, RTC, IR camera) and observatory environment (telescope control system, seeing and weather monitors, etc.) by means of Ethernet or RS-232 links. Internally, a client/server architecture allows the interactions between these subsystems and a user-operated control panel (Demailly 1996). This panel provides an overview of the system set-up and status parameters as well as relevant data for evaluating the current performance of the instrument. The AIS performs most of the optimization and control tasks, an important example of which being the modal control optimization tool which evaluates the best modal control matrix in relation to

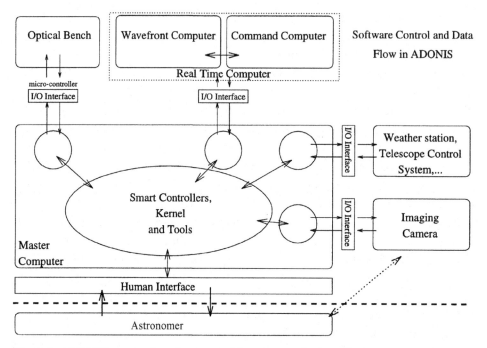

Fig. 8.6. ADONIS control software architecture and data flow. Reproduced from Demailly *et al.* (1994).

the prevailing atmospheric turbulence characteristics and the wave-front sensing noise (Gendron and Léna 1994, 1995).

8.2.4 Focal plane instrumentation

Two dedicated infrared cameras are available to the astronomer at the ADONIS $f/45$ output focus in addition to the possibility of installing visitor equipment, which could be either a different camera or a complementary observing mode such as a coronograph, a polarimeter, a 3-D spectrograph, etc.

8.2.4.1 Infrared cameras

The first camera, an improved version of SHARP II, built by MPIE in Garching (Hofmann *et al.* 1995; ESO Adaptive Optics Group *et al.* 1995), is based on a 256×256 NICMOS-3 HgCdTe detector array, sensitive to the $1-2.5\ \mu m$ spectral range. It features the standard J, H, K, and K$'$ photometric filters as well as narrow band filters centered on the spectral lines of FeII, HeI, H_2, OII, Pβ, Pγ. A low spectral resolution imaging mode (R $\sim$ 70) is also provided by a circular variable filter (CVF) in the 1.3−2.38 μm range. A set of exchangeable objectives provide three different image scales: 0.035, 0.05, and 0.1 arcsec/pixel. The differential atmospheric dispersion preventing the achievement of diffraction-limited observations at high zenithal distances in broad-band imaging, especially in J and H bands, an **A**tmospheric **D**ispersion **C**orrector (ADC) can be inserted in front of the camera in order to limit the wavelength dependent elongation to a maximum value of 0.005″.

The second camera, called COMIC (**COME-ON-PLUS** **I**nfrared **C**amera), based on a 128 × 128 HgCdTe/CCD focal plane array from the French CEA−LETI/LIR manufacturer, was developed by Observatoire de Paris and Observatoire de Grenoble (Feautrier *et al.* 1994; Lacombe *et al.* 1997). The array covers the 1−5 μm spectral range but is particularly optimized for the 3−5 μm region due to its very high storage capacity of 6×10^6 electrons for a total read-out noise of about 1000 electrons. Two different image scales can be selected, depending on the wavelength of observation: 0.035 arcsec/pixel for the J, H, and K bands, leading to a 4.5 × 4.5 arcsec FOV, and 0.1 arcsec/pixel for the L and M bands leading to a 12.8 × 12.8 arcsec FOV. Standard broadband photometric filters (J, H, K, short K, L, L′, and M) are provided, as well as exchangeable narrow-band and continuum filter doublets dedicated to specific astronomical targets (spectral lines of HeI, Brγ, H_2O, PAH, H_{3+} and Brα). Two CVFs, covering the 1.34−4.52 μm range, allow low spectral resolution imaging with R ∼ 80−120.

8.2.4.2 *Complementary modes*

Two additional modes, provided by MPIE together with the SHARP II camera, are already available and can be used with either IR camera (Hofmann *et al.* 1995): a wire grid polarizer unit working in the 1−5 μm range which allows polarimetric measurements for any linear polarizer position angle and two Fabry-Perot etalons covering the K band with typical respective spectral resolutions of 1000 and 2200 and corresponding finesses of 42 and 46.

A stellar coronograph dedicated to the COME-ON-PLUS/ADONIS systems has been developed in a collaboration between Observatoire de Paris and Observatoire de Grenoble (Beuzit *et al.* 1997). By using an occulting mask to block the flux of a bright object, coronographic techniques allow exploration of its close environment to search for faint sources such as stellar or substellar companions or disks. The coupling with AO greatly improves the efficiency of coronographic imaging: the AO system concentrates the flux from the bright source and reduces the wings of the PSF, therefore allowing the use of smaller occulting masks and consequently the exploration of a region very close to the central object. A first attempt to obtain coronographic observations with the COME-ON prototype has been described by Malbet (1996). First astrophysical results have been obtained with the ADONIS coronograph on the β Pictoris circumstellar disk (Mouillet *et al.* 1997; see also Chapter 15 in this volume). New fields will soon benefit from this coronographic mode: search for brown dwarfs around nearby stars, study of AGB outflows, etc.

Furthermore, a dedicated integral field spectrometer called GraF has been developed at Observatoire de Grenoble (Chalabaev and Le Coarer 1994). Tested on the sky in 1997, regular observations began in May 1998. GraF allows both the spectral distribution of energy as well as the spatial distribution over the sky field to be recorded simultaneously for a given source therefore providing a 3-D capability. This ensures that all monochromatic images are recorded with the same PSF. It combines a grating spectrograph with a Fabry-Perot interferometer. When used at 2.2 μm with a plate scale of 50 milliarcsec/pixel GraF will offer simutaneous imaging of a $1 \times 12''$ field in 12 narrow-band spectral channels. The resolution will range from 10 000 to 30 000.

8.3 System performance

The very first results obtained in 1989 on the OHP 1.52-m telescope were reported by Rousset *et al.* (1990a), Kern *et al.* (1990), and Merkle *et al.* (1990) and demonstrated the applicability of adaptive optics in astronomy. The observations performed showed two important features in the images:

- The gain in angular resolution leads to nearly diffraction-limited images made of a central coherent core surrounded by a broad residual halo at and above a critical imaging wavelength (2.2 μm for COME-ON under the encountered seeing conditions). The angular resolution is given by the full width at half-maximum (fwhm) of the image, mainly enforced by the central core.
- For objects smaller than the seeing disk size, the gain in energy concentration in the central core significantly improves the SNR of the IR images (whether it is background- or detector-noise limited). The energy concentration is expressed by the Strehl ratio R, that is the ratio of the peak intensity in the recorded long exposure point-spread function (PSF) over that in the theoretical diffraction-limited PSF.

A large number of results obtained on the ESO 3.6-m telescope have been reported with detailed analysis of the correction performance, firstly for COME-ON by Rigaut *et al.* (1991a,b; 1992c), Fontanella *et al.* (1991) and Rousset (1992) and secondly for COME-ON-PLUS by Rousset *et al.* (1993; 1994), Gendron and Léna (1995), and Beuzit (1995). The results presented in the next two sections are mainly selected from these publications. The observations consisted of taking long exposure IR images with and without AO compensation and simultaneously recording sets of WF measurements at the WFS frame rate and about one minute long. From the WFS data, WF phases are computed and their spatial and temporal properties deduced. Hence, observing conditions can be derived as values of $r_0(0.5 \, \mu\text{m})$, average wind speed $\bar{v}$, measurement noise, etc. The fwhm and R are obtained from the IR images recorded at several wavelengths. Residual static aberrations, like the triangular coma for instance, can also be identified in the images.

To illustrate the astronomical performance of AO, Fig. 8.7 presents images of double stars taken with correction (top) and without correction (bottom). Fig. 8.7(a) shows at $\lambda = 3.87 \, \mu\text{m}$ (L' band) the double star HR 6658 with a separation of 0.38″. Broken diffraction rings are visible around each component. Figure 8.7(b) shows at $\lambda = 2.23 \, \mu\text{m}$ (K band) another double star HR 5089 with a separation of 0.22″. In spite of the halo around the central cores due to the partial correction, the components are obviously resolved. A detailed discussion of astronomical results is given in Chapter 15.

8.3.1 Wavefront residuals

The correction brought by the AO system can be directly analyzed on the corrected WFs. Let us consider the WF residual error as measured by the WFS

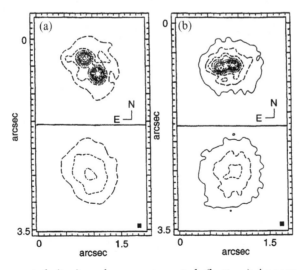

Fig. 8.7. Compensated (top) and uncompensated (bottom) images of double stars. Contour levels are 1 to 90% of the maximum. (a) HR 6658 at 3.87 μm, separation 0.38″, $r_0(0.5 \, \mu\text{m}) = 15$ cm. (b) HR 5089 at 2.23 μm, separation 0.22 ″, $r_0(0.5 \, \mu\text{m}) = 12$ cm. COME-ON, $f_{\text{bw}} = 9$ Hz. Reproduced from Rigaut *et al.* (1991a).

and compare it to the turbulence-induced distortion. The phase $\varphi(x, y)$ is expanded on the Zernike polynomials $Z_i(x, y)$ as:

$$\varphi(x, y) = \sum_{i=2}^{i=N} a_i Z_i(x, y) \tag{8.1}$$

where the a_i's are the expansion coefficients.

Figure 8.8 displays the variance $\langle a_i^2 \rangle$ of the coefficients of the first $N = 21$ Zernike polynomials (expressed at 0.5 µm) as observed in uncorrected and corrected sets of WFs (observations made with the COME-ON system on a bright star). Crosses are two samples of turbulent WFs. As theoretically predicted (Noll 1976), the low order polynomials are dominant in the disturbance, especially the two first polynomials: the tilts Z_2 and Z_3. Closed symbols (circle and square) are two samples of residual errors. The attenuation brought by AO on the two tilt variances is about 200 to 400 and about 100 on that of the three second-degree polynomials defocus and astigmatisms (Z_4 to Z_6). The low orders are much more attenuated than the higher orders. These residuals are due to the effect of the finite temporal bandwidth of the system (here $f_{bw} = 25$ Hz in open loop at 0 dB) being the same for all the modes. In Fig. 8.8, the measured WF variance $\sum_{i=2}^{21} \langle a_i^2 \rangle$ is reduced from 450 and 230 rad² for the turbulence to a residual WF error of 4.5 rad² at 0.5 µm. Note that $r_0(0.5 \text{ µm})$ is also deduced from the uncorrected measurements and allows the evaluation of the fitting error $\sum_{i=22}^{\infty} \langle a_i^2 \rangle$ given by Noll (1976): for Fig. 8.8, about 8 rad² for $r_0(0.5 \text{ µm}) \simeq 10$ cm. The total residual error variance is then about 13 rad².

The outer scale of turbulence L_0 can also be deduced from the WF

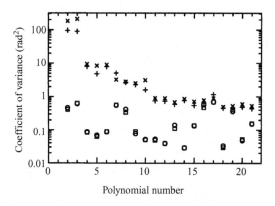

Fig. 8.8. Variance of the Zernike coefficients $\langle a_i^2 \rangle$ expressed at 0.5 µm versus the polynomial number i for uncorrected ($+$, $\times$) and AO corrected ($\bigcirc$, $\square$) sets of measured wave fronts. COME-ON, April 1991, $f_{bw} = 25$ Hz open loop 0 dB, $r_0(0.5 \text{ µm}) \simeq 10$ cm, $\bar{v} = 5$ m/s. Reproduced from Rigaut *et al.* (1991b).

expansion. Indeed, both r_0 and L_0 can be estimated by fitting the coefficient variances with theoretical values derived from a Kolmogorov spectrum modified by the outer-scale (Winker 1991). L_0 is essentially determined by the attenuation of the two tilt variances. We found L_0 varying between 10 m and infinity depending on the observing conditions (Rigaut *et al.* 1991a; Rousset *et al.* 1991). These values may be affected by any problem of telescope tracking.

The compensation by an AO system can be understood as a filtering process reducing the temporal power spectral density (PSD) of the turbulent WF (see Chapter 6). The servo-loop bandwidth f_{bw} represents the maximum frequency at which the AO system still compensates for the turbulence distortions. Figure 8.9 is an example of two measured PSDs of the defocus aberration term (Z_4) without and with correction by COME-ON. Note that the turbulence knee frequency is lower than 2 Hz and the system attenuates the turbulent PSD at low frequencies up to 20 Hz. In fact, the corrected PSD is the product of the uncorrected PSD by the error transfer function of the servo-loop. The corresponding attenuation in variance is around 100 (same data as Fig. 8.8). Note that a lower bandwidth f_{bw} would increase the turbulence residuals as was the case on COME-ON in 1990 (Rigaut *et al.* 1991a). The measurements of the turbulence knee frequencies on the Zernike PSDs have been used to estimate the average wind speed $\bar{v}$ from the expression derived by Conan *et al.* (1995). With the data used in Figs. 8.8 and 8.9, $\bar{v}$ is estimated to be 5 m/s.

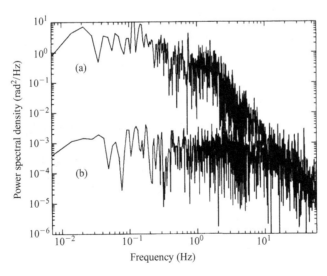

Fig. 8.9. Measured power spectral densities of the Zernike coefficient a_4 (defocus), (a) from an uncorrected WF set, (b) from a corrected set. COME-ON, April 1991, $f_{bw} = 25$ Hz open loop 0 dB, $r_0(0.5 \ \mu m) \simeq 10$ cm, $\bar{v} = 5$ m/s. Reproduced from Rigaut *et al.* (1991b).

The previous data were recorded on bright stars. For faint stars, the estimation of the noise level in the WF measurement is of great importance in the optimization of the correction. Figure 8.10 presents measured PSDs of Zernike polynomials (tilt Z_3 and 5th order Z_{20}) without any correction for high or low

(a)

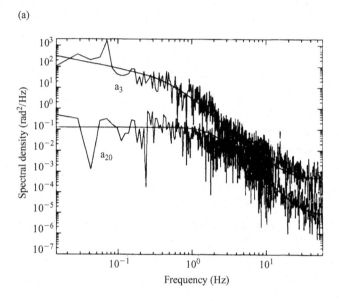

(b)

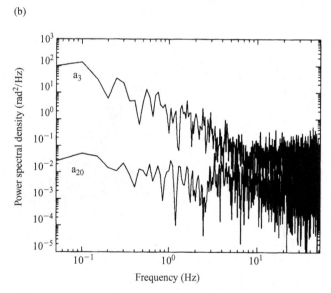

Fig. 8.10. Measured power spectral densities of two Zernike polynomials, Z_3 and Z_{20}. (a) for a bright star: $r_0(0.5\ \mu m) \simeq 9$ cm, $\bar{v} = 4$ m/s. (b) for a faint star: nine detected photons per subaperture and per exposure time, $r_0(0.5\ \mu m) \simeq 12$ cm. Reproduced from Rousset (1992).

light levels. At a low light level, an average of 9 photons per subaperture and per exposure time were detected. The SNRs of the Zernike coefficients are 9 for the tilt and 0.5 for Z_{20}. The low order polynomials (or modes) have better SNR than the higher orders, as already pointed out in Section 8.2.2. This shows the importance of the optimized modal control developed for the COME-ON-PLUS system. Indeed, the turbulence and noise contributions to the PSD are not distinguishable for Z_{20} in Fig. 8.10. Therefore, the servo-loop bandwidth for this mode must be much lower than for the tilt.

Depending on the seeing and the guide star magnitude, the mode SNRs can significantly change. For SNR < 1, the modes are not very efficient in the correction. Hence, increasing the guide star magnitude limits the degree of correction of the system. Figure 8.11 shows the number of corrected modes versus the magnitude for the COME-ON-PLUS system. Here, a mode is considered as corrected when the ratio of its corrected variance to its uncorrected one is below a threshold of 0.5. The points spread widely around the average value. These deviations depend on the observed guide star characteristics and the turbulence conditions. Figure 8.11 shows that the correction steadily decreases as the magnitude m_V increases from 8 to 13.

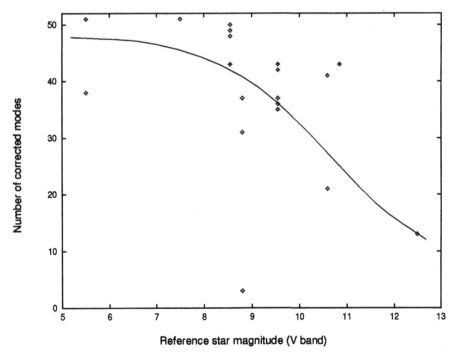

Fig. 8.11. Number of modes for which the corrected variance is less than 0.5 times the uncorrected one, versus the magnitude of the guide star. Reproduced from Gendron and Léna (1995).

8.3.2 Image characteristics

We analyze here long exposure IR images of an unresolved star, i.e. the PSF of the system. The exposure time ranges from a few seconds to several minutes. The two parameters of interest are the R and the fwhm. As an example of an image, Fig. 8.12 displays the compensated and uncompensated images of a star at $\lambda = 1.68\,\mu\text{m}$ (H band) obtained with COME-ON. For the compensated image $R = 0.24$ and fwhm $= 0.12''$. For the uncompensated one $R \sim 0.02$ and fwhm $\sim 0.65''$, the telescope aberrations being removed by the deformable mirror. The compensated image consists of a central core on top of a halo. The gain in R is of the order of 10 and the core fwhm reaches the diffraction limit of $0.096''$.

Figure 8.13 shows the IR image Strehl ratio R versus r_0 calculated at the imaging wavelength λ, for different observing conditions with COME-ON-PLUS. Let us first consider the results for bright stars ($6 \leqslant m_V \leqslant 10$) (square and diamond). R values are measured in I, J, H, K, and L$'$ bands. r_0 ($0.5\,\mu\text{m}$) ranges from 11 to 15 cm and $\bar{v}$ from 4 to 20 m/s. At $2.23\,\mu\text{m}$ (K), $0.57 \leqslant R \leqslant 0.8$ with $r_0(2.23\,\mu\text{m}) \sim 80$ cm. At $1.68\,\mu\text{m}$ (H), $0.26 \leqslant R \leqslant 0.47$ with $r_0(1.68\,\mu\text{m}) \sim 50$ cm. At $1.25\,\mu\text{m}$ (J), $0.1 \leqslant R \leqslant 0.17$ with $r_0(1.25\,\mu\text{m}) \sim 35$ cm. The given R values are lower than the values expected from the WFS data. We suspect a static aberration in the imaging path limiting the image quality. A triangular pattern is clearly seen on the J and H images. Note that the scatter of the results can be explained by the seeing variations and the influence of the parameter $\bar{v}$ and of the SNR in the WF sensing. Theoretical

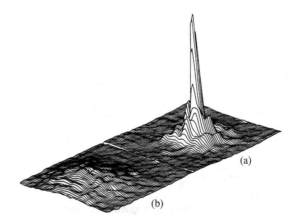

Fig. 8.12. Compensated (a) and uncompensated (b) images of an unresolved star at $\lambda = 1.68\,\mu\text{m}$. COME-ON, $f_{\text{bw}} = 25$ Hz, $r_0(0.5\,\mu\text{m}) \simeq 12$ cm, $\bar{v} = 4$ m/s. Reproduced from Rousset (1992).

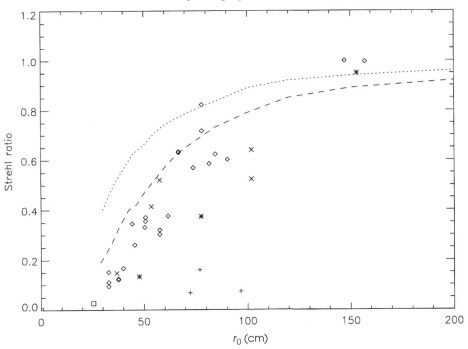

Fig. 8.13. Strehl ratio R versus $r_0(\lambda)$ for the COME-ON-PLUS system. Bright stars $6 \leq m_V \leq 10$: ($\Diamond$) values measured in J, H, K, and L′ bands, ($\Box$) in I band, $11 \leq r_0(0.5\ \mu m) \leq 15$ cm, $4 \leq \bar{v} \leq 20$ m/s. Faint stars: ($\times$) $11 \leq m_V \leq 12$, (+) $12 \leq m_V \leq 14$, ($*$) $14 \leq m_V$, $6 \leq r_0(0.5\ \mu m) \leq 10$ cm, $4 \leq \bar{v} \leq 25$ m/s. Upper and lower curves: 52 degrees of correction, respectively for an infinite and a 25 Hz bandwidth. Reproduced from Rousset *et al.* (1993).

behaviors are also reported in Fig. 8.13 for 52 degrees of correction at infinite SNR with infinite and 25 Hz bandwidth.

Let us now consider the results for fainter stars ($11 \leq m_V \leq 15$) ($\times$, + and asterisk). Strehl ratios R are measured in K and L′ bands with r_0 (0.5 μm) ranging from 6 to 10 cm and $\bar{v}$ from 4 to 25 m/s. The results dispersion is mainly due to very different turbulence conditions during the observations. For the lowest magnitudes and under bad seeing conditions, the results follow relatively well the behavior of the results obtained on bright stars. This is a demonstration of the quality of the optimized modal control. Considering the increase of the magnitude, a decrease of the correction quality is observed.

Figure 8.14 displays the fwhm measured on the same IR images as Fig. 8.13 versus the imaging wavelength λ. For bright stars, the diffraction limit is reached down to 1.68 μm (H), i.e. 0.096″. In the J band, fwhm is close to the diffraction limit. Note that a 0.2″ resolution is achieved in the I band. For faint

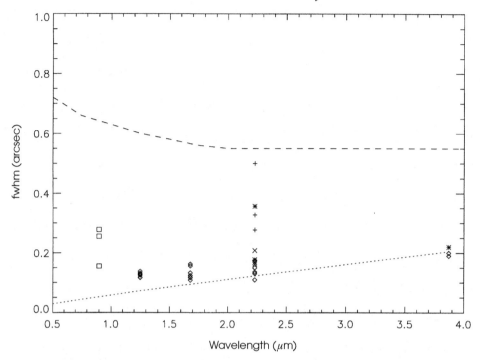

Fig. 8.14. Image fwhm versus imaging wavelength for the COME-ON-PLUS system. Same experimental data and symbols as in the previous figure. Lower curve: diffraction limit. Upper curve: seeing fwhm for $r_0(0.5\ \mu m) = 15$ cm. Reproduced from Rousset *et al.* (1993).

stars, both the magnitude and the turbulence conditions affect the fwhm. In the K band, fwhm ranges from 0.13 to 0.51″.

Figure 8.15 presents the distribution of R versus fwhm for the calibration PSFs recorded in one night at J, H, and K bands on the COME-ON-PLUS system. The exposure time was 1 s in H and K, 20 s in J. The image in K is well corrected, fwhm being around 0.15″ and R mainly between 0.2 and 0.35. In H, R drops below 10%, a key value below which the effects of the turbulence induce large variations of fwhm between 0.13 and 0.25″. It demonstrates a sensitive threshold effect on the efficiency of AO. Note that for some images the narrower diffraction core in H produces a smaller fwhm compared to K in spite of a lower R. The distribution of the points in Fig. 8.15 illustrates how the PSF varies as the turbulence conditions continuously change during the observations. When the fwhm is rescaled by the factor λ/D, the diffraction limit points clearly gather along a single curve. As underlined by this curve, R describes the PSF quality very well when R is above 10%, but only poorly for the lower values. In this latter case, fwhm is a better characteristic of the PSF quality.

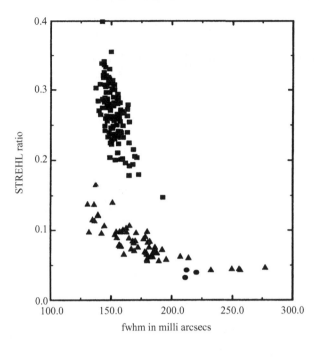

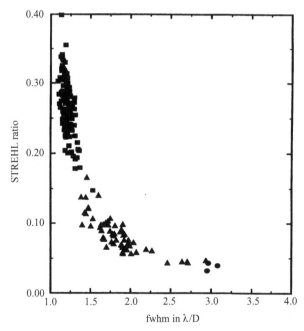

Fig. 8.15. *R* versus fwhm for individual compensated PSFs at J, H, and K bands, obtained with COME-ON-PLUS in one night. Top: fwhm in milliarcsec. Bottom: same data but fwhm in λ/D units. Reproduced from Tessier *et al.* (1994). ●, J band; ▲, H band; ■, K band.

As an example, Fig. 8.16 shows a typical long exposure image profile taken at $\lambda = 1.68$ μm with the COME-ON system. A common feature observed in a number of images is the presence of two components in the profile: namely a sharp diffraction-limited core on top of a broad halo as foreseen by simulations (Smithson *et al.* 1988). A fit of the halo is superimposed on the image profile. The central core in the long exposure image is in fact fully coherent while the halo is an incoherent superimposition of residual speckles. The fraction of energy in the core is 16% while 84% is spread around in the halo. The fwhm of the halo is 0.36″ while fwhm of the uncompensated image is around 0.5″. Even if the amount of energy concentrated in the central core is relatively small, the available coherence is of paramount importance for the interferometric combination of large telescopes (Roddier and Léna 1984; Merkle and Léna 1986; Chapter 14 in this volume).

Partial correction may range from quite poor correction (in the visible) to very good correction (at L′ band) with large variation in R. The obtained image profile, as in Fig. 8.16, results directly from how adaptive optics correct the WF disturbances. A way to characterize this correction is the analysis of the optical transfer function (OTF). In long exposures, the OTF after AO correction can be expressed as the product of the aberration-free telescope OTF by the term $\exp[-\frac{1}{2}D_\varphi(\lambda f)]$ where f is the spatial frequency and $D_\varphi(\rho = \lambda f)$ is the structure function of the WF residual phase (Conan *et al.* 1992). This function is wavelength independent when expressed in terms of the optical path

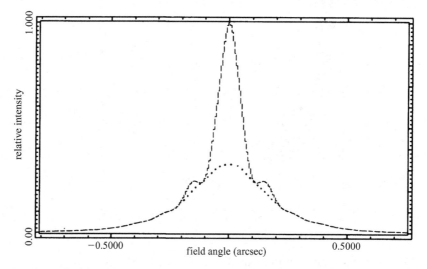

Fig. 8.16. *Dashed line:* long exposure profile of a compensated image at $\lambda = 1.68$ μm with the COME-ON system. *Dotted line:* the fitted halo. $f_{bw} = 25$ Hz, $r_0(0.5$ μm$) \simeq 18$ cm, $\bar{v} = 6$ m/s. Reproduced from Rousset (1992).

difference. Figure 8.17 shows examples of structure functions of the residual WF, derived from a set of images at J, H, and K bands and illustrates the basic effect of the correction on the WF. Because the deformable mirror compensates for the low orders (low spatial frequencies) of the WF distortion, D_φ saturates for large separations in the pupil ($\rho > 0.9$ m, the inter-actuator distance), i.e. the high spatial frequencies of the image. Indeed for large separation, the phase φ is decorrelated and as a result $D_\varphi = 2\sigma_\varphi^2$, where σ_φ^2 is the variance of the residual phase. The saturation of D_φ produces the central core of the image, and the fraction of energy in this core is $\exp(-\sigma_\varphi^2)$. Figure 8.17 shows that for small separations in the pupil, D_φ increases with ρ producing the broad halo in the image. The growth is slightly slower than the $\rho^{5/3}$ law of turbulence. In Fig. 8.17 at the largest separations in the pupil, the saturation of D_φ is disturbed by the effects of the WF low order residuals due to the finite temporal bandwidth of the system and to the measurement noise.

In Figure 8.18, the normalized profiles of two long exposure compensated images obtained for the components of the double star D177 at $\lambda = 2.2$ μm are plotted. Here, the maximum central intensities represent directly the Strehl ratios. In this experiment with the COME-ON system, the WFS guide star was the on-axis component of D177 and the IR images were recorded sequentially on the on-axis and off-axis components. The companion separation is $27''$. These two profiles illustrate the anisoplanatism effect in the AO compensation. Off-axis, we measure a loss in R of the order of 30%. Using the structure functions derived from these images, an anisoplanatism WF error of the order of 0.3 μm (rms) is found. Images recorded at $\lambda = 1.68$ μm during the same night lead also to a WF error of the order of 0.3 μm. These

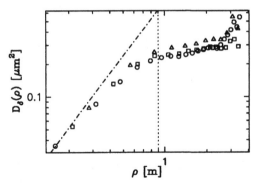

Fig. 8.17. Structure function of the WF residual phase derived from the compensated images at J, H, and K bands, versus the separation ρ in the pupil. COME-ON, April 1991. The vertical dotted line marks the inter-actuator distance. The dashed line gives the turbulence law in $\rho^{5/3}$. $\bigcirc$, J; $\square$, H; $\triangle$, K. Reproduced from Rigaut *et al.* (1992c).

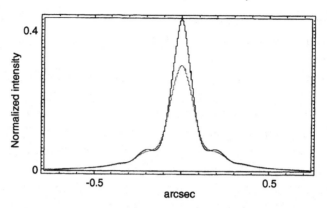

Fig. 8.18. Normalized long exposure compensated image profiles of the two components of the double star D177 (separation: 27″) at $\lambda = 2.2$ μm with the COME-ON system. Continuous line: on-axis component, $SR \simeq 0.43$. Dotted line: off-axis component, $SR \simeq 0.3$. $f_{bw} = 25$ Hz, $r_0(0.5$ μm$) \simeq 12$ cm, $\bar{v} = 3$ m/s. Images produced by F. Rigaut.

are relatively small decorrelations of the wave front for an angular distance of 27″. Note also the reduction of the central core for the off-axis image while its halo is smoothed. Indeed, the central core is much more sensitive to any additional WF error. The observed low degradation of the off-axis images when compared to conventional theoretical evaluation (Fried 1982), can be explained by the presence of important turbulence close to the telescope (dome seeing and boundary layer).

Although COME-ON was the first AO system in astronomy, since 1989 a number of new systems have been built for the 2- to 4-m class telescopes. The next step is a new generation of very large telescopes such as the VLT of ESO. The experience gained on the smaller ones will be very helpful to properly specify and design these systems in order to meet the requirements of the astronomers. The application of the Laser Guide Star concept to astronomical observations is the next challenge.

8.4 Acknowledgements

The research described here was carried out under ESO contracts and received significant financial support from the Ministère de la Recherche and the Ministère de l'Education Nationale (France). Early developments of AO components and systems at CILAS and ONERA were independently supported by the Direction des Recherches, Etudes et Techniques, Ministère de la Défense (France). We are grateful to our numerous colleagues from the different institutes and companies who were involved in the realization of the COME-ON, COME-ON-PLUS and ADONIS programs. Additional thanks are extended to the staff of the European Southern Observatory (ESO) in Garching and La Silla for their active help during the integration and tests at the telescope. We are also grateful to our colleagues from the Max Planck Institut für Extraterrestrische Physik (MPIE) in Garching who implemented and operated their SHARP II infrared camera on COME-ON-PLUS.

References

Acton, D. S. and Smithson, R. C. (1992) Solar imaging with a segmented adaptive mirror. *Appl. Opt.* **31**, 3161–69.

Barr, L. D. (ed.) (1986) Advanced technology optical telescopes III. Proc. SPIE 628. Bellingham: The Society of Photo-optical Instrumentation Engineers.

Beckers, J. M., Roddier, F. J., Eisenhardt, P. R., Goad, L. E. and Shu, K. L. (1986) In: *Advanced Technology Optical Telescopes III*, ed. L. D. Barr, Proc. SPIE 628, pp. 290–97.

Beuzit, J.-L., Hubin, N., Gendron, E., Demailly, L., Gigan, P., Lacombe, F., *et al.* (1994) ADONIS: a user-friendly adaptive optics system for the ESO 3.6-m telescope. In: *Adaptive Optics in Astronomy*, eds M. A. Ealey, F. Merkle, Proc. SPIE 2201, pp. 955–61.

Beuzit, J.-L., Hubin, N., Demailly, L., Gendron, E., Gigan, P., Lacombe, F., *et al.* (1995) ADONIS: a user-friendly adaptive optics system for the ESO 3.6 m telescope. In: *Adaptive Optics*, OSA/ESO Topical Meeting, ESO Conference Proceedings, No. 54, ed. M. Cullum, pp. 57–62. Garching: European Southern Observatory.

Beuzit, J.-L. (1995) Thèse de Doctorat. Paris. Grenoble: Université Pierre et Marie Curie.

Beuzit, J.-L., Mouillet D., Lagrange A.-M. and Paufique J. (1997) A stellar coronograph for the COME-ON-PLUS adaptive optics system. I. Description and performance. *Astron. Astrosphys. Suppl. Ser.* **125**, 175–82.

Boyer, C., Michau, V. and Rousset, G. (1990) Adaptive optics: interaction matrix measurements and real-time control algorithms for the COME-ON project. In: *Amplitude and Intensity Spatial Interferometry*, ed. J. B. Breckinridge, Proc. SPIE 1237, pp. 406–15.

Chalabaev, A. and Le Coarer, E. (1994) GraF for ADONIS: proposal of high resolution spectro-imaging option for the ESO adaptive optics instrument ADONIS. *Rapport Technique de l'Observatoire de Grenoble*, **1**.

Conan, J.-M., Madec, P.-Y. and Rousset, G. (1992) Evaluation of image quality with adative optics partial correction. In: *Progress in Telescope and Instrumentation Technologies*, ESO Conference Proceedings, No. 42, ed. M. H. Ulrich, pp. 475–78. Garching: European Southern Observatory.

Conan, J.-M., Rousset, G. and Madec, P.-Y. (1995) Wavefront temporal spectra in high resolution imaging through turbulence. *J. Opt. Soc. Am. A* **12**, 1559–70.

Cuby, J.-G., Richard, J.-C. and Lemonier, M. (1990) Electron bombarded CCD: first results with a prototype tube. In: *Instrumentation in Astronomy*, ed. D. L. Crawford, Proc. SPIE 1235, pp. 294–304.

Demailly, L. (1996) Thèse de Doctorat. Caen: Université de Caen.

Demailly, L., Gendron, E., Beuzit, J.-L., Lacombe, F. and Hubin, N. (1994) Artificial Intelligence system and Optimized Modal control for the ADONIS adaptive optics instrument. In: *Adaptive Optics in Astronomy*, ed. M. A. Ealey, F. Merkle, Proc. SPIE 2201, pp. 867–78.

Doel, A. P., Dunlop, C. N., Major, J. V., Myers, R. M., Purvis, A. and Thompson, M. G. (1990) Stellar image stabilization using piezo-driven active mirrors. In: *Advanced Technology Optical Telescopes IV*, ed. L. D. Barr, Proc. SPIE 1236, pp. 179–92.

ESO Adaptive Optics Group, Lacombe, F., Marco, O., Eisenhauer, F. and Hofmann, R. (1995) First light on COMIC and SHARP II+. *The Messenger* **82**, 16–17.

Feautrier, P., Geoffray, H., Petmezakis, P., Monin, J.-L., Le Coarer, E. and Audaire, L. (1994) The 1–5 micron imaging detector for the ADONIS adaptive optics system. In: *Infrared Spaceborne Remote Sensing II*, ed. M. Scholl, Proc. SPIE 2268, pp. 386–97.

Fontanella, J.-C. (1985) Analyse de surface d'onde, déconvolution et optique active. *J. Optics (Paris)* **16**, 257–68.

Fontanella, J.-C. Rousset, G. and Léna, P. (1991) L'optique adaptative, un élément-clé du très grand télescope européen VLT. *J. Optics (Paris)* **22**, 99–111.

Fried, D. L. (1965) Statistics of geometric representation of wavefront distortion. *J. Opt. Soc. Am.* **55**, 1427–35.

Fried, D. L. (1982) Anisoplanatism in adaptive optics. *J. Opt. Soc. Am.* **72**, 52–61.

Gaffard, J.-P., Touraut, C. and De Miscault, J.-C. (1984) Adaptive optics modal control. In: Conference on *Lasers and Electro-Optics* held in Anaheim, Proc. OSA, paper TH15. Optical Society of America, Washington.

Gendron, E., Cuby, J.-G., Rigaut, F., Léna, P., Fontanella, J-C., Rousset, G., *et al.* (1991) The COME-ON-PLUS project: an upgrade of the COME-ON adaptive optics prototype system. In: *Active and Adaptive Optical Systems*, ed. M. A. Ealey, Proc. SPIE 1542, pp. 297–306.

Gendron, E. (1993) In: *Active and Adaptive Optics*, ICO 16 Satellite Conference, ESO Conference Proceedings, No. 48, ed. F. Merkle, pp. 187–92. Garching: European Southern Observatory.

Gendron, E. and Léna, P. (1994) Astronomical adaptive optics I, modal control optimization. *Astron. Astrophys.* **291**, 337–47.

Gendron, E. and Léna, P. (1995) Astronomical adaptive optics II, experimental results of an optimized modal control. *Astron. Astrophys. Suppl. Ser.* **111**, 153–67.

Greenwood, D. P. (1977) Bandwidth specification for adaptive optics systems. *J. Opt. Soc. Am.* **67**, 390–93.

Hardy, J. W., Lefebvre, J. E. and Koliopoulos, C. L. (1977) Real-time atmospheric compensation. *J. Opt. Soc. Am.* **67**, 360–69.

Hofmann, B., Brandl, B., Eckart, A., Eisenhauer, F. and Tacconi-Garman, L. (1995) High-angular-resolution NIR astronomy with large arrays (SHARPI and SHARPII) In: *Infrared Detectors and Instrumentation for Astronomy*, ed. A.M. Fowler, Proc. SPIE 2475, pp. 192–202.

Hubin, N., Beuzit, J.-L., Gigan, P., Léna, P., Madec, P.-Y., Rousset, G., *et al.* (1992) New adaptive optics prototype system for the ESO 3.6-m telescope: COME-ON-PLUS. In: International Symposium on *Lens and Optical Systems Design*, ed. H. Zuegge, Proc. SPIE 1780, pp. 850–61.

Jagourel, P. and Gaffard, J.-P. (1991) Adaptive optics components. In: Laserdot. *Active and Adaptive Optical Components*, ed. M. A. Ealey, Proc. SPIE 1543, pp. 76–87.

Kern, P., Léna, P., Rousset, G., Fontanella, J.-C., Merkle, F. and Gaffard, J.-P. (1988) Prototype of an adaptive optics system for infrared astronomy. In: *Very Large Telescopes and their Instrumentation*, ESO Conference Proceedings, No. 30, ed. M. H. Ulrich, pp. 657–65. Garching: European Southern Observatory.

Kern, P., Rigaut, F., Léna, P., Merkle, F. and Rousset, G. (1990) Adaptive optics prototype system for IR astronomy II: first observing results. In: *Amplitude and Intensity Spatial Interferometry*, ed. J. B. Breckinridge, Proc. SPIE 1237, pp. 345–55.

Lacombe, F., Tiphène, D., Rouan, D., Léna, P. and Combes, M. (1989) Imagery with infrared arrays I: ground-based and astronomical performances. *Astron. Astrophys.* **215**, 211–17.

Lacombe, F., Marco, O., Geoffray, H., Beuzit, J.-L., Monin, J.-L., Gigan, P., *et al.* (1997) 1–5 micron Adaptive Optics Imaging on Large Telescopes. I. The COMIC Camera for ADONIS. *Pub. Astr. Soc. Pac.* in press.

Léna, P. (1995a) Astronomy with adaptive optics. In: *Adaptive Optics*, OSA/ESO Topical Meeting, ESO Conference Proceedings, No. 54, ed. M. Cullum, pp. 317–22. European Southern Observatory, Garching.

Léna, P. (1995b) From planets to galaxies: adaptive optics revolution and VLT interferometer. In: *Science with the VLT*, eds J. R. Walsh and I. J. Danziger, Springer, pp. 425–435.

Malbet, F., Rigaut, F., Léna, P. and Bertout, C. (1993) Detection of a 400 AU disk-like structure surrounding the young stellar object Z CMa. *Astron. Astrophys.* **271**, L9–12.

Malbet, F. (1996) High angular resolution coronography for adaptive optics. *Astron. Astrosphys. Suppl. Ser.* **115**, 161–74.

Merkle, F. and Léna, P. (1986) Spatial interferometry with the European VLT. In: *Advanced Technology Optical Telescopes III*, ed. L. D. Barr, Proc. SPIE 628, pp. 261–72.

Merkle, F., Rousset, G., Kern, P. and Gaffard, J.-P. (1990) First diffraction-limited astronomical images with adaptive optics. In: *Advanced Technology Optical Telescopes IV*, ed. L. D. Barr, Proc. SPIE 1236, pp. 193–202.

Mouillet, D., Lagrange, A.-M., Beuzit, J.-L. and Renaud, N. (1997) A stellar coronograph for the COME-ON-PLUS adaptive optics system. II. First astronomical results. *Astron. Astrosphys.* **324**, 1083–90.

Noll, R. J. (1976) Zernike polynomials and atmospheric turbulence. *J. Opt. Soc. Am.* **66**, 207–11.

Parenti, R. R. (1988) Recent advances in adaptive optics methods and technology. In: *Laser Wavefront Control*, ed. R. F. Reintjes, Proc. SPIE 1000, pp. 101–9.

Pearson, J. E. (1979) The whither and whether of adaptive optics. In: *Adaptive Optical Components II*, ed. S. Holly, Proc. SPIE 179, pp. 2–10.

Richard, J.-C., Bergonzi, L. and Lemonier, M. (1990) A 604 × 288 Electron-Bombarded CCD image tube for 2-D photon counting. In: *Optoelectronic Devices and Applications*, ed. S. Sriram, Proc. SPIE 1338, pp. 241–54.

Rigaut, F., Rousset, G., Kern, P., Fontanella, J.-C., Gaffard, J.-P., Merkle, F., *et al.* (1991a) Adaptive optics on a 3.6-m telescope: results and performance. *Astron. Astrophys.* **250**, 280–90.

Rigaut, F., Gendron, E., Léna, P., Madec, P.-Y., Couvée, P. and Rousset, G. (1991b) Partial correction with the adaptive optics system COME-ON. In: *High Resolution Imaging by Interferometry II*, ESO Conference Proceedings, No. 39, eds J. M. Beckers, F. Merkle, pp. 1105–12. European Southern Observatory, Garching.

Rigaut, F., Combes, M., Dougados, C., Léna, P., Mariotti, J-M., Saint-Pé, O., *et al.* (1992a) Astrophysical results with COME-ON. In: *Progress in Telescope and Instrumentation Technologies*, ESO Conference Proceedings, No. 42, ed. M. H. Ulrich, pp. 479–84. European Southern Observatory, Garching.

Rigaut, F., Cuby, J.-G., Caes, M., Monin, J.-L., Vittot, M., Richard, J.-C., *et al.* (1992b) Visible and infrared wavefront sensing for astronomical adaptive optics. *Astron. Astrosphys.* **259**, L57–60.

Rigaut, F., Léna, P., Madec, P.-Y., Rousset, G., Gendron, E. and Merkle, F. (1992c) Latest results of the COME-ON experiment. In: *Progress in Telescope and Instrumentation Technologies*, ESO Conference Proceedings, No. 42, ed. M. H. Ulrich, pp. 399–402. Garching: European Southern Observatory.

Roddier, F. and Léna, P. (1984) Long baseline Michelson Interferometry with large ground-based telescopes operating at optical wavelength. *J. Optics* (Paris) **15**, 171–182 and 363–374.

Rousset, G. (1992) Implementation of adaptive optics. In: *Wave Propagation in Random Media (Scintillation)*, ed. V. I. Tatarski, A. Ishimaru, V. U. Zavorotny, pp. 216–34. Bellingham: The Society of Photo-optical Instrumentation Engineers.

Rousset, G., Fontanella, J.-C., Kern, P., Gigan, P., Rigaut, F., Léna, P., *et al.* (1990a) First diffraction-limited astronomical images with adaptive optics. *Astron. Astrosphys.* **230**, L29–32.

Rousset, G., Fontanella, J.-C., Kern, P., Léna, P., Gigan, P., Rigaut, F., *et al.* (1990b) Adaptive optics prototype system for infrared astronomy I: system description. In: *Amplitude and Intensity Spatial Interferometry*, ed. J. B. Breckinridge Proc. SPIE 1237, pp. 336–44.

Rousset, G., Madec, P.-Y. and Rigaut, F. (1991) Temporal analysis of turbulent wavefronts sensed by adaptive optics. In: *Atmospheric, Volume and Surface Scattering and Propagation*, ICO Topical Meeting, ed. A. Consortini, pp. 77–80. International Commission for Optics.

Rousset, G., Madec, P.-Y., Beuzit, J.-L., Cuby, J-G., Gigan, P., Léna, P., *et al.* (1992) The COME-ON-PLUS project: an adaptive optics system for a 4-meter class telescope. In: *Progress in Telescope and Instrumentation Technologies*, ESO Conference Proceedings, No. 42, ed. M. H. Ulrich, pp. 403–6. Garching: European Southern Observatory.

Rousset, G., Beuzit, J.-L., Hubin, N., Gendron, E., Boyer, C., Madec, P.-Y., *et al.* (1993) The COME-ON-PLUS adaptive optics system: results and performance. In: *Active and Adaptive Optics, ICO 16 Satellite Conference*, ESO Conference Proceedings, No. 48, ed. F. Merkle, pp. 65–70. European Southern Observatory, Garching.

Rousset, G., Beuzit, J.-L., Hubin, N., Gendron, E., Madec, P.-Y., Boyer, C., *et al.* (1994) Performance and results of the COME-ON+ adaptive optics system at the ESO 3.6-meter telescope. In: *Adaptive Optics in Astronomy*, eds M. A. Ealey, F. Merkle, Proc. SPIE 2201, pp. 1088–98.

Saint-Pé, O., Combes, M., Rigaut, F., Tomasko, M. and Fulchignoni, M. (1993a) Demonstration of adaptive optics for resolved imagery of solar system objects: preliminary results on Pallas and Titan. *Icarus* **105**, 263–70.

Saint-Pé, O., Combes, M. and Rigaut, F. (1993b) Ceres Surface Properties by High-Resolution Imaging from Earth. *Icarus* **105**, 271–81.

Smithson, R. C., Peri, M. L. and Benson, R. S. (1988) Quantitative simulation of image correction for astronomy with a segmented active mirror. *Appl. Opt.* **27**, 1615–20.

Tessier, E., Bouvier, J., Beuzit, J.-L. and Brandner, W. (1994) COME-ON+ adaptive optics images of the pre-main sequence binary NX pup. *The Messenger* **78** 35–40.

Véran, J.-P., Rigaut, F., Maître, H. and Rouan, D. (1997) Estimation of the adaptive optics long exposure point-spread function using control loop data. *J. Opt. Soc. Am. A* **14**, 3057–69.

Winker, D. M. (1991) Effect of finite outer scale on the Zernike decomposition of atmospheric optical turbulence. *J. Opt. Soc. Am. A* **8**, 1568–73.

9

The UH–CFHT systems

FRANÇOIS RODDIER

Institute for Astronomy, University of Hawaii

FRANÇOIS RIGAUT

Canada–France–Hawaii Telescope

9.1 The birth of a new concept

The astronomical AO system developed at the University of Hawaii (UH) and its offspring, the AO user instrument of the Canada–France–Hawaii telescope (CFHT) are members of a new breed of AO systems based on the concept of wave-front curvature sensing and compensation (Roddier 1988). The concept emerged in the late 1980s at the Advanced Development Program (ADP) division of the US National Optical Astronomical Observatories (NOAO), as an output of a research program led by J. Beckers on the application of adaptive optics to astronomy.

Given the success of AO in defense applications, particularly surveillance systems, it was natural to seek components developed by the defense industry. The main difficulty was to obtain a good deformable mirror at a reasonable price. The technology being classified, one had no access to the latest developments. Commercially available mirrors were either monolithic mirrors, or first generation piezostack mirrors (see Chapter 4). Whereas monolithic mirrors of good optical quality could be purchased, their stroke was insufficient for the envisioned use on large astronomical telescopes. On the other hand, piezostack mirrors had enough stroke but were of poor optical quality and aged poorly. Most of all, the cost of a deformable mirror with suitable power supplies vastly exceeded budgets normally available for astronomical instrumentation.

It soon became clear that for surveillance applications one of the cost drivers was the high frequency response needed to follow the motion of satellites through the atmosphere. For astronomical applications, the requirement could be relaxed by roughly an order of magnitude. It was therefore worth looking into other potentially less expensive technologies which had been proposed but not developed owing to their lower frequency response. A particularly attractive approach was the use of bimorph mirrors as proposed by Steinhaus and

Lipson (1979). Unlike the mirrors considered above, the stroke of a bimorph mirror depends upon the scale of the deformation. It decreases as the square of its spatial frequency. This was considered an additional drawback for defense applications which require high spatial frequency corrections for observations in the visible. On the other hand, their large stroke at low spatial frequencies becomes an advantage for astronomical observations with large telescopes in the infrared, where a large stroke is needed at low spatial frequency.

Another advantage of bimorph mirrors becomes apparent when the wave-front sensor is considered. Ideally, a wave front sensor should be sensitive to any possible deformation of the flexible mirror, otherwise deformations to which the sensor is insensitive will occasionally occur and stay uncompensated, degrading the final image quality (Roddier, 1991, 1994b). Conventional wave-front slope sensors developed for defense applications are not particularly well matched to sense the deformation of their associated mirrors. To compensate for that, the number of wave-front measurements must significantly exceed the number of degrees of freedom of the deformable mirror. Applying a voltage to a bimorph electrode changes the total curvature (Laplacian) of the mirror integrated over the electrode area. Hence the idea of sensing directly the wave-front Laplacian (second order derivatives) rather than the wave-front slopes (first order derivatives). It happens that this is not only feasible but quite easy. The technique described in Chapter 5 consists of subtracting the illumination of two oppositely defocused pupil images. By matching the subapertures areas with that of the bimorph electrodes, one obtains a perfectly matched sensor with a nearly one to one relationship between the sensor elements and the bimorph actuators. The AO systems described in this chapter are based on this concept.

9.2 The experimental UH system

9.2.1 Choice of the detector

In the early 1990s, conventional systems developed for defense applications used either photomultipliers or intensified detector arrays to sense the wave front. They used the observed object itself (or bright glints on the object) as guide sources and were limited to point sources brighter than magnitude 10, or even less depending on seeing conditions. Only the brightest astronomical sources, or those that happened to be close enough to a bright star, could potentially be observed with these systems. This lack of sensitivity was considerably limiting their application to astronomy and the general consensus among US researchers was that only laser guide source systems would be useful to astronomy. This led the National Science Foundation (NSF) to request and obtain the declassification of defense research in this area and to finance the development of laser aided systems (see Chapter 1).

Meanwhile, European researchers felt sufficiently confident they could improve the sensitivity of conventional systems, to develop a natural guide star system for ESO. The gain in sensitivity was obtained by limiting the observations to long wavelengths (mainly 2.2 μm or longer) and by using a new type of detector array the EBCCD (see Chapter 8). This brought the limiting magnitude of the guide star to about 14, a significant improvement. Theoretical considerations developed in Section 3.5, show that this is by no means an ultimate limit set by the quantum nature of light, but rather a practical limit set by the performance of the detector. The University of Hawaii (UH) team therefore proposed to build a system based on a new type of detector that was just becoming commercially available, the photon counting avalanche photodiode or APD (Roddier *et al.* 1991a,b). Compared to previously used detectors, APDs have both a higher quantum efficiency and an extended sensitivity in the red, hence a larger band width. One should note that these characteristics are also shared by bare CCDs. However, until recently, the read-out noise of the CCDs was too high at the operating speed of a wave front sensor. Even today, this is a limitation of the CCDs. The drawback of APDs is that they do not form arrays, and each piece is expensive. It was therefore important to build a system that would efficiently use a minimum number of detectors. As we have seen, the concept of curvature sensing and compensation is ideal for that purpose. As currently implemented, it requires a single detector per subaperture instead of at least four for a Shack–Hartmann sensor.

The UH team was the first to build an AO system based on this concept. The number of subapertures was chosen to be only 13. This number was dictated by the desire to build an experiment that would demonstrate the feasibility of the proposed technique at minimum cost, but would also sufficiently improve image quality to produce useful astronomical results. At that time, the CFHT was already equipped with a fast tip/tilt compensation system (HR Cam). Using four photomultipliers with S-20 photocathodes in a quadrant detector, HR Cam was already producing a significant image improvement on objects brighter than mag. 17. Replacing the photomultipliers with APDs brought the limiting magnitude to 19. The UH proposal was to build an AO system that would have a similar performance on faint sources, but would also further improve image quality on brighter sources. The use of a deformable mirror would also allow compensation for telescope aberrations which were found to limit HR Cam performance under good seeing conditions. In retrospect, 15 or 16 subapertures would have been a better choice because 13 subapertures do not properly sample all the orientations of the triangular coma, an aberration often produced by mirror and lens supports. The construction of the UH instrument required the development of both a new sensor and a new mirror technology.

9.2.2 A new type of wave-front sensor

As we have seen, curvature sensing is done by measuring the difference in illumination between two oppositely defocused pupil images. Since APDs are expensive, it is desirable to use the same detector for both measurements. This has the additional advantage of avoiding the need for an accurate calibration of the detector sensitivities. The problem was to find a means to defocus a pupil image back and forth at a few kHz rate. The solution was found in the use of a vibrating membrane. The membrane has a metal coating and acts as a mirror. As it vibrates its shape alternates between a concave and a convex surface. To avoid any change in magnification of the pupil image, the membrane must be located on a guide star image. Beyond the membrane, a converging lens or a concave mirror forms an image of the telescope pupil onto the detector array which is described below. At rest the membrane is flat and the pupil image is sharp. When the membrane vibrates, the pupil image is defocused back and forth producing a modulation of the illumination related to local wave front curvatures (Fig. 9.1).

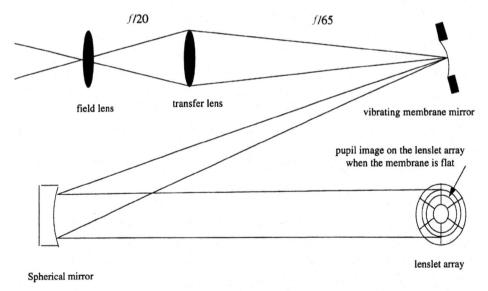

Fig. 9.1. Wave-front curvature sensing with a membrane modulator. As it vibrates the membrane defocuses back and forth the pupil image on the detector array.

The membrane modulator of the UH system was developed by J. E. Graves (Graves *et al*, 1991, 1994). Nitrocellulose (collodion) membranes were found to be very robust and of sufficient optical quality. Since the membrane is located in the image plane it does not itself produce any wave-front aberration. The membrane is driven by air pressure from an acoustic resonant cavity. It must be small enough (a few millimeters) and sufficiently stretched for its own resonance frequency to be higher than that of the cavity. As it vibrates, the minimum radius of curvature of the membrane can be as small as 6 cm without altering its optical quality. At maximum stroke it reimages to infinity a cross section of the beam located at a minimum distance $l = 3$ cm from the star image. As shown in Chapter 5 (Eq. (5.13)), the minimum useful value of l is

$$l_{\min} = \theta_b \frac{f^2}{d}, \tag{9.1}$$

where θ_b is the angular width of the source, f the focal length, and d the subaperture diameter. Let $N_s = (D/d)^2$ be the total number of subapertures. Equation (9.1) can be written

$$l_{\min} = \theta_b \sqrt{N_s} \frac{f^2}{D}, \tag{9.2}$$

which gives the required f-ratio as a function of $l_{\min}$

$$\left(\frac{f}{D}\right)^2 = \frac{l_{\min}}{\theta_b D \sqrt{N_s}}. \tag{9.3}$$

As an example, the first AO system of this type built at UH had $N_s = 13$ subapertures, and was used at the CFHT which has an aperture diameter $D = 360$ cm. Under exceptional seeing conditions a closed-loop star image can be as small as $\theta_b = 0.1''$ or 5×10^{-6} radian, even at the sensor wavelength. With $l_{\min} = 3$ cm, Equation (9.3) gives $f/D = 68$. Hence to operate properly, the membrane requires the use of a slow beam, that is a highly magnified image. Equation (9.1) shows that the distance $l_{\min}$ increases with the angular width θ_b of the source. An important advantage of the vibrating membrane is the ease with which $l_{\min}$ can be adapted to the angular size of the guide source. In practice, the membrane stroke can be modified during the observations until optimum compensation is achieved.

To feed 100 or 200 μm size APDs, one has to build a lenslet array. The subaperture geometry of the lenslet array and the electrode pattern of the deformable mirror were both optimized by simulating the whole system on a computer. The lenslet array forms a bull's-eye pattern, with seven lenslets inside the telescope aperture to sample the wave front Laplacian, and six lenslets over the aperture edge to sample the edge slopes. The latter must be large enough to collect most of the light of a highly defocused pupil image. The lenslet optics was designed by Guy Monnet, drawing from his experience with imaging spectrographs. Each lenslet is a doublet made of a converging PSK3 lens followed by a diverging LaSFN3 lens with a flat back surface. Each lenslet is cut according to the subaperture geometry and its back surface glued onto the flat front section of a LaSFN3 glass cylinder. The cylinder has a 18.5 mm diameter and a 26 mm length. All the lenslets must be made together with exactly the same focal length. Once assembled and glued on the glass cylinder, all the foci must fall on the flat rear section of the cylinder. Optical fibers 100 μm in diameter are centered with a 10μm accuracy on each individual sub-image and held in contact with the cylinder rear surface through index-matching optical grease. Each fiber is coupled to an APD though a graded index lens supplied by the manufacturer. This design has the advantage of avoiding any loss of light due to reflections in optical interfaces except on the lenslet entrance surface which has an anti-reflection (AR) coating. The total transmission from the lenslet front surface to the fiber output was measured to be 90%.

Compared to conventional wave-front sensors, the new wave-sensing technique described here has a number of advantages worth summarizing:

- it matches the deformations of a bimorph mirror,
- it has a continuously tunable sensitivity,
- it is self-referencing (no calibration is required with a plane wave),
- it utilizes a small array of high throughput detectors,
- it allows fast parallel read-out,
- data processing is reduced to a simple synchronous detection.

9.2.3 A new type of deformable mirror

The construction of the UH system also required the development of a new deformable mirror technology. This was done in cooperation with LASERDOT (now CILAS), a French company specializing in piezo-stack mirrors. UH was in charge of designing the electrode pattern and testing the mirrors, whereas LASERDOT was in charge of the actual bimorph fabrication. Following the theoretical considerations of Chapter 3, the electrode pattern was optimized to best compensate the low order Karhunen–Loève modes. This led to a bull's-eye type of pattern with seven electrodes inside the telescope aperture area to compensate the wave-front curvature errors, and six electrodes outside to provide the proper edge slopes. Because of the edge electrodes, the total mirror diameter is about twice the diameter of the telescope pupil image. Depending upon the availibility of piezoelectric material, a typical bimorph mirror consists of two 1-mm thick and 60- to 75-mm diameter piezo-wafers glued together. The diameter of the telescope pupil image is in the 30 to 40 mm range. Compared to other deformable mirrors, it has a unique advantage of being able to compensate atmospheric tip/tilt errors with a remarkably high frequency response, but a limited amplitude. Additional tip/tilt correction is still needed to compensate for telescope drive errors, but with a lower frequency response. For that purpose, advantage was taken of the mirror's light weight (50 g) by supporting it on a tip/tilt platform, thus avoiding the use of an additional tip/tilt mirror.

The mirror fabrication required the solution of a number of practical problems. Double lap polishing of the two surfaces at the same time produced satisfactory results. However, piezo-materials being porous, their polished surfaces scatter light. A workable solution was found by epoxy replication of a master plane surface onto the piezo-wafer. It is a difficult task, owing to the high (35:1) aspect ratio of the wafer. When successfully accomplished, it provides a highly reflective surface with an excellent optical quality, identical to that of the master. Another problem was to connect wires to the electrode pattern located between the two wafers. This was done by drilling holes through the back wafer. Unfortunately it was found to degrade the mirror optical quality significantly. A solution was found by moving the connections outside the telescope pupil image. Laplacian Optics, a company founded by members of the UH team, is now marketing deformable mirrors of a slightly different kind with no holes at all. Proper mirror support was also found to be a problem. It must be stiff enough to avoid spurious low frequency resonances, while still allowing the mirror to deform freely. Satisfactory results were obtained by supporting the mirror at the edge with three 'V' grooves at 120° from each other.

Unlike monolithic or piezo-stack mirrors, the stroke of a bimorph mirror is defined as the Laplacian ∇^2_{max} of the mirror surface when a maximum voltage is applied. It is related to the minimum radius of curvature R_{min} of the mirror by the relation $\nabla^2_{max} = 2/R_{min}$. It is easily determined by measuring the minimum focal length $R_{min}/2$ of the mirror and taking the inverse of it. Typical values of ∇^2_{max} are of the order of 5 to 10×10^{-5} per mm. Under a maximum voltage of 400 volts. The following formula gives the worse seeing (expressed as the full width at half-maximum of an uncompensated stellar image) that a given bimorph can compensate:

$$\alpha_{max} = 8.8 \times 10^4 \lambda^{-1/5} D^{-1} \Phi^{12/5} N^{-7/5} (\nabla^2_{max})^{6/5} \qquad (9.4)$$

where λ is the wavelength (in mm) at which the seeing angle is estimated, D is the telescope aperture diameter (in mm), Φ is the diameter (in mm) of the pupil image on the bimorph, and N is the number of electrodes along a pupil diameter. Equation (9.4) assumes that saturation may occasionally occur when the curvature of the random atmospheric wave front exceeds three times its standard deviation. It should be noted that the saturation of a bimorph mirror is often barely noticeable. This is because the high order aberration terms saturate first. Hence saturation only lowers the degree of compensation. This is in sharp contrast with the saturation of a piezo-stack mirror which affects the low order terms first such as defocus and astigmatism, and immediately degrades the quality of the compensation.

Compared with monolithic or piezo-stack mirrors, bimorph mirrors were found to have a number of advantages worth listing here:

- Their large stroke at low spatial frequencies make them ideal for near-infrared observations on large astronomical telescopes.
- The good match they provide with curvature sensors allows efficient compensation to be obtained with a small number of high performance detectors.
- They can be fabricated at low cost using standard production techniques.
- Optical quality can be excellent. Epoxy replication provides a low scatter surface. In general, open-loop aberrations can be compensated with a few percent of the total stroke. The compensation of low order terms introduces a minimum of uncompensated higher terms, that is, the compensation efficiency is high. The compensated point-spread function is clean.
- The actuator spacing can be as small as a few millimeters, allowing the size of the pupil image (that is the over all system size) to be small.
- The mirrors are robust and reliable: they are nearly insensitive to temperature changes, and age well. Under poor seeing conditions, mirror saturation has only minimal effects. In case of a connection failure the bimorph interpolates the wave front with a zero-Laplacian surface, minimizing image degradation.

9.2.4 The development of the experimental system

A step by step approach was used to develop such an experimental system, following the availability of components as they were developed. It started as a laboratory set-up on an optical bench. The first wave-front sensor was equipped with a small array of silicon diodes. This was deemed acceptable for a proof-of-concept experiment. A custom-made array was used with a bull's-eye geometry as described above. It avoided the need for a lenslet array as well as the expense of high cost APDs. Moreover, the sensitivity of the array was found to be sufficient for the sensor to work on bright stars (up to mag. 4) allowing measurements to be made at the telescope under various seeing conditions. Since bimorph mirrors were still in the development stage, the wave-front sensor was first used to control a tip/tilt mirror. Having successfully demonstrated the technique in the laboratory, telescope tests started in April 1991 as a series of short observing runs (Graves *et al.* 1991). Each time the optical bench was transported and installed at the coudé focus of the CFH telescope. Tip/tilt compensation was done by stopping the telescope aperture down to 1 m, for maximum Strehl ratio improvement. An improvement by a factor of five was observed, which is close to the theoretical maximum limit for pure tip/tilt compensation, a performance that had never been achieved before (Graves *et al.* 1992a,b; Roddier 1992). This is because the 13-channel sensor was able to measure and reject any alias due to random atmospheric coma components.

The control system also evolved step by step. Analog lock-in amplifiers were first used to detect the sensor signals. Later voltage-to-frequency converters were used to feed counters, and the synchronous demodulation was performed digitally. An ordinary 386 PC-type computer was initially used to control the feedback loop, but could not simultaneously provide real time diagnostic information. However, open-loop wave-front errors were recorded and statistically analyzed, providing invaluable information on the characteristics of the aberrations that had to be compensated (Roddier *et al.* 1993). Meanwhile LASERDOT was trying to produce bimorph mirrors of good optical quality. The first useful mirrors became available in 1992. They were tested at the telescope on July 1992. Images were recorded at 0.85 μm with a CCD camera using the full telescope aperture. This run benefited from excellent seeing conditions (0.4″ uncompensated seeing) and compensated images of Arcturus with a full width at half-maximum (fwhm) of 0.08″ were recorded. At that time, it was the sharpest long exposure image ever recorded on a ground-based telescope (Graves *et al.* 1993, Roddier 1994a).

The next step was to build a new detector array by coupling fiber-fed APDs to a lenslet array. The same counters were used to count photon pulses, but a new VME-based control system was built and later installed allowing real time diagnostics to be made. These modifications were tested at the telescope in a series of observing runs extending from March to December 1993. In December, a new generation bimorph mirror was mounted on the bench, and satisfactory results were finally obtained (Roddier *et al.* 1994). Not only was the theoretically expected performance achieved, but interesting science results were obtained, such as the discovery of a binary core in the Frosty Leo nebula (middle image on book cover). A description of these early results can be found in Roddier *et al.* (1994, 1995).

9.3 The first UH Cassegrain system

After a run at the United Kingdom Infrared Telescope (UKIRT) in January 1994, the experimental system was dismantled and the construction of a Cassegrain focus system was undertaken. A sketch of this new system is shown in Fig. 9.2. To enable the system to continue to evolve, a custom-made sturdy optical board is used. A Serrurier-type mount is attached to the board at four strong points, and serves as an interface to the telescope. The system is designed to operate at an $f/35$ focus either on the CFHT or on the UH 88″ telescope. This choice has the advantage of simplifying the transfer optics. It is a 1:1 transfer with two off-axis reflections on the same parabolic mirror as in a classical Ebert–Fastie spectrograph, the grating being replaced with a bimorph mirror. The 1024×1024 pixel HgCdTe infrared camera, newly built at the Institute for Astronomy, can be attached on a focusing mechanism provided under the board. It is fed through a hole in the bench. A dichroic beam splitter transmits 85% of the light beyond 1 μm to the infrared camera and reflects 95% of the light below 1 μm to the wave-front sensor. Part of the light sent to the sensor can be diverted to feed a CCD camera. This is done by means of additional beam splitters mounted on a wheel. The total system transmission is about 70% for the infrared camera and 50% for the sensor.

A filter wheel is mounted in front of the CCD camera. Another filter wheel is mounted before the wave-front sensor detector with various density filters to avoid saturation of the APDs on bright guide sources. Although the IR camera

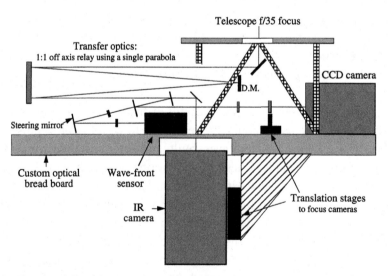

Fig. 9.2. Sketch of the UH experimental AO system. Some of the optical components actually on the bread-board are shown above for clarity.

has its own filter wheels inside the Dewar, an external filter slide is provided for additional narrow band filters. The pixel size is chosen to provide Nyquist sampling at the shortest wavelength. It gives a $36'' \times 36''$ field of view (FOV) for the infrared camera, and a $25'' \times 25''$ FOV for the CCD camera. An offset steering mirror allows the observer to pick up a guide source as far as $30''$ away from the center of the field, and move the image on the cameras. Two calibration light sources are provided using a single mode fiber fed either with a laser diode or a white light source. These sources are used for the internal focusing of the cameras and for the calibration of the control matrix. A parallel glass plate can be inserted in the beam, introducing a known defocus for wave-front sensor calibration. An optional rotatable wave plate has been installed in front of the system together with a polarizer in the IR camera for polarization measurements. All the key elements are remotely controlled. This includes the focusing of the cameras, the motion of all the filter wheels and filter slide, the tilting of mirrors for pupil alignment and offset guiding, the introduction and removal of the calibration light sources and glass plate, and the rotation of the wave plate. Every time an exposure is taken, all the parameters are automatically logged by the system.

The adaptive optics control system is basically that of the experimental system in its latest version (Anuskiewicz *et al.* 1994). The main processor consists of a VME backplane with two Force CPU-2CE SPARC single board computers. One is used as a loop processor, the other as a control/status processor. The loop processor is primarily dedicated to the feedback loop tasks which include reading the wave-front signals from the counters, doing the feedback loop calculations, and sending the output to the bimorph mirror (D/A converters). The input and output signals are conveyed through a fiber-optics interface. The control/status processor is used for managing control parameters and system status data flow. The operating system is VxWorks. A 1 Mb section of the loop processor memory is mapped onto the VME bus and is accessible for reading or writing from the control/status processor. All communications between the processors is accomplished through this shared memory. An Ethernet link is used for communication between the control/status processor and a workstation. The whole system (adaptive optics and cameras) is controlled from the same workstation through a graphical user interface. The architecture allows diagnostic data to be read while the loop is running. Several diagnostic programs can be run simultaneously. These include a pupil monitor for alignment, a sensor/drive monitor which displays the sensor a.c. or d.c. signals, or the mirror drive signals, a power spectrum monitor which analyzes and displays the power spectra of a variety of signals, and a seeing monitor which estimates the seeing condition.

The Cassegrain system was first operated in December 1994 at the CFHT, and performed as expected. It has since been regularly scheduled for astronomical observations both at the CFHT and the UH 88″ telescope. Some of the results are described in Chapter 15. At the time of writing it is being upgraded to a 36-channel system (Graves 1996). In the next section, we describe the simultaneous development of the CFHT $f/8$ user instrument which followed closely the technical developments at UH. Since both systems have similar performance, we only describe the performance of the CFHT user instrument here.

9.4 The PUEO system

The construction of a user instrument for the CFHT was considered as early as the fall of 1990, when a group was set up to advise the CFH Corporation on adaptive optics. At that time, the first compensated images had just been obtained at ESO with the COME-ON system (see Chapter 8). One possibility was to use the same technical approach. However, the expected gain in sensitivity brought by the use of APDs was sufficiently attractive for the group to contemplate from the start the use of a wave-front curvature sensor. A phase-A study ended in August 1992 with a project definition. It consisted of a 19-subaperture curvature sensor coupled to a 52-actuator piezo-stack mirror from LASERDOT. Because aberrations in the deformable mirror at its actuator spatial frequency could not be properly sampled by the curvature sensor, provision was made for a higher order Shack–Hartmann sensor to sense the mirror deformation and flatten it with an additional loop on an internal light source. Such a provision significantly increased the cost of the system. In April 1993, as bimorph mirrors produced by LASERDOT were succesfully tested by the UH team, the decision was taken to use instead a 19-actuator bimorph, and the construction of the instrument was undertaken (Arsenault *et al.* 1994).

The CFHT Adaptive Optics Bonnette (AOB) is now a facility instrument mounted at the $f/8$ Cassegrain focus of the 3.6-m CFH telescope. The 'bonnette' (adaptor in French), is also called PUEO after the sharp-sighted Hawaiian owl and is meant to '**P**robe the **U**niverse with **E**nhanced **O**ptics'. It is the result of a collaborative effort between several institutes: The CFHT (managing the project and designing the general user interface); the Dominion Astrophysical Observatory (DAO, Canada) who designed and fabricated the opto-mechanical bench, the wave-front curvature sensor and its electronics; the company CILAS (LASERDOT, France) who provided the deformable curvature mirror and the Real Time Computer and software, including a high level maintenance interface; the Observatoire de Paris-Meudon (OPM, France) who manufactured the separate tip/tilt mirror and was in charge of the final integration, testing, and calibration of the instrument. The UH adaptive optics team acted as consultants and provided guidance throughout the project. The system was commissioned at CFHT during three runs in the first semester of 1996. In the following sections, the instrument is briefly described and its performance is presented mostly in terms of image improvement. The properties of the compensated images are also discussed.

9.4.1 Instrument description and laboratory tests

The main characteristics of PUEO are summarized in Table 9.1. It has only a few optical parts, mainly reflecting ones. Making use of off-axis parabolic mirrors allows for a compact instrument with small optical components, favoring reduced flexures (see Fig 9.3). The beam, which can pass straight through the bonnette, is normally diverted by a flat mirror – on a moving slide – to the AO system, allowing it to switch rapidly from the $f/19.6$ corrected beam to the direct $f/8$ beam, if required.

The optical design (Richardson 1994) includes an $f/8$ off-axis parabola that collimates the beam and reimage the telescope pupil on the 19 electrode curvature mirror. A $f/19.6$ off-axis parabola, mounted on a fast tip/tilt platform, directs the beam to the astronomical instrument at the 'science focus'. Prior to this focus, a beam splitter reflects part of the light to the visible wave-front curvature sensor. Optionally, an atmospheric dispersion compensator can be inserted in the collimated beam for observation at visible wavelength.

The geometry (19 electrodes/subapertures divided up into two rings plus a central electrode) is well suited to circular pupils; the inner ring and the central electrode provide the wave curvature over the pupil while the outer ring provides the boundary radial slopes (Roddier 1988, see Chapter 5). Such a system, with few degrees of freedom but a high bandwidth, is particularly well suited to the Mauna Kea seeing conditions where turbulence is usually weak yet occasionally fast.

Modal control and mode gain optimization (Gendron & Léna 1994; Rigaut et al. 1994) maximize the instrument performance according to the state of turbulence and the guide star magnitude. Using the deformable mirror (DM) command covariance and the wave-front sensor (WFS) measurement covariance computed in closed-loop, the gains are optimized and updated during the closed-loop operation, allowing the system to track seeing variations.

The system has been tested in the laboratory at 0 °C and 20 °C for flexures, optical quality, and bandwidth (Lai 1996).

Table 9.1. *Characteristics of the CFHT Adaptive Optics Bonnette*

Optomechanics

Total number of mirrors in science train	5 + 1 beam splitter (in transmission)
Total number of mirrors in WFS train	9 + 1 beam splitter (in reflection)
Transmission of science train	70% (V) excluding beam splitter
	75% (H), 70% (K) including dichroic
Input/output F-ratios	8/19.6
Overall Bonnette dimension	Diameter 120 cm, thickness 28 cm
Flexures	Approximately 15 μm/hour at the $f/20$ focal plane
Optical quality	$\lambda/20$ rms at 0.5 μm with DM flattened
Instrument clear field-of-view	90″ diameter

Wave-front sensor

Type	Curvature
Number of subapertures	19
Detectors	APDs (45% peak QE, $\approx$ 20e-/s dark current)
Field of view	1–2″ depending on optical gain

Tip/tilt mirror

Type	Voice coils
Stroke	± 4.6″
Resolution	0.002″ (on the sky)
Bandwidth	>800 Hz for both axes (closed loop at −3 db)
Phase shift at 100 Hz	15° (see control below)
Diameter	60-mm (55 mm clear aperture)

Deformable mirror

Type	Bimorph
Number of electrodes	19
Stroke	Approximately ± 10 μm
First mechanical resonance	> 2-kHz
Overall dimension	80 mm
Conjugation	Telescope pupil
Pupil image size	42 mm

Control

Sampling/command frequency	Selectable (1000 Hz, 500 Hz, 250 Hz, . . .)
Max. bandwidth 0dB rejection	105 Hz
Max. bandwidth −3dB closed-loop	275 Hz
Control scheme	Modal, 18 mirror modes controlled. Closed-loop mode gains optimization
Tip/tilt control	Tip/tilt mirror operates in a nested loop. High frequency tip/tilt errors are corrected by the bimorph mirror.

Instrumentation
Visible and near-IR imagers
Integral field spectrograph (1997)

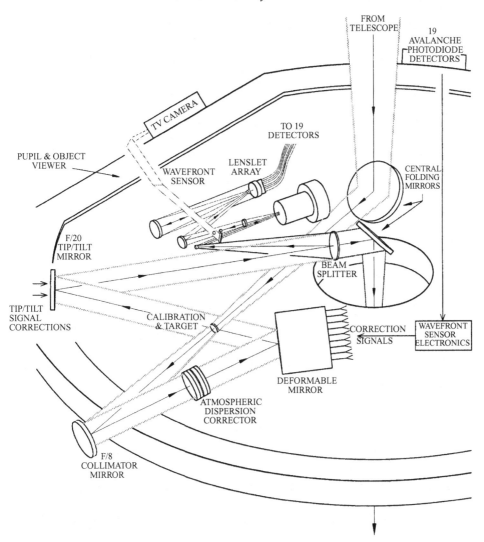

Fig. 9.3. Optical path of the instrument. The central folding mirrors are on a movable slide, so that the direct and the corrected focus are co-incident. The wave-front sensor is remotely controlled along three axes and allows the selection of a reference star different from the science object.

9.4.2 Performance on the sky

PUEO was extensively tested during three runs in the first half of 1996. The performance was evaluated in both the visible and the near-infrared wavebands, using a 2k × 2k CCD and a 256 × 256 Nicmos array (loaned from the University of Montreal). The emphasis was put on the performance in the near IR, where the instrument was expected to show its full potential. The commissioning included pure engineering tests, performance evaluation tests, and scientific programs. The latter were intended to test in 'real life' what could be achieved by the instrument, and to set up the data acquisition and reduction procedures. The engineering tests were to check that all the functions performed as expected and to make the necessary calibrations (wave-front sensor motions, atmospheric dispersion compensator calibrations, etc). In this section, we will report only on the results of the performance evaluation, mostly in terms of image quality and characteristics.

9.4.2.1. System behavior

One of the main goals in designing this system was to make a user-friendly, robust interface. The user is presented with a limited choice, simple interface. Basically, a one-button 'start/stop compensation'. This turned out to be achievable, and efficient both in terms of system operation and performance. It covers all cases, from the brightest to the dimmest objects ($m_R = 17$), thanks to the closed-loop optimized modal control. In turn, the overheads of the system are very small: the set up on an object, including the mode gains optimization automatic procedure, takes less than 1 min. Because the instrument focus is taken care of by the adaptive compensation, the overhead is actually less than for a standard, bare CCD imager.

9.4.2.2 Turbulence characterization and PSF files

In addition to the system capabilities described above, and directly associated with the modal control, a tool has been implemented that allows one to determine *a posteriori* the system state during a science exposure (Véran 1995, 1997): the wave-front sensor measurement covariance, the deformable mirror command covariance, and other parameters are computed synchronously with each exposure, and stored in PSF files. All kinds of diagnoses can be made from these data. Particularly, one can estimate D/r_0 and the system PSF for the exposure. A discussion of the method used to retrieve the PSF from the system data is beyond the scope of this paper. We refer the reader to Véran (1996, 1997) for a detailed discussion. The usefulness of having a PSF synchronous with – versus having to image a PSF calibrator after and/or before – the science exposure is many-fold: the gain of time and the assurance that the atmosphere has exactly the same behavior as during the science exposure (by definition) are the two major advantages. The D/r_0 values derived from the PSF files are the ones we use for all figures in the following discussion. D/r_0 was determined by fitting the actual variance of the system modes (excluding tip/tilt) with their theoretical Kolmogorov value, corrected for noise and spatial aliasing. The distribution of measured D/r_0 (750 measures in a dozen nights spread over a period from March to September 1996) is well represented by a log-normal distribution with a mean r_0 value of 15.5 cm at 500 nm. This corresponds to a seeing disk of 0.67″. This determination includes both free atmosphere seeing, dome seeing, and mirror seeing (although these last ones have short to very short outer scales and may not be accounted for properly). Because tip/tilt is excluded in the D/r_0 calculation, it excludes any telescope jitter or free atmosphere outer scale effect. These values were computed for the actual direction of observation and not corrected to the zenith. If we assume an average zenith distance of 30 degrees, the median seeing at zenith becomes 0.58″. This compares well with values derived from other data sets for the same Mauna Kea site (Roddier *et al.* 1990). This r_0 determination was checked against open loop exposures. The error on D/r_0 is a few percent (2–5%). Something worth noting is that the atmosphere showed most of the time a very good match with a Kolmogorov-type turbulence. On some occasions (15–20% of the time), we have noted deviations that may be attributed to dome or mirror seeing.

9.4.2.3 System performance

All images (IR and visible) were reduced following standard image reduction procedures. All exposures for which Strehl and full width at half-maximum (fwhm) were extracted are long exposure images. Most of the images have integration times of 15 s or longer, to get statistically meaningful data.

Figure 9.4 is a plot of the Strehl ratio R versus Fried's parameter r_0 at the image wavelength. In this plot only the Strehl ratios derived from images of 'bright' stars of R-magnitude lower than 13.5 are reported. The lower solid line is the Strehl ratio of the seeing limited image. Strehl ratios in this plot, *as well as in the rest of this section*, have been corrected from the system static aberrations, i.e. actual images were compared not to fully diffraction limited images, but to the images obtained using the optical bench artificial source with no turbulence. The Strehl ratio of the later 'static' images are 0.50, 0.65, 0.75, 0.77, 0.84, 0.90, and 0.93 in V, R, I, J, H, H2, and K, respectively. The cloud of points exhibits little scatter in this plot, owing mostly to the simultaneous estimation of the r_0, as reported in Section 9.4.2.2 and to the fact that the system bandwidth on bright guide stars is, in most cases, several times larger than the Greenwood frequency. The points are aligned along a theoretical curve (dotted line) which corresponds to the full compensation of eight Zernike modes (in between the $n = 2$ and the $n = 3$ curves in Fig. 3.2). Note that it is not very far from our expectation of ten Zernike modes fully compensated

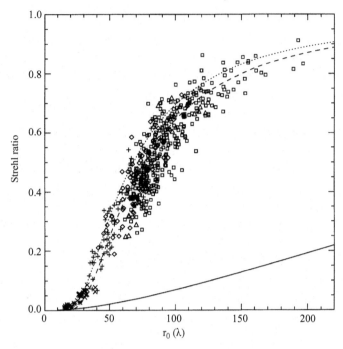

Fig. 9.4. Strehl ratio versus the Fried parameter r_0 at the image wavelength. $*$, B and V band; $\times$, R and I band; $+$, J band; $\diamond$, H band; $\triangle$, H2 band; $\square$, K band.

(dotted line in Fig. 9.4), as derived from numerical simulations (Rigaut *et al.* 1994). The difference between ten modes (expected) and eight modes (achieved) is most probably due to telescope vibrations, uncorrected high spatial frequency aberrations of our mirror train (particularly the telescope primary mirror), and mirror seeing, which is mostly made of local, high spatial frequency aberrations at the primary mirror surface. Note that this performance of eight Zernike modes fully compensated for 19 actuators is comparable, although slightly better, to that derived for the COME-ON system, the earlier version of the current ADONIS system which also had 19 actuators (see Rigaut *et al.* 1991 and Chapter 8). The difference between the number 19 of modes controllable by the system and the computed figure of 10 comes from (a) the piston mode which is not measurable, and therefore not corrected; it has no effect on image quality; (b) four of our modes which are higher order and therefore not very efficient in terms of phase variance reduction; (c) spatial aliasing, a feature intrinsic to any system (see Chapter 3); (d) finite temporal bandwidth, and finally (e) noise which is always present.

By taking the ratio of the achieved Strehl ratio (corrected from the static aberrations of the optical bench) to the uncompensated image Strehl ratio (theoretical expression, therefore also assuming no further degradation by optical aberrations, telescope jitter, etc.), one gets the gain in peak intensity, reported in Fig. 9.5, versus D/r_0. This translates into a sensitivity gain on unresolved source (up to 2.5 magnitude in J and H). The upper solid line in this figure is the limit imposed by the diffraction: if the image is fully diffraction limited, the Strehl ratio improvement is equal to one over the Strehl ratio of the seeing-limited image.

Note that the Strehl improvement peaks at an r_0 of approximately 60 cm ($D/r_0 = 6$). This is to be compared to the characteristics length associated to the correction and resulting from the geometry of the system: our mirror has 19 electrodes and the average distance between two electrodes is approximatively $d = 90$ cm, therefore $D/d = 4$.

In terms of fwhm, the images are basically diffraction limited in H and K for median seeing conditions. A fwhm of around $0.10''$ is maintained down to the I band in most of the seeing conditions encountered. The fwhms in the visible region (B, V, and R) still show a substantial gain with respect to the uncompensated image fwhms (see below). The properties of the corrected images, in terms of morphology, are discussed in the next section.

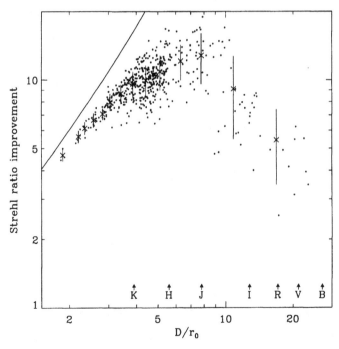

Fig. 9.5. Strehl ratio improvement versus D/r_0.

Figure 9.6 presents the fwhm *improvement* brought by the AO compensation. The gain is hereby defined as the ratio between the seeing limited image fwhm, λ/r_0, and the fwhm of the compensated image. In the figure, the solid line is the maximum gain set by the diffraction limit $((\lambda/r_0)/(\lambda/D) = D/r_0$ against r_0). To avoid mixing too many parameters in the analysis, these values are reported for stars brighter than $m_R = 13.5$.

The maximum gain in fwhm is obtained for an r_0 of approximately 40 cm ($D/r_0 = 9$). It is worth noting that the maximum gain in Strehl takes place at a larger r_0 value (60 cm or $D/r_0 = 6$), as mentioned above. The r_0 value at which the maximum resolution in terms of fwhm is obtained has been called the critical r_0 value (Rousset *et al.* 1990), which, for median seeing conditions, corresponds to a given wavelength that has been called the critical wavelength of the system. For PUEO, this critical wavelength is 1 μm. Referring to Fig. 9.4, one can see that the largest gain in resolution is obtained for images with Strehl between 10 and 15%. At $r_0 = 20$ cm ($D/r_0 = 18$), which corresponds approximatively to the typical value of r_0 in the V band, the gain in fwhm is still 2–3.

A different way to present the same fwhm results, which is useful in practice, is shown in Fig. 9.7. The normalized fwhm of the compensated images is

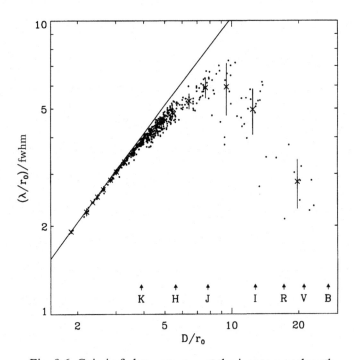

Fig. 9.6. Gain in fwhm versus r_0 at the image wavelength.

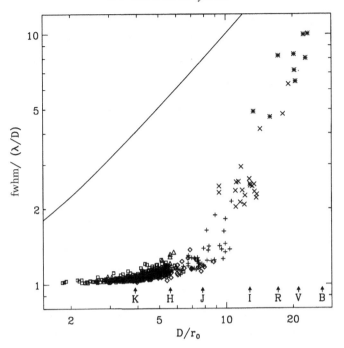

Fig. 9.7. Normalized image fwhm versus r_0 at the image wavelength. $*$, B and V band; $\times$, R and I band; $+$, J band; $\diamond$, H band; $\triangle$, H2 band; $\square$, K band.

plotted against r_0 at the image wavelength, again for stars brighter than $m_R = 13.5$. The normalized fwhm is the fwhm of the image in units of λ/D at the image wavelength. In turn, the normalized fwhm lower limit is 1. In the plot, the upper solid line is the fwhm of the seeing limited image. As seen on this figure, the normalized fwhm, as the Strehl ratio, is a characteristic of the system which depends only upon D/r_0 and not upon the image wavelength (other than the r_0 dependence). Two regimes, with a very clear cut-off, are revealed in this plot. The first regime, up to $D/r_0 = 7$, is characterized by high Strehl ratios ($> 20\%$) and diffraction limited images in terms of fwhm (normalized fwhm ≈ 1). Above $D/r_0 = 7$ lies a regime of more partial correction, with low Strehl ratio images ($< 20\%$) and fwhm strongly dependent upon the turbulence conditions. However, as reported above, the fwhm's gain is still quite attractive in this domain. This is particularly true for astronomical direct imaging. In the visible, compensated images with fwhm 0.1–$0.2''$ are commonly obtained at CFHT. Even these modest resolution gains can make a huge difference in the feasibility/efficiency of a lot of astronomical programs.

Another key issue is how the performance degrades with the guide star magnitude m_{GS}. This is shown Fig. 9.8, in terms of Strehl ratio attenuation versus the guide star R magnitude. The Strehl ratio attenuation is merely the attenuation with respect to Strehl ratio values obtained on bright guide stars, in the same turbulence conditions. This curve was computed using the results presented in Fig. 9.4: Strehl ratio at H band on dim guide stars were divided by the expected Strehl value for bright guide stars in the given r_0 conditions, binned by magnitude and plotted against m_{GS}. These points were fitted with a function of the form:

$$S_{\mathrm{att}} = \exp(-\sigma^2_{\mathrm{noise}}) \quad \text{with} \quad \sigma^2_{\mathrm{noise}} \propto \frac{1}{N_{\mathrm{ph}}} \qquad (9.5)$$

where N_{ph} is the number of photons detected by the WFS. This function was then extrapolated to other wavelengths using a dependence $\sigma^2_{\mathrm{noise}} \propto \lambda^{-2}$. The magnitude for which the Strehl ratio attenuation is 50% is 15.7 for K band images, 15.0 for H, and 14.4 for J. A direct extrapolation to R, less meaningful at this wavelength where low Strehl ratio are usually obtained, gives 13.0. These values should not really be considered as limiting magnitudes, for which there is no satisfactory or unambiguous definition. In a real life situation, the actual limit depends more on the scientific goal one wants to achieve, coupled

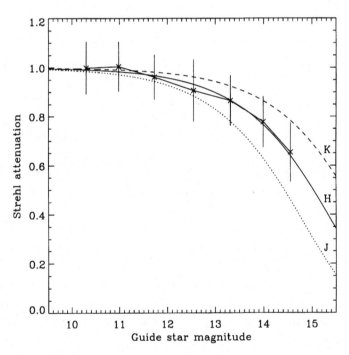

Fig. 9.8. Strehl attenuation with respect to the bright guide star case versus the guide star magnitude.

Table 9.2. *Performance summary in median seeing conditions*

Waveband	V	R	I	J	H	K
Wavelength [μm]	0.54	0.65	0.83	1.25	1.65	2.23
Median $r_0(\lambda)$ [cm]	17	21	28	46	65	93
D/r_0	21.3	17.1	12.7	7.8	5.6	3.9
Strehl ratio (%)	1	2	5	21	41	61
fwhm (arcsec)	0.24	0.19	0.12	0.09	0.11	0.14
Gain$_{Strehl}$	4.0	5.0	7.0	12.5	11.6	9.5
Gain$_{FWHM}$	2.6	3.2	4.8	5.9	5.0	3.6

with the turbulence conditions. As an illustration, we have achieved 0.17″ fwhm in the K band on a R magnitude 17 guide star, under good seeing conditions (0.38″ seeing).

In real observing situations, from the knowledge of r_0, one can determine the performance (Strehl and fwhm) at any wavelength and any guide star magnitude using Figs 9.4 to 9.8.

To conclude this section, Table 9.2 summarizes the compensated images Strehl ratio and fwhm at various wavelengths for median seeing conditions. Also reported are the gains in Strehl ratio and fwhm over the uncompensated case, as defined above.

9.4.2.4 Image properties

We now discuss global properties of compensated images. Most of the material in this section belongs as well to the previous section. However, the properties we attempt to derive here are not only relevant to PUEO but also to any general adaptive optics system.

The partial correction image profile, with a coherent core – broadened by tip/tilt residuals – on top of a diffuse halo, is well known. Examples of such point-spread functions can be found in other chapters. Here, we will go one step further and explain why the PSF has such a shape.

A very educational and global way to understand the effect of the compensation by an AO system is to consider the phase itself. The phase structure function D_φ is a powerful tool to investigate the phase properties and image characteristics. With some limitations, it is possible to derive the phase structure function from the point-spread function. In this work, we used the following estimator based on Eq. (2.36) of Chapter 2:

$$D_\varphi = -2.\log\left[\frac{|FT(\text{image})| - \text{noise}(|FT(\text{image})|)}{|FT(\text{psf\_stat})| - \text{noise}(|FT(\text{psf\_stat})|)}\right], \qquad (9.6)$$

averaged azimuthally. Here *FT* stands for Fourier transform. The function 'psf_stat' is the point-spread function acquired on the internal artificial source that includes all uncorrected non-common path aberration (mostly the imaging camera). The average noise level is determined from the spatial frequency domain lying outside the telescope cut-off frequency. The noise on the Fourier transforms usually prevents an accurate determination of the structure functions at separation greater than approximately $0.8D$, corresponding to spatial frequencies for which the amplitude of the TF drops down to the noise level, i.e. close to the cut-off frequency D/λ. Using 18 images recorded at different wavelengths (J, H, H2), under various seeing conditions during two consecutive nights, it was possible to compute a *characteristic phase structure function* of the system, plotted in Fig. 9.9. This latter function, plotted as a solid line, is the average of the structure functions obtained on the 18 images, normalized by D/r_0 at the image wavelength. The error bars were computed simply as the rms deviation of this ensemble of curves. The structure function for a Kolmogorov-type turbulence is plotted as a dashed line for a $D/r_0 = 1$. The dashed-dotted line is the phase structure function corresponding to a theoretically ideal correction of the PUEO mirror modes (i.e. where only the fitting

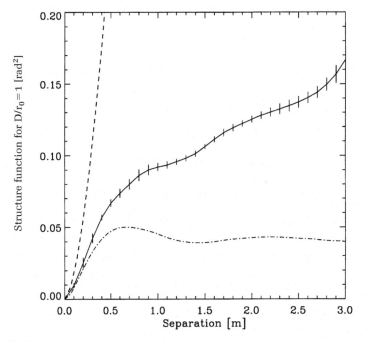

Fig. 9.9. PUEO characteristic phase structure functions. Experimental (solid line + error bars), turbulent (dashed line), and high order phase residual alone (dashed-dotted line).

error is taken into account). We also call this latter function the 'Noll structure function', by analogy with the Noll residual. It has been computed using Monte-Carlo realizations of turbulent wave fronts, from which the contribution of the system/mirror modes was removed. It is therefore the structure function of the compensated phase in absence of any errors such as noise, spatial aliasing, and servo-lag error.

This figure requires several remarks:

- Contrary to the uncompensated structure function, the *Noll phase structure function* saturates and forms a 'plateau' over most of the separation domain. This ensures the coherence – more precisely the partial coherence – over the whole telescope pupil, allowing the formation of an image coherent core. The saturation takes place for separations for which the phase becomes *uncorrelated*. Therefore the phase structure function at these separations can be expressed as

$$D_\varphi = \langle |\varphi(\mathbf{r}) - \varphi(\mathbf{r} + \boldsymbol{\rho})|^2 \rangle = 2\langle \varphi(\mathbf{r})^2 \rangle = 2\sigma_\varphi^2 \qquad (9.7)$$

in other words, the structure function saturation value is twice the phase variance over the pupil. In addition, we know that the Strehl ratio is approximately given by $\exp(-\sigma_\varphi^2)$ (see Section 2.3). Hence, the higher the 'plateau', the smaller the coherence and the smaller the Strehl ratio (assuming we identify at first order the coherent energy to the Strehl ratio).
- In the Noll structure function, the saturation takes place at a separation of approximately 70 cm, which is roughly the 'inter-actuator' distance d. This is not a coincidence: only the phase corrugations of scale larger than the distance between two actuators can naturally be corrected.
- The *phase structure function achieved by the system* (solid line) is larger than the Noll structure function at all separations. The 'plateau' is destroyed partially. The rising of the function at scales larger than the inter-actuator distance can only be explained by the presence of low-spatial frequency aberrations, such as tip/tilt, defocus, etc., which have not been fully corrected by the system. A more detailed analysis shows that, when using bright guide stars, these low order modes are principally the result of spatial aliasing (noise and servo-lag errors are small because of the large number of photons available on these particular examples and because the system bandwidth (70–100 Hz) is several times larger than the Greenwood frequency). Overall, both our numerical simulations and experimental results show that for curvature systems, the spatial aliasing induce a phase error which is comparable in amplitude to the mirror fitting error. As far as images are concerned, the difference between the Noll structure function and the system structure function means a smaller Strehl ratio, and a not-fully diffraction limited core, slightly larger than the Airy pattern.
- The decorrelation at small scales in the system structure function is of the same type (although quantitatively different, see below) as the loss of coherence induced by

the seeing. This is what forms the compensated image 'halo'. However, the system structure function increases more slowly than the uncompensated one at a small scale. The cause of this lies in the fact that large scale perturbations play a non-negligible role in the phase decorrelation at small scales (especially because the power in these large scale perturbations is so large), therefore, correction of perturbations of scale $> d$ affect the phase structure function at scale $< d$. The consequence of this is that the coherence of the wave front is increased everywhere, and in particular at small scales. The net effect is equivalent to having a larger r_0 value for the behavior of the structure function at a small scale. The 'halo', directly linked to the wave-front coherence at small scales, will therefore have properties different than the long exposure image: it will be narrower (because the coherence length is increased).

As shown above, all the observed properties of the AO compensated images can be explained by the consideration of the phase structure function. In addition the phase structure function expresses the correlation of the phase independently from the image formation process. *The knowledge of this function characterizes entirely a given system performance* (in the bright guide star regime) and allows the computation of the image at *any* wavelength.

Figure 9.10 displays the image Strehl ratio versus the image normalized

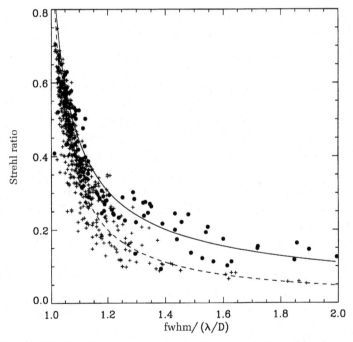

Fig. 9.10. Image Strehl ratio versus normalized fwhm for bright stars (crosses) and dim stars (filled circles).

fwhm, as defined earlier. In this figure, the crosses refer to guide stars brighter than $m_R = 13.5$ and the filled circles to guide stars fainter than $m_R = 13.5$. These curves show that for bright guide stars, there is a *very* tight relation between the normalized fwhm and the Strehl ratio (Tessier 1995). To any given Strehl ratio corresponds a well-defined normalized fwhm, therefore a given image shape, *whatever the wavelength and the atmospheric conditions*. This is a natural conclusion if one considers the interpretation of the AO compensation in terms of structure function as described above.

Use of faint guide stars modifies the relationship between the Strehl ratio and the normalized fwhm. In terms of modal decomposition, this can be easily understood if one considers that the effect of noise propagation on the corrected modes is different from that of errors when noise does *not* dominate. For instance, it is known that noise propagates in a very large part on tip/tilt (especially for curvature systems), and therefore the error on tip/tilt, *proportionally to the other modes*, will be larger when dominated by noise measurement error, broadening the image further, and modifying the Strehl/fwhm relation.

As a conclusion for this section, we report in Fig. 9.11 some data on anisoplanatism. The globular cluster M71 was observed at visible wavelengths (B, V, and I) with a field-of-view of $40 \times 40''$ squared. Strehl ratios are reported here against the distance off center. The angle for which the Strehl ratio drops by a relative 30% is of the order of $12''$ in the I band, which would translate in $18''$ in the J band, and $37''$ in the K band. Those are typical values as quoted by other authors for Mauna Kea (Northcott *et al.* private communication).

9.4.2.5 Scientific programs

The exploitation of the AOB for regular scientific programs started in August 1996. It has already included a survey of multiple stars in the Pleiades, and imaging of the comet Hale–Bopp, the nucleus of M31, the galactic center, Seyfert galaxies, QSO hosts galaxies, and high redshift galaxies. Some of these early results are presented in Chapter 15.

9.5 Acknowledgements

The development of the UH system was funded by the National Science Foundation (US). ONERA (France) supported the development of bimorph mirrors by LASERDOT. D. Rouan, J.-P. Véran and O. Lai are to be thanked for their help in writing the section in PUEO.

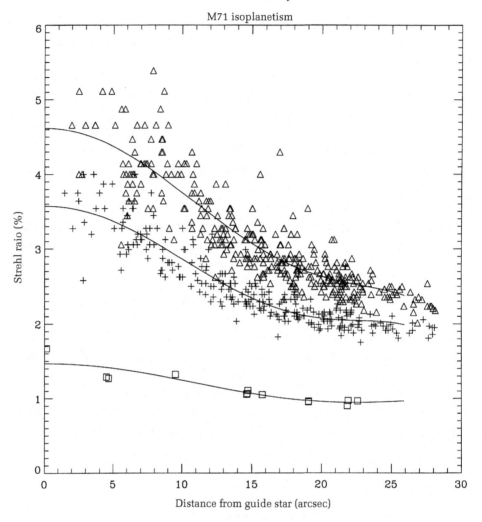

Fig. 9.11. Strehl ratio recorded as a function of the angular distance to the guide star. △, I band; +, V band; □, B band.

References

Anuskiewicz, J., Northcott, M. J. and Graves, J. E. (1994) Adaptive Optics at the University of Hawaii II: Control system with real-time diagnostics. In: *Adaptive Optics in Astronomy*, Kona, March 17–18, 1994. eds. M. A. Ealey and F. Merkle, Proc. SPIE 2201, pp. 879–888.

Arsenault, R., Salmon, D. A., Kerr, J., Rigaut, F., Crampton, D. and Grundmann, W. A. (1994) PUEO: The Canada–France–Hawaii Telescope adaptive optics bonnette I: system description. In: *Adaptive Optics in Astronomy*, Kona, March 17–18, 1994, eds. M. A. Ealey and F. Merkle, Proc. SPIE 2201, pp. 833–42.

Gendron, E. and Léna, P. (1994) Astronomical adaptive optics I, modal control optimization. *Astron. Astrophys.* **291**, pp. 337–47.

Graves, J. E. and McKenna, D. (1991) University of Hawaii adaptive optics system III. Wave-front curvature sensor. In: *Active and Adaptive Optics*, San Diego (California), July 22–24, 1991. ed. M. A. Ealey, Proc. SPIE 1542, pp. 262–72.

Graves, J. E., Roddier, F., McKenna, D. and Northcott, M. (1992a) Latest results from the University of Hawaii prototype adaptive optics system. In: Proc. workshop *Laser Guide Star Adaptive Optics* at Albuquerque, New Mexico, March 10–12, 1992, ed. R. Q. Fugate, Starfire Optical Range, Phillips Lab./LITE, Kirtland AFB, N. M. 87117.

Graves, J. E., McKenna, D. L., Northcott, M. J. and Roddier, F. (1992b) Recent results of the UH adaptive optics system. In: Proc. OSA Meeting, *Adaptive Optics for Large Telescopes*, Lahaina (Maui Hawaii), Aug. 17–21, 1992. Tech. Digest Series, pp. 59–62.

Graves, J. E., Northcott, M. J., Roddier, C., Roddier, F., Anuskiewicz, J. Monnet, G., *et al.* (1993) The UH experimental adaptive optics system: first telescope results with a photon counting avalanche photodiode array. In: Proc. ICO-16 Satellite Conference, *Active and Adaptive Optics*, Garching (Germany), Aug. 2–5, 1993. ed. F. Merkle, pp. 47–52. ESO, Garching.

Graves, J. E., Roddier, F. J., Northcott, M. J. and Anuskiewicz, J. (1994) University of Hawaii adaptive optics system IV: a photon-counting curvature wave-front sensor. In: *Adaptive Optics in Astronomy*, Kona, on March 17–18, 1994, eds M. A. Ealey and F. Merkle, Proc. SPIE 2201, pp. 502–7.

Graves, J. E. (1996) Future directions of the University-of Hawai Adaptive Optics Program. In: Proc. OSA Meeting, *Adaptive Optics*, Maui (Hawaii), July 8–12, 1996. Tech. Digest Series, pp. 49–52.

Lai O. (1996) PhD Thesis, Université Paris VII.

Richardson, E.H. (1994) Integrated adaptive optics systems. In: *Adaptive Optics for Astronomy*, eds D. M. Alloin and J. M. Mariotti, NATO ASI Series. pp. 227–36. Kluwer, Dordrecht.

Rigaut, F., Rousset, G., Kern, P., Fontanella, J-C., Gaffard, J-P., Merkle F., *et al.* (1991) Adaptive optics on a 3.6-m telescope: results and performance. *Astron. Astrosphys.* **250**, 280–90.

Rigaut, F., Arsenault, R., Kerr, J., Salmon, D. A., Northcott, M. J., Dutil, Y., *et al.* (1994) PUEO: Canada-France-Hawaii Telescope adaptive optics bonnette II: simulations and control. In: *Adaptive Optics in Astronomy*, Kona, March 17–18, 1994, eds M. A. Ealey and F. Merkle, Proc. SPIE 2201, pp. 149–60.

Roddier, F. (1988) Curvature sensing and compensation: a new concept in adaptive optics. *Appl. Opt.* **27**, 1223–5.

Roddier, F. (1991) Wave-front curvature sensing and compensation methods in adaptive optics. Proc. SPIE 1487, pp. 123–8.

Roddier, F. (1992) The University of Hawaii Adaptive Optics System. CFHT Bull. No 27 (2nd semester 1992), pp. 7–8.

Roddier, F. (1994a) Adaptive Optics: performance and limitations In IAU Symposium No 158 on *Very High Angular Resolution Imaging* held in Sydney (Australia) on Jan. 11–15, 1993, eds J. G. Robertson and W. J. Tango, pp. 273–81. Kluwer, Dordrecht.

Roddier, F. (1994b) The problematic of adaptive optics design. In: *Adaptive Optics in Astronomy*, eds D. M. Alloin and J. M. Mariotti, pp. 89–111. Kluwer, Dordrecht.

Roddier, F., Cowie, L., Graves, J. E., Songaila, A., McKenna, D., Vernin. J., *et al.*

(1990) Seeing at Mauna Kea: a joint UH-UN NOAO-CFHT study. Proc. SPIE 1236, pp. 485–91.

Roddier, F., Nortcott, M. and Graves, J. E. (1991a) A simple low-order adaptive optics system for near-infrared applications. *Pub. Astr. Soc. Pac.* **103**, 131–49.

Roddier, F. J., Graves, J. E., McKenna, D. and Northcott, M. J. (1991b) The University of Hawaii adaptive optics system: I. General approach. Proc SPIE 1524, p. 248.

Roddier, F., Northcott, M. J., Graves, J. E. and McKenna, D. L. (1993) One-dimensional spectra of turbulence-induced Zernike aberrations: time-delay and isoplanicity error in partial adaptive compensation *J. Opt. Soc. Am. A* **10**, 957–65.

Roddier, F., Anuskiewicz, J., Graves, J. E., Northcott, M. J., Roddier, C., Surace, J., *et al.* (1994) First astronomical observations with the University-of–Hawaii experimental adaptive optics system. In Proc. Conf. *Instrumentation for Large Telescopes in the 21st Century. Scientific and Engineering Frontiers for 8–10 m Telescopes*, Tokyo (Japan), October 4–6, 1994. eds M. Iye and T. Nishimura, pp. 313–9. Univ. Acad. Press, Tokyo.

Roddier, F., Roddier, C., Graves, J. E., Jim, K. and Northcott, M. J. (1995) Adaptive Optics Imaging at the CFHT. In: Proc. 4th CFHT User's Meeting Lyon (France), May 15–17, 1995. ed. M. Azzopardi, pp. 125–131. CFHT.

Rousset, G., Fontanella, J-C., Kern, P., Gigan, P., Rigaut, F., Léna, P., *et al.* (1990). First diffraction-limited astronomical images with adaptive optics. *Astron. Astrosphys.* **230**, L29–32.

Steinhaus, E. and Lipson, S. G. (1979) Bimorph piezoelectric flexible mirror. *J. Opt. Soc. Am.* **69**, 181–7.

Tessier, E. (1995) Analysis and calibration of natural guide star adaptive optics data In: *Adaptive Optics*, OSA/ESO Topical Meeting, Tech. Digest Series, No. 23, 257–9.

Véran, J.-P., Rigaut, F., Rouan, D. and Maître, H. (1996) Adaptive optics long exposure point spread function retrieval from wave-front sensor measurements. In: *Adaptive Optics*, OSA/ESO Topical Meeting, Tech. Digest Series, pp. 171–5.

Véran, J.-P., Rigaut, F., Maître, H. and Rouan D. (1997) Estimation of the adaptive optics long exposure point spread function using control loop data, *J. Opt. Soc. Am. A* **14**, 3057–69.

10

Adaptive optics in solar astronomy

JACQUES M. BECKERS

National Solar Observatory/NOAO[a]
Tucson, AZ 85718

10.1 Introduction

Much of the early experimentation on astronomical adaptive optics was done in the 1970s on the Vacuum Tower Telescope (VTT) at Sacramento Peak (Buffington *et al.* 1977; Hardy 1981, 1987) either on stellar objects or on the sun itself. That telescope, although only 76 cm in aperture, was ideally suited for such experimentation because of its attractive environment for instrumentation. Diffraction limited imaging at visible wavelengths on both stars and the sun was achieved by Hardy. The solar results then clearly demonstrated the limitations on adaptive-optics-aided solar research resulting from the small isoplanatic patch size (a few arcseconds). Since then a few other efforts have been mounted to achieve diffraction limited imaging in solar observations.

Solar adaptive optics systems differ in a number of significant aspects from systems developed for night-time astronomy. Specifically:

(i) since the sun is an extended object, wave-front sensing on point-like objects as is done mostly in night-time adaptive optics systems is not an option.

(ii) solar seeing is generally worse than night-time seeing because the zenith angle/air mass at which the sun is being viewed is large in the early morning when night-time seeing still prevails and because seeing caused by ground heating by sunlight becomes severe later in the day when the sun is seen at greater elevations.

(iii) solar telescopes have generally much smaller apertures than night-time telescopes none, except for the 150-cm aperture McMath–Pierce facility on Kitt Peak, exceeding 1 meter in diameter.

(iv) wave-front sensing methods using solar surface structure are not photon starved as is often the case for night-time adaptive optics.

[a] The National Optical Astronomy Observatories (NOAO) are operated by the Association of Universities for Research in Astronomy (AURA) under a Cooperative Agreement with the National Science Foundation.

235

Emphasis for solar adaptive optics has so far been on visible light observations so that the complexity for the short wavelength (500 nm), poorer seeing, $r_0(0.5 \; \mu m) = 8$ cm, solar adaptive optics systems for the 0.76-meter aperture Sac Peak telescope is similar to, or worse than that of night-time systems where the K spectral band is generally specified under better seeing conditions. For example, for the 10-meter Keck telescope a typical $r_0(2.2 \; \mu m) = 120$ cm requires as many adaptive elements but at a control time constant which is an order of magnitude shorter because of the visible wavelengths used and because of the poorer daytime seeing.

In contrast to the many efforts being pursued in the development of night-time adaptive optics, the pursuit of solar adaptive optics is limited both because of the smaller community, and hence smaller resources, involved and because other methods have been very effective for diffraction limited imaging. The sun, with its abundant photon flux, allows the very short exposure, relatively low noise observations required for broadband imaging. Hence post-detection image reconstruction and image selection techniques have been successfully applied to solar observations. They will be described in the next section. Nonetheless, narrow spectral bandwidths (0.01 nm) and high spectral resolutions ($> 10^5$) are often needed in solar observations. In those cases the use of pre-detection image restoration with adaptive optics and high duty cycles becomes especially attractive.

In the following section pre-detection and post-detection techniques will be briefly reviewed to put the use of solar adaptive optics in a broader perspective. Then the methods used for wave-front sensing in solar adaptive optics are reviewed. They, and the higher control bandwidth, are the prime factors distinguishing it from other systems described in this monograph. Finally a description is given of the two solar adaptive optics systems currently under development.

10.2 High-resolution imaging in solar research

10.2.1 Image selection techniques

Although r_0 results in a seeing angle for long exposure images of the order of λ / r_0, short exposure images vary greatly in quality because of the statistical nature of the wave-front disturbances caused by atmospheric turbulence. The time constant τ involved could be as short as r_0 / V_{wind} where V_{wind} corresponds to the typical wave-front translation velocity. With $V_{wind} = 10$ m/s and $r_0 = 10$ cm, $\tau = 10$ ms. Broadband observations of the solar disk indeed allow exposures of this length. It is therefore possible to obtain images significantly

Table 10.1. *Gain in angular resolution for short exposure images using image selection*

D/r_0	3	4	5	7	10	15	20	50	100
76 percentile	2	2	1.9	1.7	1.5	1.4	1.3	1.2	1.1
10 percentile	2.7	3.1	3.1	2.8	2.4	2.1	1.9	1.5	1.4
1 percentile	2.9	3.4	3.6	3.4	3.0	2.6	2.3	1.7	1.6
0.1 percentile	3.0	3.6	4.0	4.1	3.8	3.2	3.0	2.1	1.8

better (and also worse!) than λ/r_0 in quality by selecting the right moment. Fried (1978) first calculated the probability of getting such a 'lucky short exposure image'. Hecquet and Coupinot (1985) further elaborated on this concept. Table 10.1 is taken from their publication. It gives the gain in angular resolution of short exposure images over that of a long exposure image for images selected for different upper seeing percentiles.

Image selection thus results in major gains in angular resolution especially in the $D/r_0 = 3$ to 30 range. Image selection is being used at a number of solar observatories for solar surface structure imaging. On days of sub-arcsecond seeing it has, for example, resulted in diffraction limited images at visible wavelengths at the 50-cm aperture Swedish Solar Telescope on La Palma and at the 76-cm aperture Vacuum Tower Telescope at Sacramento Peak (see Fig. 10.1). It is expected to give diffraction limited images at 1.6 μm in the near infrared with the 150-cm aperture McMath–Pierce Telescope on Kitt Peak if the present efforts to improve the image quality at that telescope are successful. As was pointed out by Beckers (1988b) image selection obviously only works well if the telescope optical quality is significantly better than that of the atmosphere above it. Because the atmospheric wave-front disturbances can occasionally compensate the optical aberrations, it is, however, not necessary that the telescope is fully diffraction limited!

Image selection techniques have an additional advantage over other techniques, like adaptive optics, that the 'lucky short exposure image' not only has a high angular resolution in the long exposure isoplanatic patch, but also that the size of the isoplanatic patch is increased as well. Short exposure solar images actually tend to have high image quality over areas well exceeding in size the few arcseconds normally mentioned as isoplanatic areas for visible adaptive optics. Part of that is undoubtedly due to the image selection technique, part may be due to the fact that a relatively large percentage of the daytime seeing occurs close to the telescope.

The use of image selection techniques to achieve diffraction limited imaging

Fig. 10.1. Example of diffraction-limited image taken with the Sacramento Peak Vacuum Tower Telescope at 460 nm. Exposure time was 16 ms. (Courtesy T. Rimmele).

is of course limited to observations where photon fluxes are high and for D/r_0
smaller than about 5. For photon-starved narrow band filter imaging and high
resolution spectroscopy adaptive optics will be needed. Adaptive optics is also
desired for diffraction limited imaging in the future with very large aperture
solar telescopes $(D > 5r_0)$, although post-detection image restoration may
remain an option for broadband imaging. The combination of low order adaptive
optics with image selection is an interesting future possibility for broadband
imaging with very large solar telescopes. Not only would it decrease the
number of required adaptive subapertures by more than an order of magnitude,
it would also presumably benefit from the increased isoplanatic patch.

10.2.2 Post-detection image restoration techniques

Speckle image reconstruction methods have been used extensively in high
resolution solar imaging. As is the case for image selection techniques, it is not
within the context of this monograph to review their methodology and application in detail. Instead we refer to other reviews like the one by von der Lühe
(1992) for a more detailed description. It is, however, of interest here to review
the relation of these techniques to the use of adaptive optics. Solar and night-
time applications both have used Knox–Thompson and Bispectrum/Triple

Correlation/Speckle Masking techniques extensively to obtain close to diffraction limited images from many objects of interest. Recently phase-diversity techniques have found interesting applications in solar astronomy (Löfdahl and Scharmer 1993; Paxman *et al.* 1992; Seldin and Paxman, 1994). As shown in Fig. 10.2, it generally uses two simultaneous, short exposure images taken at different focus/defocus positions near the focal plane. The two images have a known phase aberration difference in the incoming wave front referred to as 'phase diversity'. In the case of different focus/defocus images it is quadratic with the distance to pupil center. From this known phase diversity it is then possible to recover both the unaberrated image as well as the aberrations of the optical system, including the atmosphere. Following conventional speckle imaging, the use of many such focus/defocus positions to enhance the quality of the final image is referred to as phase-diverse speckle imaging.

In solar applications, problems associated with photon noise are minimized again by using broadband imaging. Some degree of image selection is common to minimize the degree of image restoration needed. Differential methods have been used in which the image restoration function (or the atmosphere/telescope aberration) is determined from low noise broadband observations and then applied to the restoration of noisier narrow-band observations (Keller *et al.* 1991a; Keller *et al.* 1991b).

Post-detection image restoration techniques in solar astronomy presently compete successfully with the gains anticipated from the use of adaptive optics. They are, however, very computer intensive and are generally used only for specific research programs. The introduction of adaptive optics in solar astronomy will reduce the need for this. In addition adaptive optics will be necessary for spectroscopy and probably for photon-starved imaging with very large aperture solar telescopes in the future.

10.2.3 Adaptive optics techniques

As is evident from the discussions above, the implementation of adaptive optics in solar astronomy is desirable for a number of reasons including:

(i) *Spectroscopy*, where the short exposure times and the one-dimensional spatial coverage preclude the other methods for high resolution observations to achieve the required sensitivities.

(ii) *Narrow-band imaging*, where photon noise in the short exposures prevents image selection techniques and where the sensitivities may be compromised in post-detection image restoration techniques. Adaptive optics allows longer exposures and full time coverage.

(iii) *Large aperture telescopes*, where image selection techniques start breaking down

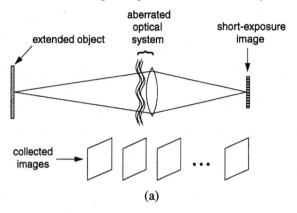

(a)

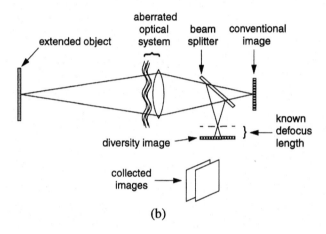

(b)

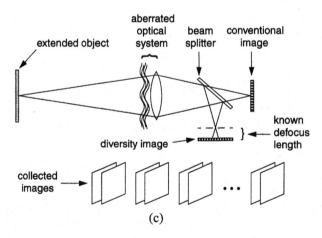

(c)

Fig. 10.2. Schematic of conventional speckle image observations (a), phase diversity imaging (b), and phase-diverse speckle imaging (c). Courtesy Seldin and Paxman (1994).

for $D/r_0 > 5$ and where the speckle transfer functions become very small at high spatial frequencies. In addition one expects the lifetime for the smallest solar features to decrease more or less linearly with feature size and hence with the angular resolution of the telescope (proportional to D^{-1}). Since the light flux per resolution element in a diffraction limited telescope is independent of its diameter D, the required decrease of exposure times works against image selection and reconstruction techniques.

(iv) *Routine observing*, where the observation of the highly time variable solar atmosphere and the need to obtain maximum coverage of solar active regions requires the best image all the time, and where routine computer intensive image reconstructions are unacceptable.

(v) *Coronal observing*, where the absence of a sufficiently bright surface structure requires the use of laser guide star technology.

As has already been pointed out for image selection techniques, a hybrid of adaptive optics, image selection, and post-detection techniques will be very useful in many cases. Image restoration techniques will enable the removal of residual image deterioration effects resulting from, for example, a not fully effective adaptive optics system and from anisoplanicity effects.

10.3 Wave-front sensing techniques on extended objects

Non-solar adaptive optics systems use point or point-like objects for wave-front sensing. Solar adaptive optics differs from these primarily in the need to do wave-front sensing on an extended, structured object (see Fig. 10.1). The solar surface structure most suitable is the solar granulation which is present anywhere on the solar disk, has a typical size of 1.5", is visible in broadband images and has a (wavelength-dependent) contrast of 5–10%. In solar active regions sunspots and their small scale, sub-arcsecond structure can be used as well. Sometimes very small sunspots with sizes near one arcsecond are available, so-called pores. In that limited case non-solar wave-front sensing techniques can be used with the pore acting as the (inverse) point-like object. In general, however, it is necessary to develop extended object wave-front sensing techniques. This section describes the concepts used in different efforts towards developing these.

10.3.1 Tip/tilt sensing using correlation trackers

Very low order solar adaptive optics systems correcting for wave-front tilt variations only have been in operation for a number of years. The earlier ones relied on tip/tilt sensing using pores (Tarbell and Smithson 1981). Recent systems rely on sensing using granulation patterns (Ballesteros *et al.* 1993, 1996; Rimmele *et al.* 1991, 1993; von der Lühe *et al.* 1989). The sensors are called 'correlation trackers' since they use correlation techniques to measure the shifts of the images with time. High speed CCD array detectors with typically 32×32 pixels, each pixel 1/4 to 1/2 arcsec in size, continuously record the image of a small area of the solar disk, and the shift between the images is determined using high speed correlation techniques. The correlation technique used should be optimized to reflect the motion of the granulation pattern, be insensitive to other effects like changes in the large scale intensity gradients across images and be as sensitive as possible considering the low residual contrasts (1–2%) in the granulation patterns encountered under mediocre seeing conditions.

The vacuum solar telescopes on Sacramento Peak, Tenerife, and La Palma all use correlation tracking for their tip/tilt correction systems. In addition to their use for these low order adaptive optics systems, the sensors also deliver a feature contrast measure which is often used to trigger exposures in the image selection mode. Correlation trackers of course, only work well over durations short compared to the lifetimes of the solar structures used. For solar granulation this lifetime is of the order of 500 s, for pores it is longer. Over longer lifetimes the telescopes have to rely on the solar limb autoguiding taking into account the solar rotation effects (amounting to $1''$ in 6 min at the solar equator).

10.3.2 Shack–Hartmann sensors

A natural extension of the correlation tracker technique for measuring image shifts is the Shack–Hartmann wave-front sensor similar to the one frequently used in night-time adaptive optics. In it a small area of the solar image is selected by a field-stop in the focal plane and re-imaged by the lenslet array onto a larger format CCD array where the differential motions between the images are measured by correlation techniques as done in a correlation tracker. The outline of the field-stop can be used as the zero position reference for the Shack–Hartmann images. The size of the pupil subapertures for a solar Shack–Hartmann sensor, of course, has to be large enough to maintain a significant amount of the 1 to 2 arcsec-sized granulation contrasts. This effectively sets a lower limit to the subpupil size of about 10 cm.

Solar Shack–Hartmann sensing is being used for sensing wave-front aberrations (Rousset *et al.* 1987; Denker *et al.* 1993; Michau *et al.* 1993; Owner-Peterson, 1993; Owner-Peterson *et al.* 1993; Rimmele *et al.* 1997). However, currently their speed is insufficient for use in adaptive optics systems. These speeds are limited by a combination of the time needed for read-out of the detector and by the computational efforts required to determine the image shifts from correlation analysis. Neither of these limitations are unsolvable with presently existing digital capabilities since almost all processing involved is parallel in character. The future use of this relatively straightforward form of wave-front sensing based on existing correlation tracker techniques is therefore likely.

10.3.3 The liquid crystal multi-knife-edge sensor

Foucault or knife-edge testing is frequently used to assess wave-front quality in optical systems. Commonly they use point sources as light input, but any object with significant intensity gradients can be used. In solar telescopes knife-edge testing often uses the solar limb resulting in an estimate of wave-front tilt variations in the direction at right angles to the knife-edge/solar limb direction. Von der Lühe (1987a, 1988) describes a wave-front sensor concept using the intensity gradients present in the solar granulation pattern. Experiments using this concept are being carried out for the US National Solar Observatory adaptive optics system (Dunn *et al.* 1988; Dunn 1990).

The concept uses a LCD, a liquid crystal, computer programmable 2-D transmission mask for the knife-edge pattern. To program its transmission, an image of the small area of the solar image being used for wave-front sensing is analyzed to determine the locations where the largest positive and negative intensity gradients in two orthogonal x- and y-directions occur. At least two LCD masks are needed as in common knife-edge testing, one each for the two orthogonal x- and y-directions. For maximum sensitivity two masks are desirable for each of these directions, one programmed for the optimum response to the positive intensity gradients, the other to the negative gradients. This results in four LCD masks. The wave-front system thus includes as basic components:

(i) a beam splitter which creates four images of the area on the sun being measured to be used with the four masks,
(ii) the LCD mask in each of these images programmed appropriately for the orthogonal directions and for the intensity gradients,
(iii) optics to form the four images of the telescope pupil,
(iv) an array of photo sensors corresponding in configuration to the spatial resolution pattern desired for the wave-front sensing on the telescope pupil.

Combination of the negative and positive intensity gradient results makes the wave-front sensing less sensitive to temporal intensity changes, combining the results from the two orthogonal directions allows the reconstruction of the full wave-front pattern. For further description of variants of this form of wave-front sensing as well as of some of the aspects of its implementation, e.g. the removal of the diffraction patterns caused by the LCD pixel masks, we refer to von der Lühe 1987a, 1988; Dunn 1990; Dunn *et al.* 1988. Although this is a very promising wave-front sensing technique, a full demonstration still has to be done.

10.3.4 *The shearing interferometer sensor*

In the adaptive optics systems developed at ITEK by Hardy shearing inter-ferometers were used as wave-front sensors. Hardy in his experimentation at the Sacramento Peak telescope applied the same technique successfully on solar disk structures (Hardy 1980, 1987). Since then shearing interferometry has not been used for solar wave-front structure analysis.

10.3.5 *The curvature sensor*

The successful implementation of wave-front curvature sensing techniques and curvature control adaptive optics has led to efforts to analyze the use of similar wave-front sensing techniques for extended sources like the sun (Kupke *et al.* 1994; Roddier *et al.* 1996). Initial results show that curvature sensing may be useful also for solar applications. No experimental results of this technique have so far been obtained or published.

10.3.6 *The phase-diversity or focal volume sensor*

As described earlier (Section 4.3.2.2) phase-diversity image reconstruction techniques result in an estimate both of the undisturbed image and of the wave-front aberrations. It, as well as possibly some other speckle reconstruction techniques, could therefore be used as wave-front estimators. Denker *et al.* (1993) demonstrated on actual solar observations that phase-diversity and Shack–Hartmann sensing gave identical results. However, the time-consuming computational efforts involved in phase-diversity sensing effectively precludes them of effective use in adaptive optics systems.

10.4 Other aspects of solar adaptive optics

Although the main item that distinguishes solar adaptive optics systems from night-time systems is the wave-front sensing, we want to comment on some of the other aspects of solar adaptive optics as well.

10.4.1 Adaptive mirrors

Segmented mirrors have been used in early solar adaptive optics systems (Smithson *et al.* 1984; Title *et al.* 1987; Peri *et al.* 1988; Acton and Smithson 1992) primarily because of the more direct coupling between the wave-front tilt measurements and the segment tilt control. The continuity of the mirror surface was assured either by using a separate interferometric test of the mirror or by high quality control of the tip/tilt piston actuators. With the advance of high speed, sophisticated wave-front reconstructors and adaptive optics controls newer systems use continuous face-plate adaptive mirrors.

10.4.2 Wave-front reconstructors

Wave-front reconstruction techniques and control loops in solar and night-time adaptive optics are identical. We refer to Chapter 5 for a discussion.

10.4.3 Multi-conjugate adaptive optics

Since the sun is an extended object it is ideally suited for the application of multi-conjugate adaptive optics for the extension of the isoplanatic patch size (Beckers 1988a, 1993). Multi-conjugate adaptive optics is the detailed correction of the atmospheric wave-front disturbances not just as they appear at the bottom of the atmosphere, but at the different heights at which they occur. It requires the determination of the wave-front variations caused by different layers in the Earth's atmosphere and the use of adaptive components located at the conjugates of these layers. For the sun the wave-front disturbances should be measurable as a function of height by 'atmospheric tomography' using the measurements of the total projected wave-front disturbance from different areas on the solar surface. For night-time adaptive optics this is normally not an option unless arrays of laser beacons are used. Multi-conjugate adaptive optics has not progressed beyond the conceptual stage and probably won't until the single conjugate systems currently under development have been made to work routinely.

10.4.4 An option for adaptive optics for solar corona observations

So far the discussion has been limited to solar adaptive optics systems for use on the solar disk where the solar surface structure can be used for wave-front sensing. For solar corona observations this form of wave-front sensing is not possible. When observing this faint envelope around the sun the radiation of the faint solar corona ($\leq$ one-millionth of the solar disk brightness) is buried in a scattered light background resulting from atmospheric Rayleigh and aerosol scattering and from scattering within the coronagraph. On excellent days this scattering is of the order of 10 millionths of the solar disk brightness. In terms of stellar magnitudes this amounts to $V = 1.9$ per squared arcseconds within which a coronal point-like source structure with a real 100% contrast would have a contrast of less than 10%. All coronal structures are much less contrasty than that, and are best visible only in the few coronal emission lines present. The low photon fluxes, the low feature contrasts and the high sky photon noise all exclude the sensing of the wave-front on the coronal structures themselves.

Sodium laser beacons provide a good alternative, however, for future coronal adaptive optics development. Using a narrow band (0.1 nm) filter to isolate the center of the sodium Fraunhofer line in the scattered light background the sky background decreases to $V = 11 - 12$ per squared arcsec, low enough to make sodium laser beacons clearly visible and usable for coronal adaptive optics. The solving of the tip/tilt ambiguity remains a problem. Once diffraction limited, it should be possible to observe a number of stars through the corona against the scattered light background. The number depends on a number of factors including telescope size, sky background, and wavelength used, but will in any case allow the observing of only a fraction of the corona. Because of the Earth's revolution around the sun these stars will move at a rate of 0.04 arcsec/sec so that total observing time is limited to a few minutes for each coronal location. Another way of solving the tip/tilt problem is the use of coronal structure itself. Its applicability depends on the properties of the coronal fine structures in the continuum or spectral lines, which is presently unknown.

10.5 Description of solar adaptive optics systems in development

Presently only the following two higher order (i.e. beyond tip/tilt correction) solar adaptive optics systems are in use and under development.

10.5.1 The Lockheed adaptive optics system

As described by Acton (1993, 1995) and Acton and Smithson (1992) the Lockheed system includes the following features:

(i) A 19 segment adaptive mirror with three actuators per segment for tip/tilt and piston control. The segments are hexagonal in shape and about 35 mm in size.
(ii) A 19 element Shack–Hartmann wave-front sensor for segment tip/tilt control using quad-cells working on pores.

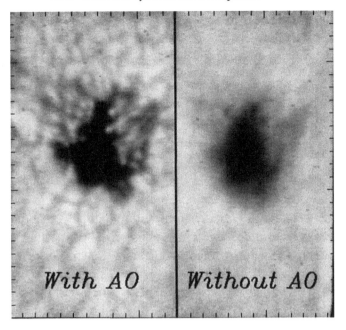

Fig. 10.3. Image of a small sunspot on the solar surface obtained with the Lockheed adaptive optics system located at the Sacramento Peak Vacuum Telescope. The wavelength used is 520 nm. The telescope aperture was masked down to 32 cm. Right: uncorrected image. Left: corrected image. The divisions at the edge of the image are 1″ apart (from Acton and Smithson 1992).

(iii) Separate control of the segment piston by means of a white light interferometer.
(iv) Analog wave-front reconstruction and control with a 300 Hz (3 dB) bandpass.

The Lockheed system has been used at both the Sacramento Peak Vacuum Telescope and at the German Vacuum Tower Telescope at the Teide Observatory in the Canary Islands. Figure 10.3 shows the results of one of the Sacramento Peak observing runs taken under mediocre seeing conditions with the telescope aperture masked down to 32 cm. Until now no images have been obtained which approached the quality obtained by image selection (see Fig. 10.1).

10.5.2 The US National Solar Observatory adaptive optics system

Dunn (1987a, 1990) and Dunn *et al.* (1988) described the Sacramento Peak/ NSO system. After a number of changes it now includes the following features:

(i) A continuous face-plate adaptive mirror 4.5 mm thick with 61 actuators separated by 32 mm.

(ii) A liquid-crystal multi-knife-edge wave-front sensor.

(iii) A digital wave-front reconstructor using matrix manipulations for either modal or zonal or other forms of reconstruction.

(iv) A rate of wave-front sensing and reconstruction of 5000 Hz.

 (v) A servo-control bandwidth of 500 Hz (3 dB).

The NSO adaptive system will be incorporated in an optical system which will allow its use with a number of different instruments being used at the Sacramento Peak Vacuum Telescope. Its operation is intended to be user friendly. In addition it is intended as a prototype for future large solar telescopes like LEST (Dunn, 1987b; Jiang Wenhan *et al.*, 1987; von der Lühe, 1987b; Owner-Peterson, 1993). A significantly modified version of this system has recently received more emphasis. It combines a 97-actuator commercial (Xinetics) adaptive mirror with a 24-element Shack-Hartmann wave-front sensor. The low temporal frequency, active optics mode of this system has been demonstrated. The full adaptive optics mode is anticipated to be functioning in the near future (Rimmele *et al.* 1997; Rimmele and Radich, 1998).

10.6 Conclusion

The small community of solar users, the complexities associated solar specific adaptive optics systems (i.e. wave-front sensing and worse seeing), and the limited resources have made progress in developing these systems slow. Even though solar physicists were among the first to start the development of astronomical adaptive optics systems, these issues and the lower amount of technology transfer that was possible from military efforts have caused solar adaptive optics development to fall behind. The smaller telescope sizes available for solar research and the rapid increase of sophistication in image selection and restoration techniques have offset to some extend the losses suffered from this slower development. As the size of ground-based solar telescopes increases in the future these alternate high resolution techniques will decrease in effectiveness and the need for solar adaptive optics will become even more apparent. In view of the implementation of future large telescopes, aimed at the study of the very small scale, energetically very important magneto-hydrodynamic structure at all layers in the solar atmosphere, the pursuit of solar adaptive optics is urgent. This pursuit should focus on the development of optimum wave-front sensing techniques unique to solar applications and on the adaption of other system components (adaptive mirrors, wave-front reconstructors and controllers, laser beacons) from other astronomical and military efforts.

References

Acton, D. S. and Smithson, R. C. (1992) Solar imaging with a segmented adaptive mirror. *Appl. Opt.* **31**, 3161.

Acton, D. S. (1993) Status of the Lockheed 19-segment solar adaptive optics system. In: *Real Time and Post Facto Solar Image Correction*, ed. R. Radick, Proceedings of the 13th NSO/SP Summer Workshop. p. 1.

Acton, D. S. (1995) Correction of static optical errors in a segmented adaptive optical system. *Appl. Opt.* **34**, 7965.

Ballesteros, E., Bonet, J. A., Martin, C., Fuentes, F. J., Lorenzo, F., Manescau, A., *et al.* (1993) A solar correlation tracker using a video motion estimation processor. In: *Real Time and Post Facto Solar Image Correction*, ed. R. Radick, Proceedings of the 13th NSO/SP Summer workshop. p. 44.

Ballesteros, E., Collados, M., Bonet, J. A., Lorenzo, F., Viera, T., Reyes, M., *et al.* (1996) *Astron. Astrophys. Suppl. Ser.* **115**, 353–65.

Beckers, J. M. (1988a) Solar image restoration by adaptive optics. In: NATO Advanced Study Workshop on Solar and Stellar Granulation, eds R. Rutten and G. Severino, p. 43. Kluwer Academic Publishers, Dordrecht.

Beckers, J. M. (1988b) Improving solar image quality by image selection. In: NATO Advanced Study Workshop on Solar and Stellar Granulation, eds. R. Rutten and G. Severino, p. 55. Kluwer Academic Publishers, Dordrecht.

Beckers, J. M. (1993) Adaptive optics for astronomy: principles, performance and application. *Ann. Rev. Astron. Astrophys.* **31**, 13.

Buffington, A., Crawford, F. S., Muller, R. A., Schemin, A. J. and Smits, R. G. (1977) Correction of atmospheric distortion with an image sharpening telescope. *J. Opt. Soc. Am.* **67**, 298.

Denker, C., Restaino, S. and Radick, R. (1993) A comparison of two wavefront sensors. In: *Real Time and Post Facto Solar Image Correction*. ed. R. Radick, Proceedings of the 13th NSO/SP Summer Workshop. p. 86.

Dunn, R. B. (1987a) Adaptive Optical System at NSO/Sac Peak. In: *Proc. Workshop on Adaptive Optics in Solar Observations*. eds. F. Merkle, O. Engvold and R. Falomo. LEST Technical Report 28, p. 87.

Dunn, R. B. (1987b) Specifications for the LEST Adaptive Optical System. In: *Proc. Workshop on Adaptive Optics in Solar Observations*. eds. F. Merkle, O. Engvold and R. Falomo. LEST Technical Report 28, p. 243.

Dunn, R. B. (1990) NSO/SP adaptive optics program. In: *Adaptive Optics and Optical Structures*, eds. J. J. Schulte-in-den-Baeumen and R. K. Tyson, Proc. SPIE 1271, p. 216.

Dunn, R. B., Streander, G. and von der Lühe, O. (1988) Adaptive Optical System at Sac Peak. In: *High Spatial Resolution Solar Observations*, ed. O. von der Lühe, p. 53. National Solar Observatory, Sunspot, NM.

Fried, D. L. (1978) Probability of getting a lucky short-exposure image through turbulence. *J. Opt. Soc. Am.* **68**, 1651.

Hardy, J. W. (1981) Solar Isoplanatic Patch Measurements, In: *Solar Instrumentation: What's Next?* ed. R. B. Dunn, p. 421. Sacramento Peak National Observatory, Sunspot, NM.

Hardy, J. W. (1987) Adaptive optics for solar telescopes. In: *Proc. Workshop on Adaptive Optics in Solar Observations*. eds. F. Merkle, O. Engvold and R. Falomo. LEST Technical Report 28, p. 137.

Hecquet, J. and Coupinot, G. (1985) Gain en résolution par superposition de poses courtes recentrées et selectionées. *J. Optics* (Paris) **16**, 21.

Jiang Wenhan, Lui Yueai, Shi Fang and Tang Guomao (1987) Study of a modal multidither image sharpening adaptive optical system for solar applications. In: *Proc. Workshop on Adaptive Optics in Solar Observations.* eds. F. Merkle, O. Engvold and R. Falomo. LEST Technical Report 28, p. 133.

Keller, C. U. and von der Lühe, O. (1991a) Solar speckle polarimetry. *Astron. Astrophys.* **261**, 321.

Keller, C. U. and von der Lühe, O. (1991b) Application of differential speckle imaging to solar polarimetry. *High-Resolution Imaging by Interferometry* II ESO Conference and Workshop Proceedings, eds. J. M. Beckers and F. Merkle, p. 451. ESO Garching, Germany.

Kupke, R., Roddier, F. and Mickey, D. L. (1994) Curvature-based wavefront sensor for use on extended patterns. In: *Adaptive Optics in Astronomy*, eds. M. A. Ealey and F. Merkle, Proc. SPIE 2201, p. 519.

Löfdahl, M. and Scharmer, G. (1993) Phase-diversity restoration of solar images. In: *Real Time and Post Facto Solar Image Correction.* ed. R. Radick, Proceedings of the 13th NSO/SP Summer Workshop. p. 89.

Michau, V., Rousset, G. and Fontanella, J. (1993) Wavefront sensing from extended sources. In: *Real Time and Post Facto Solar Image correction.* ed. R. Radick, Proceedings of the 13th NSO/SP summer Workshop. p. 124.

Owner-Peterson, M. (1993) An algorithm for computation of wavefront tilts in the LEST solar slow wavefront sensor. In: *Real Time and Post Facto Solar Image Correction.* ed. R. Radick, Proceedings of the 13th NSO/SP Summer Workshop. p. 77.

Owner-Peterson, M., Darvann, T. and Engvold, O. (1993) Design of the LEST slow wavefront sensor. In: *Real Time and Post Facto Solar Image Correction.* ed. R. Radick, Proceedings of the 13th NSO/SP Summer Workshop. p. 63.

Paxman, R. G., Schulz, T. J. and Fienup, J. R. (1992) Joint estimation of object and aberrations by using phase diversity. *J. Opt. Soc. Am.* **9**, 1072.

Peri, M. L., Smithson, R. C., Acton, D. S., Frank, Z. A. and Title, A. M. (1988) Active optics, anisoplanatism, and the correction of astronomical images. In: *NATO Advanced Study Workshop on Solar and Stellar Granulation* (eds. R. Rutten and G. Severino), p. 77. Kluwer, Dordrecht.

Rimmele, Th., von der Lühe, O., Wiborg, P., Widener, A. L., Spence, G. and Dunn, R. B. (1991) Solar Feature Correlation Tracker. Proc SPIE Conf. 1542, p. 186.

Rimmele, T., Kentischer, T. and Wiborg, P. (1993) High resolution observations with NSO/KIS correlation tracker. In: *Real Time and Post Facto Solar Image Correction.* ed. R. Radick, Proceedings of the 13th NSO/SP Summer Workshop. p. 24.

Rimmele T., Beckers, J., Dunn, R., Radick, R. and Roeser, M. (1997) High resolution solar observations from the ground. In: Proc. SAO Workshop, *Solar Atmospheric Dynamics*, Gloucester (Mass.), June 3–4 1997, eds. L. Golub and E. Deluca, (in press).

Rimmele T. R. and Radick, R. R. (1998) Solar adaptive optics at the National Solar Observatory. In: *Adaptive Optical System Technologies*, eds. D. Bonaccini and R. K. Tyson, Proc. SPIE 3353 (in print).

Roddier, F., Kupke, R. and Mickey, D. (1996) A new wave-front sensing technique for solar adaptive optics. In: Proc. Workshop, *Science with Themis*, Paris-Meudon Observatory, Nov. 14-15 1996, eds. N. Mein and S. Sahal-Bréchot, pp. 179–85.

Rousset, G., Primot, J. and Fontanella, J.-C. (1987) Visible wave-front sensor development. In: *Proc. Workshop on Adaptive Optics in Solar Observations.* eds. F. Merkle, O. Engvold and R. Falomo. LEST Technical Report 28, p.17.

Seldin, J. H. and Paxman, R. G. (1994) Phase-diverse speckle reconstruction of solar data. In: *Image Reconstruction and Restoration*, eds. T. J. Schulz and D. L. Snyder. Proc. SPIE 2302, p. 268

Smithson, R., Marshall, N., Sharbaugh, R. and Pope, T. (1984) A 57-actuator active mirror for solar astronomy. In: *Solar Instrumentation, Whats Next?* (ed. R. B. Dunn), p. 66. Sacramento Peak National Observatory, Sunspot, NM.

Tarbell, T. and Smithson, R. (1981) A simple image motion compensation system for solar observatories. In: *Solar Instrumentation: What's Next?* (ed. R. B. Dunn), p. 491. Sacramento Peak National Observatory, Sunspot, NM.

Title, A. M., Peri, M. L., Smithson, R. C. and Edwards, C. G. (1987) High resolution techniques at Lockheed solar observatory. In: *Proc. Workshop on Adaptive Optics in Solar Observations*, eds. F. Merkle, O. Engvold and R. Falomo. LEST Technical Report 28, p. 107.

von der Lühe, O. (1987a) A wavefront sensor for extended, incoherent targets. In: *Proc. Workshop on Adaptive Optics in Solar Observations*, eds. F. Merkle, O. Engvold and R. Falomo. LEST Technical Report 28, p. 155.

von der Lühe, O. (1987b) Photon noise analysis for a LEST multidither adaptive optical system. In: *Proc. Workshop on Adaptive Optics in Solar Observations*, eds. F. Merkle, O. Engvold and R. Falomo. LEST Technical Report 28, p. 255.

von der Lühe, O. (1988) Wavefront error measurement technique using extended, incoherent light sources. *Opt. Eng.* **27**, 1078.

von der Lühe, O., Widener, A. L., Rimmele, Th., Spence, G., Dunn, R. and Wiborg, P. (1989) Solar correlation tracker for ground-based telescopes. *Astron. Astrophys.* **224**, 351.

von der Lühe, O. (1992) High spatial resolution techniques. In: *Solar Observations: Techniques and Interpretation*, eds. F. Sánchez, M. Collados and M. Vázquez, p. 1. Cambridge University Press, Cambridge.

Part four

Adaptive optics with laser beacons

11

Overview of adaptive optics with laser beacons

DAVID G. SANDLER

ThermoTrex Corporation, San Diego, California, USA

11.1 Motivation for using laser beacons

11.1.1 The goal of diffraction-limited correction of large telescopes

In Chapters 8–10, we have seen that adaptive optics (AO) is a powerful tool to enhance the resolution and contrast of astronomical images. Several 2- to 4-m class astronomical telescopes now have AO user instruments. These include the ESO and Canada–France–Hawaii 3.6-m telescope, and the Mt Wilson 100-inch telescope. All are being used for scientific observations, and are producing dramatic results, fulfilling the promise of AO to overcome the problem of seeing which has plagued astronomers for centuries.

The AO systems on the above telescopes, as well as several other systems which will be operational in the near future, use the light from a field star to sense the wave-front aberrations, as originally envisaged by Babcock (1953). As discussed in Section 3.5, to produce diffraction-limited correction a bright source must be available within the isoplanatic patch. (We will quantify this requirement in the following section.) There are many applications where a natural star can be utilized, notably in applications of AO for stellar astronomy and the search for faint companions around bright stars. In addition, many extended objects will contain a bright stellar component suitable for wave-front sensing. However, the brightness requirement for field stars is quite severe, resulting in a low probability for finding a sufficiently bright "guide" star for diffraction-limited correction (within $\sim 10''$ at an imaging wavelength $\lambda_i = 1$ μm) for arbitrary program objects.

A solution to the scarcity of bright natural guide stars (NGS) is to produce an artificial star from the ground (McCall and Passner 1978; Foy and Labeyrie 1985). The method is based on the use of artificial laser guide stars (LGS), or laser beacons. The basic concept is to project a laser beam from the telescope and focus it to a spot, to produce an artificial 'star' along the line of sight to

255

any astronomical target. The back-scattered (or back-radiated) light from the spot can then be used to sense the wave front between the beacon and the telescope primary mirror. The laser beacon can be made very bright, by concentrating either continuous or pulsed laser light in an area equal to the seeing disc. The beacon will be elongated, since light will scatter on either side of the focus. The two types of laser beacon in current use will be discussed in Section 11.2.

The use of laser beacons in principle allows photon-noise errors to be made arbitrarily small, even for very fast system servo bandwidths. Measurement and temporal errors can thus be controlled. Off-axis anisoplanatism is eliminated near the center of an imaging field, and entirely for spectroscopic applications. Thus, the use of a laser allows a man-made reference star to be supplied from the ground for adaptive control of wave-front errors to very high accuracy. We will treat the expected performance of laser beacon AO in Chapter 12.

LGS AO will have a large impact on science capabilities with the new generation of large, 6- to 10-m telescopes. The sensitivity of observations with adaptive optics (in the regime where sky airglow emission is not the dominant source of background noise) increases inversely as the fourth power of the aperture size, one factor of D^2 from the light concentration in the diffraction limited core, and another factor from the collection area. As we will see in Section 11.2, laser beacons are not used for sensing overall tilt, since the combined upward and downward motion of the laser spot usually does not correlate with global image motion. Thus, a natural field star is required to sense overall tilt, and the sky coverage for LGS AO will be determined by the brightness requirement and allowed angular field for tilt stars. Maximizing sky coverage of LGS AO favors large telescopes, since very faint field stars can be used for image motion sensing, and there is an increased isoplanatic angle for tilt correction, both of which lead to a larger number of available field stars. In addition, because the required laser power depends on the size of r_0 and not on the telescope aperture, the use of lasers becomes easier to justify for large telescopes, where the number of diffraction-limited resolution elements across the isoplanatic patch will exceed 1000 even in the K band.

The use of LGS AO has caught on quickly within the astronomical community in large part because, equipped with adaptive optics operating at the diffraction limit in the near-infrared J, H, and K bands, the new 6- to 10-m telescopes will have the capability to match the angular resolution of the 2.4-m Hubble Space Telescope (HST) in the visible and to exceed its resolution in the near infrared. This will allow high-resolution imaging and spectroscopy to be performed between wavelengths $\lambda = 1.2-2.5$ μm that will supplement HST's already impressive results. Thus, while the challenges for LGS AO are great, so are the potential benefits for ground-based astronomy.

11.1.2 Brightness requirements for laser beacons

In this section, we quantify the above estimates for reference star brightness requirements for diffraction-limited correction. This will demonstrate the potential for laser beacons to increase sky coverage, and give a first order estimate of the required laser power for LGS AO.

For an operational definition of diffraction-limited correction, we take $R \geqslant 0.3$, where R is the total system Strehl ratio. This Strehl ratio is generally the minimum value which will lead to a well-defined Airy pattern PSF, with a fraction R of total energy contained within a spot of fwhm $\sim \lambda/D$, and the remaining fraction $1 - R$ distributed in a halo of extent $\sim \lambda/d$, where d is the width of the subapertures used for wave-front sensing. For Strehl ratios < 0.3, either the fwhm of the PSF begins to spread beyond the diffraction angle, or the residual halo level will become appreciable, both of which decrease the effective resolution. For Strehl ratios of 10–20%, post-processing of images can lead to recovery of a diffraction-limited component, but at a lower SNR than for diffraction-limited correction.

11.1.2.1 Simple model for NGS AO performance

We adopt a simple analytical model for AO error sources, following the treatment and notation of Chapter 2. For natural guide star AO, the residual AO error can be written

$$\sigma^2 = \sigma_{\text{fit}}^2 + \sigma_{\text{noise}}^2 + \sigma_{\text{time}}^2 + \sigma_{\text{aniso}}^2, \tag{11.1}$$

where

$$\sigma_{\text{fit}}^2 = 0.3 \left(\frac{r_s}{r_0} \right)^{5/3} \tag{11.2}$$

is the fitting error (see Eq. (2.18) and Hudgin 1977). In Eq. (11.1), the next error source is due to wave-front measurement noise, which for a Shack–Hartmann sensor is (see Eqs. (5.42) and (12.47))

$$\sigma_{\text{noise}}^2 \simeq 1.4 \frac{\pi^2}{n_{\text{ph}}}, \tag{11.3}$$

where n_{ph} is the number of photons per subaperture collected in an AO cycle time, τ. In Eq. (11.3), we have assumed a Shack–Hartmann sensor, with four pixels per subaperture to calculate local slopes from image centroids. For simplicity, we have neglected the effect of finite detector read noise (see Eq.

(12.47)). The third term in Eq. (11.1) is the temporal delay error, for which we take the simple form (see Eq. (2.26))

$$\sigma^2_{\text{time}} = \left(\frac{\tau}{\tau_0}\right)^{5/3}. \tag{11.4}$$

The final term is the anisoplanatic error, which was originally treated by Fried (1982) for full correction ($r_s \leq r_0$), and which depends on θ/θ_0 and D/r_0. We treat the case $\theta/\theta_0 = 1$ (field star at the edge of the isoplanatic patch), for which $\sigma^2_{\text{aniso}} = 0.3, 0.4,$ and 0.5 (corresponding to Strehl ratio degradations of 0.7, 0.62, and 0.55) for $D/r_0 = 5, 10,$ and 20, respectively.

The number of photons collected in a wave-front sensor subaperture with a diameter d is given by

$$n_{\text{ph}} = Fd^2\tau\eta_{\text{opt}}\eta_q, \tag{11.5}$$

where F is the stellar flux (photons/area/time). The quantities η_{opt} and η_q are optical transmission efficiency to the sensor and effective quantum efficiency of the detector, respectively. In Fig. 11.1 we have plotted n_{ph} per millisecond of

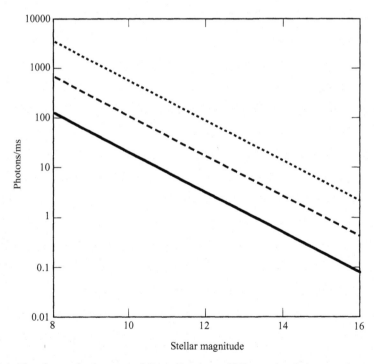

Fig. 11.1. Number of photons collected per millisecond of integration time as a function of stellar magnitude, for an r_0 sized subaperture in 0.7″ seeing. Three wavelengths are shown: 0.5 (solid line), 1 (dashed line), and 2 (dotted line) μm.

integration time, for $d = r_s = r_0(\lambda)$ at three wavelengths, $\lambda = 0.5$ μm, 1.0 μm, and 2.0 μm. We have treated the case of $0.7''$ seeing $[r_0(0.5$ μm$) = 15$ cm$]$. The flux F is taken for a mean wavelength of $\lambda = 0.65$ μm, and $\Delta\lambda = 300$ nm. The total efficiency is taken as $\eta = 0.32$, corresponding to $\eta_{opt} = 0.4$ and $\eta_q = 0.8$. From Fig. 11.1, we see that the limiting stellar magnitudes $m_0(\lambda)$ for $F = 100$ ph/msec/r_0^2 are $m_0(0.5$ μm$) = 8$, $m_0(1$ μm$) = 10$, and $m_0(2$ μm$) = 12$.

To illustrate the limiting magnitudes for diffraction-limited NGS AO, we evaluate Eq. (11.1) for $D/r_0 = 10$, and wavelengths of $\lambda_i = 0.5$, 1, and 2 μm. This corresponds to $D = 1.5$ m, 3.5 m, and 8 m, respectively, for the three wavelengths. The field star magnitudes are taken to be m$(0.5$ μm$) = 8$, m$(1$ μm$) = 12$, and m$(2$ μm$) = 16$. We take $d = r_0(\lambda)$, and an effective turbulence weighted wind speed of $v_w = 15$ m/s ($\tau_0 = 3(16)$ ms at $\lambda = 0.5(2)$ μm). The NGS is assumed to be at angular offset $\theta/\theta_0 = 1$. All other parameters are set as in the above evaluation of n_{ph}. Figure 11.2 shows the resulting Strehl ratios plotted as a function of τ. We see that for each wavelength there is an

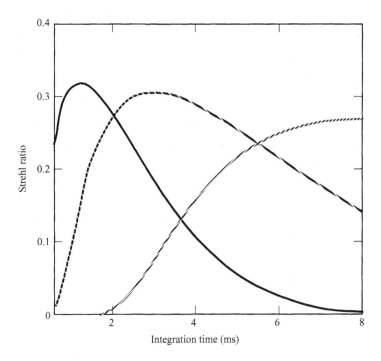

Fig. 11.2. Strehl ratio in a simple analytical model for NGS AO as a function of wave-front sensor integration time for each update. Results correspond to $D/r_0 = 10$ and subapertures of width $d = r_0$. Curves are shown for three wavelengths: 0.5, 1, and 2 μm. All curves correspond to a field star of magnitude m at the outer edge of the isoplanatic patch. Solid line, 0.5 μm, $m = 8$ star; dashed line, 1 μm, $m = 12$ star; shaded line, 2 μm, $m = 16$ star.

optimum Strehl ratio corresponding to value of τ which maximizes the Strehl. For the stellar magnitudes chosen, the optimum Strehl ratios are ~ 0.3, corresponding to the lower limit for our definition of diffraction-limited correction.

The stellar magnitude value we have chosen for each wavelength corresponds to the limiting magnitude for diffraction-limited correction, for the telescope sizes considered and the case of field star at the outer edge of the isoplanatic patch. For larger offset angles θ, the limiting magnitudes decrease substantially. For example, for $\theta/\theta_0 = 2$, the Strehl ratios at $D/r_0 = 10$ from anisoplanatism alone decrease to 0.4, 0.3, and 0.2, respectively, for $\lambda_i = 0.5$ μm, 1 μm, and 2 μm. Thus, the non-isoplanatic system errors in Eq. (11.1) must be driven to very small values to achieve a Strehl ratio $R \geqslant 0.3$.

11.1.2.2 Estimated sky coverage for diffraction-limited NGS AO

In order to understand the severe constraints on sky coverage that the above results place on adaptive optics with natural stars, we need to estimate the density of stars. From Allen (1973), we can estimate $\rho(m)$, the mean number of stars/degree^2, to be $\rho(8) = 1$, $\rho(12) = 60$, and $\rho(16) = 2000$. To evaluate the probability $P(\lambda)$ of finding field stars for diffraction-limited correction, we take the respresentative value $\theta_0 = 3''$ at $\lambda = 0.5$ μm (Sandler et al. 1994a). (For comparison, this value agrees well with measurements at the MMT site in Arizona (Lloyd-Hart et al. 1995), and is 50% larger than nominal values on Mt Haleakala on Maui, and 100% larger than the mean at Starfire Optical Range (SOR) in Albuquerque.) From this choice, it follows that $P(0.5) \sim 7 \times 10^{-7}$, $P(1) \sim 2.2 \times 10^{-4}$, and $P(2) \sim 0.038$. These probabilities are very low, and illustrate the potential power of LGS AO, which can provide even brighter beacons than assumed above well within the isoplanatic patch. There will be times when θ_0 gets very large, of course, and there may be sites with significantly less high altitude turbulence on average. In addition, studies have been made which show that using a DM conjugate to a dominant layer can increase the isoplanatic angle by a factor of two or more. Nonetheless, our basic conclusion on the need for laser beacons to routinely achieve the diffraction limit with large sky coverage, for wavelengths down to 1 μm, still holds.

We close this section by pointing out that the actual laser power implemented at most observatories will correspond to brighter field stars than derived above for NGS AO, for several reasons. First, we have assumed a 300 nm spectral bandwidth for wave-front sensing with field stars, and the equivalent V-magnitudes are usually given for a canonical 100 nm band. Also, the requirement for n_{ph} is larger for laser beacons, because the images of laser spots are larger than the short exposure seeing disc for star images by a factor of at least $\sqrt{2}$. This is due to several factors, including laser beam quality effects, the fact that the laser spots propagate upward first and then downward to the wave-front sensor, and some beam projection techniques may produce spurious elongation of laser beacon images. Thus, more photons are needed to maintain the same SNR. The subject of wave-front sensing with laser beacons is discussed in Section 12.3.3. Finally, as shown in the next section, there is an additional error source for LGS AO which arises because the laser beacon is not at infinite range, and to maximize the Strehl ratio, subapertures of width smaller than r_0 are chosen to reduce the fitting error.

11.1.3 Special requirements for LGS AO

We will see in Section 11.2 that there is some increase in system complexity attendant with the use of laser beacons to achieve diffraction-limited perform-ance. For example, laser projection optics are required, and the optical system must now be of exceedingly high quality to limit systematic errors to a very small fraction of λ/D. We have already mentioned above that a separate sensor (along with, in some cases, a separate flat steering mirror) is needed to measure and correct for the fast image motion caused by atmospheric tip/tilt. In addition, to take advantage of the bright artificial star, high temporal band-widths and fast reconstructors are required, and for large telescopes 1024 × 1024 detectors are needed for imaging fields of one to two isoplanatic patches at diffraction-limited resolution. Further, the necessary lasers may be rather expensive; in fact, we shall show that in some cases they currently challenge the state of laser technology.

In addition to increased complexity, there is also a new fundamental error which is present in LGS AO. Called "focus anisoplanatism" (FA), or the cone effect, this error arises because the rays of light from the beacon to the telescope trace out a cone, rather than the desired cylindrical volume of air above the telescope. Thus, instead of measuring the phase across an aperture of diameter D for each turbulent layer at height h above the telescope, the photon return from a laser beacon will sample a region of the layer of diameter $D(1 - h/H)$, where H is the distance of the beacon from the telescope. If there were only one layer at a single height, the measured phase could be extrapolated to the full aperture, by scaling by a factor H/h. However, in practice, there are several turbulent layers (or a continuum), and there is no known way with a single beacon to unravel this sampling error.

The FA error sets the limiting performance for LGS AO. The mean square wave-front error is given by (Fried and Belsher 1994)

$$\sigma^2_{FA} = \left(\frac{D}{d_0}\right)^{5/3}, \tag{11.6}$$

where d_0 is a parameter which depends only on wavelength and the turbulence profile at the telescope site. In Chapter 12, we will give expressions for calculating d_0, and show that d_0 is proportional to $\lambda_i^{6/5}$. From the geometric structure of the error, it is clear that the value of d_0 will depend on the strength of mid- and high-altitude turbulence at $h = 10-20$ km, and larger values of H will reduce the cone effect. Typical values are $d_0 \sim 1(5)$ m at $\lambda_i = 0.5(2)$ μm for $H = 10-15$ km, and $d_0 = 4(20)$ m for $H = 90$ km. Thus, a key requirement will be to choose H large enough to limit the FA error, and at the same time to ensure that adequate photon return is available to limit measurement noise errors.

In this and the following two chapters, we concentrate on LGS methods for obtaining diffraction-limited correction with full sky coverage for field stars. The remainder of Chapter 11 is devoted to an overview of LGS AO. In Section 11.2 we discuss a typical architecture for LGS AO, pointing out the differences compared to AO systems which use natural field stars. Section 11.3 gives a brief history of LGS AO, tracing the origins of the concept in US Department of Defense (DoD) research programs and continuing to the present time when several LGS systems are under development by astronomical observatories around the world. Chapter 12 presents the analysis tools required to predict the expected performance of LGS AO systems for astronomy with application to optimized LGS AO design for large telescopes. Chapter 13 gives an overview of existing and planned LGS systems for astronomical telescopes. We will see that in the next few years, systems will be implemented at several sites.

11.2 General architecture of LGS AO systems

Figure 11.3 shows in schematic form the general architecture of LGS adaptive optics systems. A laser beam is projected to a focus at altitude H, forming a spot whose size and shape depend on the projection optics, atmospheric conditions, and laser beam quality. In the figure, we have illustrated the case of projection through optics behind the secondary mirror of the telescope. Other projection schemes configurations can be used, including projecting the laser using the full primary mirror and using a small projecting telescope attached to the main telescope optical axis. Two types of laser beacon, corresponding to different backscatter mechanisms, have been tested. One is a Rayleigh beacon (Gardner *et al.* 1990), which uses incoherent Rayleigh backscatter off air molecules. There is a strong 180° elastic scattering cross section, and laser

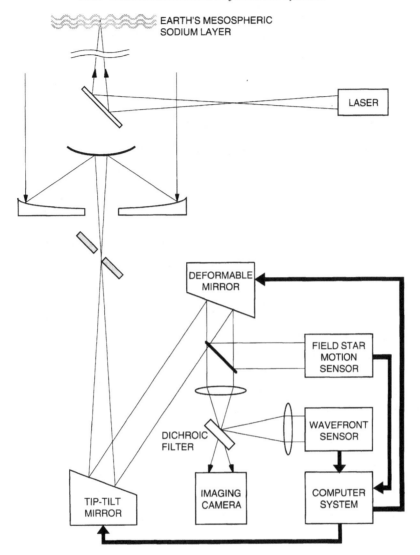

Fig. 11.3. Schematic layout of LGS AO system for an astronomical telescope.

power at visible wavelengths of ~ 100 watts has been shown to produce adequate photon return for spots at 10–20 km altitude (20 km is the limiting altitude due to the rapid fall-off in air density). Pulsed lasers must be used for Rayleigh beacons, since the pulse produces backscatter at all altitudes. Range- or time-gating of the pulse around the focus altitude is necessary, to limit the vertical extent of the beacon.

The other type of laser beacon uses resonant excitation of atoms in the

Earth's mesospheric sodium layer at 90 km (Happer *et al.* 1994; Thompson and Gardner 1987). As we shall see, sodium beacons are favored for astronomy because their higher altitude produces a better estimate of the atmospheric phase distortion in the column over the telescope in the direction of the science target. The finite, 10-km depth of the sodium layer provides automatic range-gating, so either pulsed or continuous wave (CW) lasers can be used. If a CW laser is used, the beacon must be projected in a way that prevents low altitude Rayleigh scattering from entering the science and wave-front detectors. The projection shown in Fig. 11.3 accomplishes this by blocking using the secondary mirror and forcing low altitude scatter to wide angles.

The natural field star used for image motion correction is imaged onto a quadrant detector, which measures the instantaneous atmospheric tilt to be corrected by the tip/tilt mirror (or, if enough stroke is available, by the deformable mirror). The WFS, typically of Shack–Hartmann type, forms an image of the pupil which is recorded on a fast detector. The wave-front measurements are sent to a wave-front reconstructor computer for processing and issuing of commands to the deformable mirror. A control loop is closed around the laser beacon measurements, and the long exposure image is recorded on the science detector.

Within this general architecture, the choice of optical layout and specific hardware components will vary depending on the detailed system requirements. We will see in Chapter 13 that there are several innovative designs under development.

11.3 A brief history of LGS development

Before moving on to discuss the design of LGS systems for astronomical adaptive optics, and the limiting performance expected from LGS adaptive optics, we briefly review the history leading to the first concepts and field tests. A detailed discussion of the history leading to development of artificial guide stars is given by Benedict, Breckinridge, and Fried in the Introduction to the February, 1994 special issue of *Journal of the Optical Society of America A* (*JOSA A*11, Jan–Feb 1994) devoted to atmospheric compensation technology. Here we restrict ourselves to a sketch of key developments.

11.3.1 Research and development by the US defense community

In the mid-1970s, the US Department of Defense (DoD) sponsored the development of a 100-actuator adaptive optics system for imaging of space objects using the 1.6-m AMOS telescope on Mount Haleakala. The compensated imaging system (CIS) was the first full-scale adaptive optics system, and successfully demonstrated the concept of real time adaptive control using an array of slope measurements (Hardy 1978). The CIS, installed in 1982 and still in use today, uses light from the imaging target as a source for wave-front measurements, and is restricted to rather bright objects ($m_0 \geq 6$).

By the 1980s, however, it was recognized that a method of correcting telescopes in the absence of a bright reference source was crucial to progress in both remote imaging and efficient laser beam projection through the atmosphere. Spurred by the success of CIS, various agencies in the DoD began to sponsor improved adaptive optics component development by both DoD laboratories and private industry. The emphasis of this work was on the technology required for systems with several hundred to over a thousand actuators, for correcting 3-m class telescopes at optical wavelengths. As part of this effort, new concepts were explored for artificial beacons created by Rayleigh backscatter, as first proposed by J. Feinleib of Adaptive Optics Associates in 1982. Several ingenious approaches were considered: tracking the motion of multiple thin beams during upward propagation; range-gating of a collimated laser pulse as it propagates upward, solving for phase layers from scintillation measurements; and imaging of fringes formed in the Earth's atmosphere by two interfering laser beams, solving for the phase distortion from the fringe pattern. By the mid-1980s, it was generally considered that the method of projecting a focused laser spot in the atmosphere, and using the spot as if it were an extended natural star, formed the most solid basis for further work. Performance calculations, including the dependence of Strehl ratio versus laser spot altitude were performed. For optical imaging or laser transmission using $D \geq 3$ m telescopes the residual errors caused by focus anisoplanatism were found to be too large for a single Rayleigh spot. In 1982, Happer suggested the use of the Earth's mesospheric sodium layer at 90 km for generating a high altitude artificial reference star (Happer *et al.* 1994).

Numerous successful experiments followed quickly, each demonstrating some key aspect of implementing adaptive optics. In 1983–84, two important experiments compared measured wave-front slopes obtained from a laser beacon and a bright natural star, both resulting in good agreement with theoretical predictions. Fugate and collaborators (Fugate *et al.* 1991) used a Rayleigh beacon at 10 km to measure 18 slope vectors, and Humphreys *et al.* (1991) used a sodium laser beacon to measure two slope vectors. Based on the encouraging results, several experimental efforts were begun to test the use of laser beacons for control of wave-front distortion for 1-m class telescopes. MIT/Lincoln Laboratory ($\lambda_i = 0.5$ μm) (Primmerman *et al.* 1994) and ThermoTrex Corporation ($\lambda_i = 0.35$ μm) (Sandler *et al.* 1994b) performed experiments to test Rayleigh

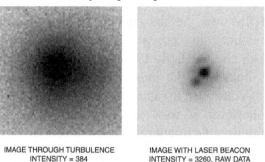

IMAGE THROUGH TURBULENCE
INTENSITY = 384

IMAGE WITH LASER BEACON
INTENSITY = 3260, RAW DATA

60 SECOND EXPOSURES, 1.66 ARCSEC FIELDS,
0.85 μm IMAGING WAVELENGTH, 1.5 m APERTURE

Fig. 11.4. One minute exposure at 0.85 μm of Beta Del taken with the laser beacon AO system at the SOR 1.5-m telescope. Left: uncorrected, fwhm = 2″. Right: LGS correction using a Rayleigh beacon at 10 km, fwhm = 0.19″.

beacon correction at short wavelengths, and both groups performed initial experiments to test the possibility of using an array of Rayleigh spots, each spot being used to sense the distortion over a limited portion of the aperture.

The most significant legacy of the LGS system experiments that began in the 1980s has been the adaptive optics system for the 1.5-m telescope at the US Air Force Starfire Optical Range (SOR). The SOR system (Fugate *et al.* 1994), now with its second generation LGS system, uses a 241-actuator continuous facesheet deformable mirror controlled at 100 Hz closed-loop bandwidth using either a Rayleigh beacon at 10–15 km or bright natural stars. The system has yielded a wealth of valuable information on the performance of LGS operation and has taught many important lessons, while producing diffraction-limited images of a number of science objects. As an example of stellar imaging, Fig. 11.4 shows 1 min. exposures of Beta Del taken with and without LGS AO at the SOR 1.5-m telescope. The Strehl ratio at $\lambda = 0.85$ μm was increased by a factor of 10. The closest of four companions to the central star is resolved in the corrected image, at a $\Delta m = 1$ and a separation of 0.6″. Since many aspects of the SOR system have guided the design of LGS adaptive optics for astronomical telescopes, the system will be discussed in more detail in Chapter 13.

11.3.2 The spread of AO within the astronomical community

Interest in laser beacons within the astronomical community began in 1985 with the publication of a paper by Foy and Labeyrie (1985), who proposed the basic laser beacon concept for high resolution astronomical imaging. Further astronomical interest in laser beacons followed Thompson and Gardner's

(1987) publication of an image of a sodium laser beacon projected through a telescope on Mauna Kea. In 1991, the DoD research summarized above was presented to astronomers at a workshop, sponsored by the Air Force Phillips Laboratory. Since that time, several astronomical groups have embarked on research and development programs for LGS AO. The first experiments comparing natural star and sodium beacon measurements at an astronomical telescope have been performed by the University of Arizona group (Lloyd-Hart *et al.* 1995), by using simultaneous laser beacon and out-of-focus star images from the six 1.8-m mirrors of the Multiple Mirror Telescope (MMT). The experiments, including the successful control of two of the mirrors using sodium backscatter (Lloyd-Hart *et al.* 1995), led to detailed estimates of both FA and natural star anisoplanatism, essentially confirming the theoretical predictions, so that one can now have faith that sodium LGS systems will indeed have tremendous value for high resolution astronomy. As of fall 1997, two groups have succeeded in obtaining corrected stellar images using closed-loop sodium LGS AO on astronomical telescopes with continuous primary mirrors.

11.3.2.1 University of Arizona at the MMT array telescope

The Arizona group has also obtained the first image of an extended field corrected using closed-loop AO with a sodium laser beacon (Lloyd-Hart *et al.* 1998). The images shown in Fig. 11.5 were taken at the MMT array telescope in January, 1996. The two panels show 100 s exposures at 2.2 μm of the central region of the nearby globular cluster M13 (NGC 6205), which orbits the nucleus of our Galaxy. On the left is an image recorded with no adaptive correction. The sodium laser beacon appeared as bright as a natural star of magnitude 10 seen through a standard *V* filter. The beacon was centered in the field-of-view of the infrared array to minimize the effect of FA. The star used to sense global tilt over the six telescope primaries is not seen here; it is 35″ north of the center of the field-of-view, and has a *V* magnitude of 15.

The adaptive correction for this first demonstration consisted of simultaneous tip/tilt control of all six MMT mirrors. Although phasing of the mirrors was not possible with this system (due to the lack of spatial coherence of the ∼ 1 arcsec sodium beacon), the spirit of full LGS AO was demonstrated since local slopes of the wave front were measured. While the increase in resolution is a modest factor of ∼ 1.5, the Strehl increase is ∼ 2 and there is considerable increase in contrast over the field. The system at the MMT array has since yielded images at the 0.3″ diffraction-limit of the individual 1.8-m MMT primary mirrors.

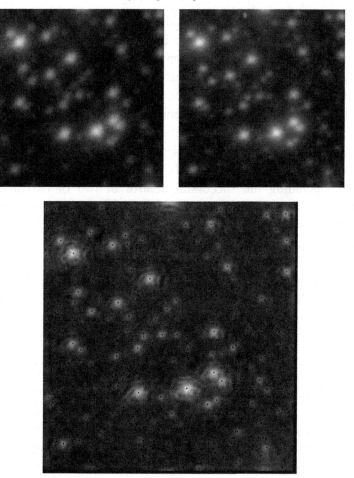

Fig. 11.5. K_s-band images before and after correction with the MMT adaptive optics system, using a sodium laser beacon as the wave-front reference source. *Upper left*: With no correction, the stellar images are seeing limited at 0.72″ in a 60 s exposure. *Upper right*: With correction on the basis of the sodium laser beacon, and global image motion corrected by reference to a natural star, the images in this 60 s exposure have been improved to 0.51″. *Bottom*: After PSF subtraction using the iterative blind deconvolution algorithm on the corrected image, the image size is further improved to 0.36″. The positions of 570 stars from a V-band image from the HST's WF/PC-1 camera are superposed as dots. All images are shown on a logarithmic gray scale to give a dynamic range of 1000, while exaggerating the halos of the bright stars. North is approximately 15° clockwise from vertical; east is left.

11.3.2.2 LLNL at the 3.5-m Lick Observatory Shane Telescope

The first successful closed-loop demonstration of sodium beacon AO for a continuous primary mirror has been by the AO research group at Lawrence Livermore National Laboratory (LLNL), working on AO development for the

Keck II Telescope with astronomers in California and Hawaii. They have obtained sharpened images of a bright star using a sodium laser, showing a factor of $\sim$ 6 improvement in Strehl ratio over uncorrected images (Max *et al.* 1997). The system developed by LLNL at Lick Observatory will be discussed in Chapter 13.

A general overview of LGS projects for astronomy is given by Thompson (1994), and these programs are reviewed in Chapter 13. At this point, we can say that while there has been much progress in demonstrating the key elements, much remains to be demonstrated to achieve the diffraction limit routinely, using the very faint field stars required to allow full sky coverage. Several observatories have set their sights on achieving this goal in the very near future.

References

Allen, C. W. (1973) *Astrophysical Quantities* (3rd edn.), p. 243. The Athlone Press, London.

Babcock, H. W. (1953) The possibility of compensating astronomical seeing. *Pub. Astr. Soc. Pac.* **65**, 229–36.

Foy, R. and Labeyrie, A. (1985) Feasibility of adaptive telescope with laser probe. *Astron. Astrophys.* **152**, 129–31.

Fried, D. L. (1982) Anisoplanatism in adaptive optics. *Astrophys. J.* **493**, 950–4.

Fried, D. L. and Belsher, J. F. (1994) Analysis of fundamental limits to artificial-guide-star adaptive optics system performance for astronomical imaging. *J. Opt. Soc. Am. A* **11**, 277–87.

Fugate, R. Q., Fried, D. L., Ameer, G. A., Boeke, B. R., Browne, S. L., Roberts, P. H., *et al.* (1991) Measurements of atmospheric wave-front distortion using scattered light from a laser guide star. *Nature* (London) **353**, 144–6.

Fugate, R. Q., Ellerbroek, B. L., Higgins, C. H., Jelonek, M. P., Lange, W. J., Slavin, A. C., *et al.* (1994) Two generations of laser-guide-star adaptive optics experiments at the Starfire Optical Range. *J. Opt. Soc. Am. A* **11**, 310–24.

Gardner, C. S., Welsh, B. M. and Thompson, L. A. (1990) Design and performance analysis of adaptive optical telescopes using laser guide stars. *Proc. IEEE* **78**, 1721–43.

Happer, W., MacDonald, G. J., Max, C. E. and Dyson, F. J. (1994) Atmospheric-turbulence compensation by resonant optical backscattering from the sodium layer in the upper atmosphere. *J. Opt. Soc. Am. A* **11**, 263–76.

Hardy, J. (1978) Active optics: a new technology for the control of light. *Proc. IEEE* **66**, 651–97.

Hudgin, R. J. (1977) Wave-front compensation error due to finite corrector-element size. *J. Opt. Soc. Am.* **67**, 393–5.

Humphreys, R. A., Primmerman, C. A., Bradley, L. C., and Herrmann, J. (1991) Atmospheric turbulence measurements using a synthetic beacon in the mesospheric sodium layer. *Opt. Lett.* **16**, 1367–9.

Lloyd-Hart, M., Angel, J. R. P., Jacobsen, B., Wittman, D., Dekany, R., McCarthy, D., *et al.* (1995) Adaptive optics experiments using sodium laser guide stars. *Astrophys. J.* **439**, 455–73.

Lloyd-Hart, M., Angel, J. R. P., Groesbeck, T., Martinez, T., Jacobsen, J., McLeod, B., *et al.* (1998) First astronomical images obtained with adaptive optics using a sodium laser guide star. *Astrophys. J.* **493**, 950–4.

McCall, S. L. and Passner, A., (1978) Adaptive optics in astronomy In: *Adaptive Optics and Short Wavelength Sources* 6, Physics of Quantum Electronics, Addison-Wesley Pub., Reading, MA.

Max, C. E., Olivier, S. S., Friedman, H. W., An, J., Avicola, K., Beeman, B. V., *et al.* (1997) Image improvement from a sodium-layer laser guide star adaptive optics system. *Science* **277**, 1649–52.

Primmerman, C. A., Murphy, D. V., Page, D. A., Zollars, B. G. and Barclay, H. T. (1994) Compensation of atmospheric optical distortion using a synthetic beacon. *Nature* (London) **353**, 141–3.

Sandler, D., Stahl, S., Angel, J. R. P., Lloyd-Hart, M. and McCarthy, D. (1994a) Adaptive optics for diffraction-limited infrared imaging with 8-m telescopes. *J. Opt. Soc. Am. A* **11**, 925–45.

Sandler, D., Cuellar, L., Lefebvre, M., Barrett, T., Arnold, R., Johnson, P., *et al.* (1994b) Shearing interferometry for laser-guide star atmospheric correction at large D/r_0. *J. Opt. Soc. Am. A* **11**, 858–73.

Thompson, L. A. (1994) Adaptive optics in astronomy. *Physics Today*, **47**, No. 12, 24–31.

Thompson L. A. and Gardner, C. S. (1987) Experiments on laser guide stars at Mauna Kea Observatory for adaptive imaging in astronomy. *Nature* (London) **328**, 229–31.

12

The design of laser beacon AO systems

ThermoTrex Corporation, San Diego, California, USA

12.1 Introduction: outline of the chapter

In this chapter, the design of LGS AO is treated, with special emphasis on analysis tools required for predicting the expected performance of laser beacon adaptive optics on astronomical telescopes. Analytical expressions are given for estimating the most important error sources encountered in LGS AO operation. Numerical examples are given to illustrate the relative importance of various errors, and to aid in the choice of design parameters.

There have been several treatments of LGS error budget analyses (for example: Gardner *et al.* 1990; Kibblewhite 1992; Gavel *et al.* 1994; Parenti and Sasiela 1994; Olivier and Gavel 1994) each emphasizing different aspects of LGS AO, and using a wide assortment of approximations. While the tools discussed in this chapter have general applicability to LGS AO, the treatment (which parallels and extends that of Sandler *et al.* 1994a) is aimed at diffraction-limited imaging on astronomical telescopes, with the goal of optimizing performance for near-infrared imaging and spectroscopy for the new large telescopes, as discussed in Chapter 11. Examples are often chosen which are relevant to this problem.

We begin in Section 12.2 by decomposing the AO system error into image motion, or tilt errors and higher order wave-front errors. In the regime of diffraction-limited correction, the tilt and high order wave front errors can be treated independently, with corresponding Strehl ratios S_θ and S_ϕ, respectively. We will see below that the quantities S_θ and S_ϕ depend on a very large number of atmospheric and system parameters, and very strongly on the imaging wavelength λ_i. For this reason, to keep from attempting to design by juggling parameters in a large design space, the design of LGS AO should follow from well-defined performance requirements and *a priori* constraints on system hardware, operation, and cost. In the remainder of Chapter 12, we will focus on choosing the design parameters from the analysis of error sources.

As set out in Chapter 11, diffraction-limited correction corresponds to $S_{\text{sys}} \geqslant 0.3$, where S_{sys} is the total system Strehl ratio. The first constraint in the LGS AO system design is to determine the minimum wavelength at which this performance is required. We will focus mainly on $\lambda_i \geqslant 1$ μm, although our analysis will allow treatment of arbitrary wavelength. A sweeping simplification of the design process can then follow by stipulating independently target Strehl ratios for image motion and high order errors. Motivated by the need for nearly full sky coverage, we take a target image motion Strehl of $S_\theta = 0.5$, and for high order correction $S_\phi = 0.6$. We will see that these values lead to a good compromise between maximizing sky coverage for large telescopes, and setting reasonable requirements for laser power, deformable mirror complexity, and control electronics.

The discussion to follow of individual error sources making up the total tilt and high order Strehl will be presented so that further simplifications in design approach (by apportioning subsystem errors) will become apparent. Section 12.3 is devoted to higher order laser guide star error sources, including FA, photon-noise errors in wave-front sensing and laser power requirements. Section 12.4 follows with a self-contained discussion of image motion correction, because of its fundamental importance. In Section 12.5 we apply the tools discussed in Sections 12.2–12.4 to performance analysis and optimization of LGS AO for large telescopes, in particular the new 6.5-m MMT.

Thus, our goal is that by the end of this chapter, we will have taken the reader through the basic concepts and formulas necessary to design LGS AO for astronomical telescopes.

12.2 System Strehl ratio: image motion and higher order errors

As discussed in Chapter 11, in LGS AO image motion (or tilt) correction is performed using a natural star and has its own control loop separate from the laser beacon correction of higher order (figure) errors. For this reason, it is convenient to treat the two types of error separately. In addition, tilt errors cause residual image jitter on the science detector, and therefore to a good approximation their effect can be treated as a smearing operation during long exposures. On the other hand, high order wave-front errors cause spreading of the PSF and scattering of light into a halo surrounding the diffraction pattern. We can write the total uncorrected wave-front distortion, φ, as a sum of tilt plus high-order phase terms:

$$\varphi(x) = \frac{2\pi}{\lambda}\theta x + \phi(x). \tag{12.1}$$

In the first term on the right hand side (rhs) of Eq. (12.1) the wave-front tilt is written in terms of the angular variable θ representing the atmospheric wedge which produces angular image motion about the optical boresight on the science detector. The phase $\phi(x)$ is the tilt-removed, or high-order wave-front distortion, corresponding to the sum of all Zernike modes $\geqslant 4$.

Let the estimate of the wedge obtained with field star photons be θ_{star}, and denote the estimated higher order wave-front obtained using the laser beacon as ϕ_{laser}. The measurements equal the true values plus uncertainties introduced because the LGS AO system is not perfect:

$$\theta_{\text{star}} = \theta + \varepsilon_\theta, \tag{12.2}$$

where ε_θ is the error in the estimate θ_{star}. Similarly,

$$\phi_{\text{laser}}(x) = \phi(x) + \epsilon_\phi(x). \tag{12.3}$$

We assume the errors ϵ_θ and ϵ_ϕ have zero mean, which means that the relative alignment of the science and tilt detectors is calibrated, and the overall piston of the wave front is removed before moving the deformable mirror actuators. Then, from the AO system parameters and hardware properties, in conjunction with Kolmogorov theory, we can estimate the expected value of the rms image motion error and the rms high order error:

$$\sigma_\theta \equiv \langle \epsilon_\theta^2 \rangle^{1/2} \tag{12.4}$$

$$\sigma_\phi \equiv \left[\frac{4}{\pi D^2} \int w(x)\epsilon^2_{\phi}(x)\,dx \right]^{1/2}. \tag{12.5}$$

The total system Strehl ratio is written as a product of image motion and high-order Strehl ratios:

$$S_{\text{sys}} = S_\theta S_\phi. \tag{12.6}$$

In the following sections, we will decompose ϵ_θ and ϵ_ϕ into a number of separate errors which can arise during LGS AO operation. Except where otherwise indicated, we will assume the various errors are uncorrelated, which has been shown to be a good approximation to estimate the Strehl ratio in the regime of full correction. In addition, we will often use the Marechal approximation to estimate S_ϕ, since the various contributions to σ_ϕ must be $\ll 1$ radian to meet the goal of diffraction-limited correction.

12.3 High-order wavefront errors for LGS AO

12.3.1 Focus anisoplanatism

12.3.1.1 The parameter d_0

As discussed in Chapter 11, the magnitude of focus anisoplanetism (FA) error is described by the parameter d_0, as $\sigma_{FA}^2 = (D/d_0)^{5/3}$, where d_0 is independent of aperture size. In this section, we present working formulas for d_0. We start by defining the spatial FA error profile over the aperture as

$$\epsilon_{FA}(\mathbf{r}) = \phi_b(\mathbf{r}) - \phi(\mathbf{r}), \tag{12.7}$$

where

$$\phi(r) = \frac{2\pi}{\lambda} \int n(\mathbf{r}, z)\, \mathrm{d}z \tag{12.8}$$

is the phase distortion across the telescope aperture caused by turbulent fluctuations, equivalent to the phase accumulated by a stellar beacon at effectively infinite range. The phase $\phi_b(\mathbf{r})$ is the wave front measured by light from a laser beacon at finite range H, and is given by

$$\phi_b(\mathbf{r}) = \frac{2\pi}{\lambda} \int n\left[\mathbf{r}(1 - \frac{z}{H}), z\right] \mathrm{d}z. \tag{12.9}$$

In Eqs. (12.8) and (12.9), $n(\mathbf{r}, z)$ is the index-of-refraction fluctuation at altitude z. The mean square FA error is then

$$\sigma_{FA}^2 = \frac{1}{A} \int w(\mathbf{r})\langle\epsilon_{FA}^2(\mathbf{r})\rangle\, \mathrm{d}\mathbf{r}, \tag{12.10}$$

where A is the area of the telescope aperture, and $w(\mathbf{r})$ is the aperture weighting function.

The above expression for the σ_{FA}^2 counts errors in sampling all modes of the distorted wave front, including both overall piston and tip/tilt. We know that piston error is irrelevant to the Strehl for a single aperture. Also, recall from Chapter 11 that a laser beacon does not measure overall tilt, so errors in sampling tilt should not be included. Nonetheless, insight into the physics of the cone effect can be obtained by first treating the case with piston and tilt included, which we will see in the next section yields values for the FA error which are often not too far from the piston- and tilt-removed values.

To simplify notation, we write Eq. (12.8) in the limit of turbulent phase distortion represented as a finite set of phase screens:

$$\phi(\mathbf{r}) = \sum \phi_k(\mathbf{r}), \tag{12.11}$$

where $\phi_k(\mathbf{r}) \equiv \phi(\mathbf{r}, z_k)$, and z_k is the altitude of the kth screen. The screens are taken to be statistically independent

$$\langle \phi_k(\mathbf{r})\phi_l(\mathbf{r}')\rangle = \sigma_\phi^2 \delta_{kl}\delta(\mathbf{r}-\mathbf{r}'). \tag{12.12}$$

With each screen we can associate a coherence length

$$(\mathbf{r}_{0,k})^{-5/3} = \frac{2.91}{6.88}\left(\frac{2\pi}{\lambda}\right)^2 C_n^2(z_k)\Delta z_k, \tag{12.13}$$

where Δz_k is the distance between screens k and $k = 1$. Then Eq. (12.10) becomes

$$\sigma_{\mathrm{FA}}^2 = \frac{4}{\pi D^2}\int w(\mathbf{r})\left\langle \left\{\sum \phi_k(\mathbf{r}) - \sum \phi_l\left[\mathbf{r}\left(1-\frac{z_l}{H}\right)\right]\right\}^2\right\rangle d\mathbf{r}. \tag{12.14}$$

Squaring the quantity in braces, we can simplify Eq. (12.20) to

$$\sigma_{\mathrm{FA}}^2 = 2\sigma_\phi^2 - \frac{8}{\pi D^2}\sum\sum\int w(\mathbf{r})\left\langle \phi_k(\mathbf{r})\phi_l\left[\mathbf{r}\left(1-\frac{z_l}{H}\right)\right]\right\rangle d\mathbf{r}. \tag{12.15}$$

Using Eq. (2.12), we can eliminate one of the sums on the rhs of Eq. (12.5), giving

$$\sigma_{\mathrm{FA}}^2 = 2\sigma_\phi^2 - \frac{8}{\pi D^2}\sum\int w(\mathbf{r})\left\langle \phi_k(\mathbf{r})\phi_k\left[\mathbf{r}\left(1-\frac{z_k}{H}\right)\right]\right\rangle d\mathbf{r}. \tag{12.16}$$

The covariance in Eq. (12.16) can be calculated directly from the Kolmogorov structure function for each phase screen:

$$\left\langle \phi_k(\mathbf{r})\phi_k\left[\mathbf{r}\left(1-\frac{z}{H}\right)\right]\right\rangle = \sigma_\phi^2 - \tfrac{1}{2}D_\phi\left(\frac{\mathbf{r}z_k}{H}\right). \tag{12.17}$$

Inserting Eq. (12.17) into Eq. (12.16), we arrive at

$$\sigma_{\mathrm{FA}}^2 = \frac{4\times 6.88}{\pi D^2}\sum \mathbf{r}_{0,k}^{-5/3}\left(\frac{z_k}{H}\right)^{5/3}\int w(\mathbf{r})\mathbf{r}^{5/3}\,d\mathbf{r}. \tag{12.18}$$

This expression can be simplified by appealing to Eq. (12.13),

$$\sigma_{\mathrm{FA}}^2 = \frac{4}{\pi D^2 H^{5/3}}\int w(\mathbf{r})\mathbf{r}^{5/3}\,d\mathbf{r}\left[(2.91)\left(\frac{2\pi}{\lambda}\right)^2\sum C_n^2(z_k)z_k^{5/3}\Delta z_k\right]. \tag{12.19}$$

In the limit of a large number of phase screens, the quantity inside the brackets on the rhs of Eq. (12.19) is recognized as $\theta_0^{-5/3}$, where θ_0 is the isoplanatic angle (see Chapter 2). The integral over the aperture can be performed, giving

$$\frac{4}{\pi D^2}\int w(\mathbf{r})\mathbf{r}^{5/3}\,d\mathbf{r} = \frac{3}{11}2^{-2/3}D^{5/3}.$$

Our final expression for the mean square (piston/tilt-included) FA error is

$$\sigma_{\text{FA}}^2 = 0.172 \left(\frac{D}{\theta_0 H} \right)^{5/3}. \tag{12.20}$$

Note that when piston and tilt errors are included, the mean square FA error is proportional to $(\theta_0 H)^{-5/3}$. The dependence on $D^{5/3}$ allows d_0 to be easily calculated as

$$\widehat{d}_0 = 2.91 \theta_0 H, \tag{12.21}$$

where the 'hat' over d_0 has been used to denote the piston/tilt-included parameter, distinguished from the full piston- and tilt-removed d_0. Also, note that the wavelength scaling $\widehat{d}_0 \sim \lambda^{6/5}$ follows from Eq. (12.21).

We can see that $\widehat{d}_0$ depends on the mid- and high-altitude turbulence in the same way as θ_0. For a site with $\theta_0 = 3''$ (~ 15 μrad) at $\lambda = 0.5$ μm, $\widehat{d}_0 = 0.87$ m for $H = 20$ km (Rayleigh beacon), and $\widehat{d}_0 = 4$ m for $H = 90$ km (sodium beacon). At $\lambda = 2.2$ μm, the values are a factor of 5.92 larger, with $\widehat{d}_0 = 5.1$ m for $H = 20$ km and $\widehat{d}_0 = 23.7$ m for $H = 90$ km.

As mentioned above, $\widehat{d}_0$ must underestimate the true value of d_0 since FA errors in estimating overall piston and tilt should not be included. To remove these components, we must replace the beacon and stellar phases in Eq. (12.7) with

$$\phi_{\text{ptt}}(\mathbf{r}) = \phi(\mathbf{r}) - \frac{4}{\pi D^2} \int \phi(\mathbf{r}') \, d\mathbf{r}' - \frac{64}{\pi D^4} \int w(\mathbf{r}') \mathbf{r} \cdot \mathbf{r}' \phi(\mathbf{r}') \, d\mathbf{r}. \tag{12.22}$$

Fried and Belsher (1994) were the first to carry out the full derivation of σ_{FA}^2 using Eq. (12.22). They took a straightforward, but very lengthy, approach. Later, Tyler (1994a) used a polynomial expansion to arrive at the closed-form:

$$d_0 = \lambda^{6/5} \cos^{3/5}(\zeta) \left[\int C_n^2(z) F\left(\frac{z}{H} \right) dz \right]^{-3/5}, \tag{12.23}$$

where ζ is the angle of the direction of the laser beacon from the zenith. The function $F(z/H)$ is defined by

$$F\left(\frac{z}{H} \right) = 2.24 \left[1 + \left(1 - \frac{z}{H} \right)^{5/3} \right] - 36.26 \left\{ \frac{6}{11} {}_2F_1 \left[-\frac{6}{11}, -\frac{5}{6}; 2; \left(1 - \frac{z}{H} \right)^2 \right] \right.$$

$$\left. - \frac{6}{11} \left(\frac{z}{H} \right)^{5/3} - \frac{10}{11} \left(1 - \frac{z}{H} \right) {}_2F_1 \left[-\frac{11}{6}, \frac{1}{6}; 3; \left(1 - \frac{z}{H} \right)^2 \right] \right\} \tag{12.24}$$

for $z > H$, and

$$F\left(\frac{z}{H} \right) = 15.82$$

for $z > H$. In Eq. (12.24), ${}_2F_1$ is a hypergeometric function, which appears often in Kolmogorov theory. We note that while the above expression for $F(z/H)$ is quite complicated, a plot of $F(z/H)$ versus $(z/H)^{5/3}$ is nearly a straight line for atmospheric turbulence models of interest to astronomy. Also, note that the contribution to d_0 from $z > H$ is just the fitting error for Kolmogorov turbulence, since the turbulence above the beacon is not compensated.

Another very useful expression for σ_{FA}^2 has been derived by Sasiela (1994), using a formulation of LGS error sources using Fourier-space filter functions. The development is elegant, but advanced, and requires a basic understanding of Mellin Transforms. Sasiela shows that the mean square FA error (for turbulence at altitudes $z \leqslant H$) can be written as

$$\sigma_{FA}^2 = \frac{4\pi^2}{\lambda^2} D^{5/3} \left[\frac{\mu_{5/3}(H)}{H^{5/3}} - 0.903 \frac{\mu_2(H)}{H^2} \right]. \tag{12.25}$$

The moment μ_n is defined as

$$\mu_n(H) = \int C_n^2(z) z^n \, \mathrm{d}z, \tag{12.26}$$

where the upper limit of integration is taken to be H. In this expression, the first term in brackets on the rhs of Eq. (12.25) is identical to Eq. (12.20) derived above. The correction from removing piston and tilt appears in the second term; the fractional correction to the first order result given in Eq. (12.20) is $\sim 0.903(\mu_2(H)/\mu_{5/3}(H))H^{-1/3}$.

Eq. (12.25) gives a simple prescription for evaluating the FA error especially when $C_n^2(z)$ is an analytic function, and the moments can be easily computed.

12.3.1.2 Model C_n^2 distributions

In this section, we present model C_n^2 distributions for use in numerical examples of LGS errors. Numerous C_n^2 profiles have been used through the years for estimating LGS system performance. Analysis for DoD systems has used models which generally correspond to $1-2''$ seeing, and values of θ_0 ranging from $1.5-2''$. Often-used parameterized models include the SLC (Parenti & Sasiela 1994) day and night models (used for modeling at the AMOS site), and members of the Hufnagel-Valley (HV) family (Ellerbroek 1994) of models (used for SOR). Obviously, it is desirable to have at hand a measured C_n^2 for the astronomical site which will be home to the adaptive optics system. Progress in this direction has been made for several sites, but often the best that one has is a statistical survey of atmospheric parameters over time. See Beland (1992) for results of several site surveys.

Given a general form for C_n^2, it is not difficult to adjust the integrated seeing by adding a ground layer or local dome seeing, or tuning the strength of one or two layers in the first several kilometers above the site. The most important remaining degrees of freedom in choosing C_n^2 are the strength and shape of higher-altitude turbulence, especially near the tropopause from 8–12 km, and the distribution of atmospheric wind speeds. These properties of C_n^2 determine the nature of isoplanatic and FA errors (which in turn determine in turn sky coverage for natural field stars and the number of LGS sources required), and the temporal bandwidths required for image motion and phase correction (and hence the laser power requirement).

In this section, we will use the parameterized profile

$$C_n^2(h) = a_1 \left(\frac{h}{h_1}\right)^{10} \exp\left(\frac{-h}{h_1}\right) + a_2 \exp\left(\frac{-h}{h_2}\right) + a_3 \exp\left(\frac{-h}{h_3}\right). \qquad (12.27)$$

The functional form Eq. (12.27) is similar to the HV form, and for this reason we will refer to it as modified HV (MHV). (Note that several MHV profiles have been used in the literature with substantially different properties than the one used here.) The parameters are taken to be $a_1 = 2.0 \times 10^{-23}$ m$^{-2/3}$, $a_2 = 1.4 \times 10^{-15}$ m$^{-2/3}$, $a_3 = 0$; and $h_1 = 1$ km, $h_2 = 1.5$ km, $h_3 = 0.1$ km. With these parameters, MHV model contains high altitude turbulence consistent with measurements made at Mauna Kea (Roddier *et al.* 1990), and thus should be representative of a very good astronomical site. In this section, we will take the observatory altitude to be 3 km, so all integrals over the vertical C_n^2 profile will range from 3 to 30 km. Then for MHV we have $r_0 = 15$ cm at $\lambda = 0.5$ µm, corresponding to 0.7″ seeing, and $\theta_0 = 2.5″$. The MHV C_n^2 profile used here also gives qualitative agreement with observations using both laser and natural guide stars at the MMT on Mt Hopkins south of Tucson, Arizona (Lloyd-Hart *et al.* 1995).

Occasionally we will refer to the models used by Sandler *et al.* (1994a). Their models M1 and M2 are plotted in Fig. 12.1, along with the continuous MHV profile. Model M1 is monotonically decreasing in h, and it is very similar to the MHV model except that turbulence has been lumped in to discrete layers, consistent with the results of data analysis given by Roddier *et al.* (1990). In model M2, the strong layer at $h = 5$ km has been moved to $h = 10$ km, so that M2 is intended to represent a very strong tropopause layer. These models give (r_0, θ_0) values at $\lambda = 0.5$ µm of M1 = (17 cm, 3″), M2 = (15 cm, 1.3″). Thus, the seeing for all three models is similar ($\sim 0.7″$).

The vertical distribution of wind speed will vary substantially with time of year and weather conditions. In particular, when the jet stream is nearby, one

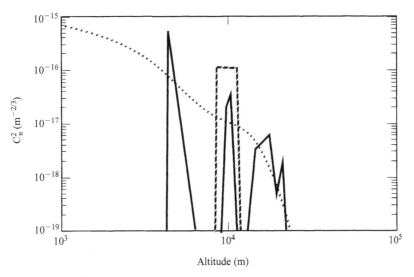

Fig. 12.1. Model C_n^2 distributions used in this section for numerical examples of LGS errors. Shown are M1, M2, and MHV models as discussed in the text. Solid line, M1; dashed line, M2; dotted line, MHV.

can expect very strong high altitude winds, and at other times the winds may be fairly calm. We will use the Bufton wind model (Bufton 1973), which was derived some years ago from analysis of astronomical observations:

$$v(h) = v_0 + v_1 \exp\left[-\left(\frac{h - 10}{5}\right)^2\right] \tag{12.28}$$

where $v(h)$ is the speed (m/sec) at altitude h (km). This wind model gives a value of $\overline{(v^{5/3})}^{3/5} = 18.5$ m/sec (where the bar stands for averaging with altitude over $C_n^2(h)$), which in turn gives $t_0 = 3$ ms at $\lambda = 0.5$ µm for 0.7″ seeing.

12.3.1.3 Numerical examples of d_0

Figure 12.2 plots the parameter d_0 at $\lambda = 0.5$, 1, and 2 µm as a function of altitude H, for the MHV turbulence model discussed above. From Fig. 12.2, we see that for the three imaging wavelengths, at $H = 90$ km, $d_0(90) = 4$ m, 9 m, and 22 m, respectively. For $H = 20$ km, the values are $d_0(20) = 1.2$ m, 3 m, and 7 m. For comparison, the values of the piston- and tilt-included FA parameter $\widehat{d_0}$ for MHV at $\lambda_i = 0.5$ µm are $d_0(90) = 3.3$ m

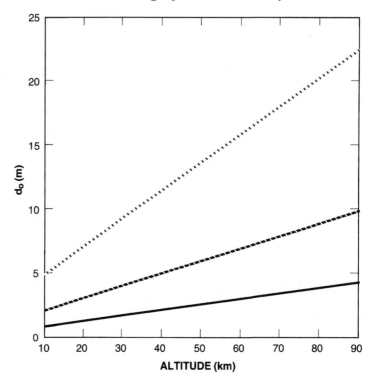

Fig. 12.2. The FA parameter d_0 plotted as a function of beacon altitude H, for the MHV turbulence model. Three wavelengths are shown: 0.5, 1, and 2 µm. Solid line, 0.5 µm; dashed line, 1 µm; dotted line, 2 µm.

and $d_0(20) = 0.77$ m which are 18 and 36% lower, respectively, than the corresponding values of d_0.

From Sandler *et al.* (1994a), for the MK model, $d_0(90) = 4.2$ m at $\lambda_i = 0.5$ µm, and for MMK $d_0(90) = 2.3$ m. Thus, the MK model leads yields a similar FA effect to the MHV, whereas the MMK model is much more severe.

Figure 12.3 plots S_{FA} as a function of wavelength, for the MHV model, for two cases: $D = 8$ m, $H = 90$ km, and $D = 4$ m, $H = 20$ km. From the figure, for $D = 8$ m and a sodium beacon, the Strehl ratio for FA alone remains > 0.8 down to H band ($\lambda = 1.6$ µm), and > 0.5 through J band ($\lambda = 1.25$ µm). The figure also illustrates that for a $D = 4$ m telescope, a single Rayleigh beacon gives a reasonable Strehl ($S_{FA} > 0.6$) in K band, but the FA Strehl begins to drop rapidly in H band. In the following two sections, we summarize experimental data on FA, and extend the above argument to estimate limiting imaging wavelengths for diffraction-limited correction.

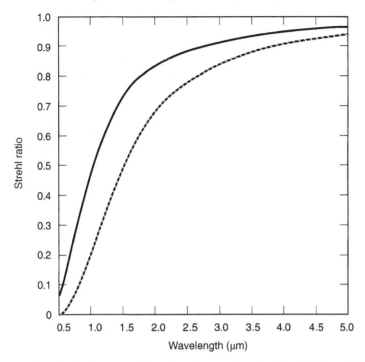

Fig. 12.3. Strehl ratio for FA as a function of wavelength, for MHV turbulence. Two cases are shown: solid line, $D = 8$ m with sodium laser beacon at 90 km, and dotted line, $D = 4$ m with Rayleigh beacon at 20 km altitude.

12.3.1.4 Experimental data on focus anisoplanatism

In spite of its fundamental importance to LGS AO, there has been limited experimental work on FA. In Chapter 11 we discussed the early measurements of d_0 in DoD programs, which offered much encouragement that FA is not strong enough to prevent diffraction-limited imaging using LGS AO. Since that time, by far the most data on LGS AO has been collected at the SOR 1.5-m telescope, where a Rayleigh LGS with $H = 10-15$ km has been used. The experimental data are consistent with a mean square FA error of $0.5-1$ rad^2 at $\lambda_i = 0.88$ μm, corresponding to $d_0 = 0.76-1.2$ m at $\lambda_i = 0.5$ μm. This indicates that FA is considerably stronger at SOR than for a site with MHV turbulence, which is to be expected since the mean SOR value of $\theta_0 \sim 1.5''$.

More important is the fact that SOR has provided a wealth of data on closed-loop LGS AO with a Rayleigh beacon. For imaging of bright stars, the magnitude of the FA Strehl can be computed directly by comparing NGS and LGS images. (Other error sources besides FA in LGS AO operation can differ appreciably from their NGS counterparts; however, the differences are small when bright stars are used for tilt control and the laser beacon brightness is high.) As an example of experimental measurement of FA, short exposure images of Arcturus at $\lambda_i = 0.88$ µm obtained at the SOR 1.5-m telescope have been analyzed, comparing Strehl ratios for NGS and LGS correction. The NGS corrected image has a Strehl ratio of 0.59, while the Strehl for the LGS corrected image is 0.48. For these images, the Strehl degradation from FA is then $S_{FA} = 0.81$ at $\lambda_i = 0.88$ µm, which scales to $d_0 = 1.9$ m at $\lambda_i = 0.5$ µm. This corresponds to a value of d_0 a factor of ~ 2 times larger than average at the SOR site.

The measurements of FA at an astronomical telescope were first performed by the Arizona group (Lloyd-Hart *et al.* 1995). They measured the FA for low-order modes (focus through trefoil) at the MMT array telescope, by comparing simultaneous images with all six 1.8-m mirrors of natural stars and sodium laser beacons. The CCD camera used to acquire the data was focused at 90 km, so all six images of the sodium beacon were in focus. The secondary mirrors for the six telescopes were intentionally defocused to separate the blurred stellar images from the sodium spot images. By comput- ing the instantaneous motion of all twelve images, the components of $Z_4 - Z_9$ were estimated; by computing the rms relative motion between sodium and natural star images, the FA corresponding to these modes was calculated. The star-beacon rms error for the six modes was found to be $(\sigma_{FA})_{4-9} = 0.27$ rad at $\lambda_i = 2.2$ µm. We can estimate the FA parameter by scaling from these six modes to the full wave front for a $D = 6.9$ m aperture, using the results of Sandler *et al.* (1994a), where fractional contribution to FA is given for low- order modes. The result of this analysis gives $d_0 = 15.4$ m at $\lambda_i = 2.2$ µm, predicting a mean square FA error of $\sigma_{FA}^2 = 0.26$ rad^2 ($S_{FA} = 0.77$). These values are in good agreement with the MK model results given in Sandler *et al.* (1994a).

A group from ONERA (Laurent *et al.* 1995) has performed detailed experiments on FA for a Rayleigh beacon and off-axis anisoplanatism at the 1.5-m telescope at CERGA observatory. They obtained simultaneous wave- front measurements from NGS and LGS sources, using a Shack–Hartmann sensor with $d = 10$ cm subaperture. They computed the modal correlation between NGS and LGS wave fronts, for bright natural stars and laser beacons from 10–15 km altitude. Their results are consistent with the expected properties of FA, showing, for instance, a drop in star-beacon correlation as the beacon altitude is lowered. The total integrated FA error over all modes is consistent with strong upper altitude turbulence, similar to that at the SOR site.

12.3.1.5 Limiting regime for a single LGS beacon

We have seen in the previous section that predictions based on simple models are in general agreement with limited experimental data on the stength of FA. In particular, data exist which indicate that for a good astronomical site, the limiting Strehl for FA should be $\geqslant 0.8$ at $\lambda_i = 2.2$ μm, for LGS AO on a large, 6- to 10-m telescope. The corresponding Strehl will decrease to $S_{FA} \sim 0.68$ at $\lambda = 1.6$ μm, and to $S_{FA} \sim 0.36$ at $\lambda = 1$ μm. Recall from Section 12.1 that our scheme for optimization of LGS AO budgets a high order Strehl ratio $S_\phi \geqslant 0.6$, including all errors. Thus, from these simple considerations it would appear that if the MMT FA results are typical, the limiting wavlength for diffraction-limited correction using a single sodium beacon is 1.6 μm. For objects with a sufficiently bright field star nearby to give high tilt Strehl ($S_\theta \sim 1$), a sodium beacon can still come close to meeting the diffraction-limited goal down to 1 μm.

The above conclusions are tentative, and further experiments at astronomical telescopes are clearly called for. In the meantime, the results of this section indicate that sound analytical models can predict FA with reasonable confidence, and there is sound evidence that use of a single sodium beacon will allow diffraction-limited infrared imaging on large telescopes in J band and longer wavelengths.

12.3.2 Multiple beacons for LGS AO

12.3.2.1 Introduction to multiple beacons

The results of the previous section illustrate that the effect of FA is to limit diffraction-limited performance for a given aperture size to wavelengths above a critical wavelength. To have $S_{FA} > 0.7$ requires $d_0 > 2D$. In the early 1980s, the small community of DoD researchers active in LGS AO realized this, quite some time before Labeyrie and Foy's paper in 1985 suggesting the use of an array of laser beacons for astronomical AO. The problem of FA was especially acute for the compensated projection of laser beams through the atmosphere, since the lasers under study all operated at wavelengths $\lambda \leqslant 1$ μm. This was made even worse by the emphasis at that time on the use of Rayleigh guide stars, since the understanding of sodium resonant excitation and the technology for building sodium lasers had not progressed far. For a $D = 3$ m class telescope and $d_0 = 1-2$ m for Rayleigh beacons at 10–20 km, $\sigma_{FA}^2 \gg 1$ rad^2, and this was recognized as a major impediment to achieving diffraction-limited correction.

A focused effort by the DoD workers produced an enormous number of predictions of enhanced performance for multiple laser beacons, as well as initial laboratory and field experiments by ThermoTrex and MIT/Lincoln Laboratory. A small portion of the theoretical work is summarized in the papers by Parenti and Sasiela (1994), Sasiela (1994), and Tyler (1994b); unfortunately, much of the work remains unpublished in the open literature. To do justice to the concepts and detailed results of this work would require a long Chapter devoted to this subject, and is beyond the scope of this book. Moreover, as shown above, FA becomes a major impediment to diffraction-limited correction for astronomical imaging only at wavelengths below J band, even for large telescopes. The restriction to a single sodium beacon is thus not holding up progress for astronomical AO. Therefore, we will restrict the discussion of multiple beacons to an overview of the basic concepts and problems encountered in implementing multiple beacon schemes for astronomy. We will also suggest areas for future research that, even after many years of work, still remain uncertain and could benefit from innovation.

12.3.2.2 Potential reduction in focus anisoplanatism

The major incentive for using more than one beacon can be understood from the defining equation $\sigma^2_{FA} = (D/d_0)^{5/3}$. If the aperture is broken up into several smaller contiguous sections of width s, each with its own laser beacon for wave-front sensing, then the FA error will reduce to

$$\sigma^2_{FA} = \left(\frac{s}{d_0}\right)^{5/3}. \tag{12.29}$$

Since the area of each section is $a = s^2$, the number of beacons is $n_b \sim (D/s)^2$, and the mean square FA error is thus reduced by $n_b^{-5/6}$. For five laser beacons, theoretically the FA can be reduced to $1/4$ its value for a single beacon.

Consider the situation for a $D = 8$ m telescope imaging at optical wavelengths, with site conditions corresponding to a sodium beacon value of $d_0 = 20$ m at $\lambda = 2.2$ μm. Then at $\lambda = 0.7$ μm, $d_0 = 5.7$ m or $\sigma^2_{FA} = 1.8$ rad^2 ($S_{FA} = 0.17$). To meet our goal of $S_{FA} = 0.8$ for diffraction-limited imaging, we must reduce the FA error by a factor of 9, to 0.2 rad^2. In this case, a total of $n_b = 15$ sodium laser beacons would be required, with each beacon covering a section $s \sim 1.8$ m. For Rayleigh beacons at $H = 20$ km, the FA must be reduced by a factor of 60, so $n_b \sim 100$ beacons are needed, with $s = 0.7$ m.

There are two major problems with this scenario. First, such a scheme could be very expensive and difficult to implement, especially if each section had its own laser projector. But novel schemes to overcome the laser power requirement and limit complexity have been conceived, and offer some hope of potential implementation. These are discussed below. The second problem, however, is more fundamental. In going to more than one laser beacon, additional error sources are introduced, and substantially reduce the $n_b^{-5/6}$ performance improvement. These error sources are of basic nature, in that they involve errors in sampling the Kolmogorov wave front.

12.3.2.3 Performance limits of multiple beacon AO

The r_0^* effect. To understand the errors which limit the performance of multiple beacon LGS AO, let us go to the limit of a very dense array of beacons, so that s becomes small, but the density of beacons n_b/D^2 remains finite. The situation is pictured in Fig. 12.4, for Rayleigh beacons (a) and sodium beacons (b). The dominant turbulence layers end at 20–25 km, so the sodium beacons

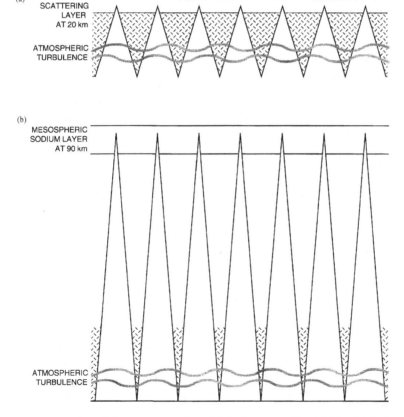

Fig. 12.4. Schematic of dense array of multiple beacons for (a) Rayleigh beacons at 20 km, and (b) sodium beacons at 90 km.

are far above all layers $(1 - z_{max}/H = 0.8)$, while the Rayleigh beacons are embedded within the turbulence $(1 - z_{max}/H = 0)$. Observe in Fig. 12.4 that even in the limit of a dense array of laser spots, half of the turbulence below the beacons is not sampled (shown as the shaded regions in the figure).

B. Spivey of ThermoTrex first identified this error in 1984, and labeled it the 'r_0^*' effect, referring to the effective coherence length for the unsampled turbulence. From the geometry shown in Fig. 12.4, we can derive the expression for r_0^* in analogy with the definition for r_0. The result is

$$(r_0^*)^{-5/3} = \frac{2.91}{6.88} \left(\frac{2\pi}{\lambda}\right)^2 \int \left(\frac{z}{H}\right)^2 C_n^2(z)\, dz. \tag{12.30}$$

The mean square error can then be calculated from the piston- and tilt-removed Kolmogorov fitting error:

$$\sigma_{r_0^*}^2 = 0.134 \left(\frac{D}{r_0^*}\right)^{5/3}, \tag{12.31}$$

which gives

$$\sigma_{r_0^*}^2 = \frac{2.17}{\lambda^2} D^{5/3} \int \left(\frac{z}{H}\right)^2 C_n^2(z)\, dz. \tag{12.32}$$

For the MHV model and $\lambda = 0.7\ \mu m$, Eq. (12.32) yields $r_0^* = 1.4\ m$ for $H = 20\ km$ and $r_0^* = 8.5\ m$ for $H = 90\ km$. Fig. 12.5 plots the corresponding Strehl ratios as a function of D. We see that for $D = 4\ m$ and Rayleigh beacons at 20 km, the limiting Strehl is $S_{r_0^*} \sim 0.4$, dropping to $S_{r_0^*} \sim 0.1$ for an 8-m telescope.

For sodium beacons, the corresponding Strehls remain very high, with $S_{r_0^*} \sim 0.9$ for $D = 8\ m$.

From these results, we conclude multiple Rayleigh beacons are precluded as a means of overcoming FA as a limitation to diffraction-limited imaging at optical wavelengths. (However, we will see below that the use of two layers of Rayleigh beacons can in principle overcome this performance limit, although at the cost of even more complexity.) On the other hand, the limiting performance from the r_0^* effect is not severe for multiple sodium beacons.

Beacon wander. The second new error source for multiple LGS AO is the beacon wander error. The ideal aimpoint for each laser beacon is the midpoint of its section on the primary mirror projected to altitude H. If each beacon is broadcast from the subaperture corresponding to its section, then it will wander on the way up, by an angle given by the tilt across its section. Each beacon is then projected through uncommon turbulence, resulting in nearly random displacements in beacon position from one beacon to the next. However, by

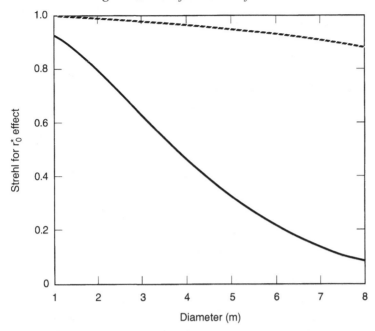

Fig. 12.5. Strehl ratios for the r_0^* effect, as a function of aperture diameter D. The wavelength is taken to be 0.7 μm. Solid line, Rayleigh spots at 20 km; dotted line, sodium spots at 90 km.

reciprocity, the overall image motion recorded on a tilt sensor will show no average tilt across each section, since the beacons wander back through the same turbulence. Thus, the average tilts across each section will register as zero, instead of the true atmospheric section tilt.

The magnitude of the angular spot-wander (SW) error is given by the rms value of the gradient (edge-to-edge) tilt:

$$\sigma_{SW}^2 = 0.339 \left(\frac{D}{r_0}\right)^{5/3} \left(\frac{\lambda}{s}\right)^2. \tag{12.33}$$

Generally, the section size (or distance between spots) s is large compared to r_0, where r_0 is evaluated at the laser wavelength, which we may take here to be $\lambda_b \sim 0.5$ μm. From Eq. (12.33), we see that the relative SW error can be very large. For example, for $s = 2$ m and $r_0 = 15$ cm, $\sigma_{SW} = 0.8''$, larger than the short exposure fwhm beacon size, $\theta_{spot} \sim 0.5''$. This amounts to many waves of random tilt error across each section. Thus, on top of the residual FA error across each section, a much larger error will result from integrating the local slopes from each beacon. The tilts over each section will be in error by σ_{SW}, and these errors will reconstruct to a gross low frequency error across the aperture.

To attempt to overcome the effect of SW, several groups studied methods to 'stitch' the wave front. For instance, a scheme was studied in which sensors near the edge of each section look at neighboring spots, as well as the spot above the section. However, this was found to be very sensitive to noise, and cumbersome to implement, possibly requiring laser spots of slightly different color to discriminate signals at overlapping sensor points.

The important simplifying step was conceived again by Spivey in 1984, when he showed that the use of full-aperture observation (FAO) of the spots can effectively eliminate SW error. To implement FAO, the focused beacon array at altitude H is imaged through the full telescope aperture onto a detector, and the centroids of each spot are computed to pin down its instantaneous position in the sky. This position is then taken to be the tilt for the corresponding section of the aperture. Another way to view FAO is from the point of view of its converse, full-aperture broadcast (FAB), in which each spot is effectively projected through the full aperture. The SW error is drastically reduced for both FAO and FAB, since the beacons are imaged (projected) through turbulence which is highly common for the entire array. This is especially true if the dominant turbulent layers are near the ground, in which case FAO and FAB eliminate SW error completely. For realistic profiles, however, there will be small, residual SW errors, due to anisoplanatism introduced by imaging (projecting) through non-overlapping regions of the upper atmosphere.

The magnitude of the residual SW error for FAO/FAB has been computed by Ellerbroek (1984). He finds that the error in determining the positions of two spots separated by distance s, when they are both viewed by the same aperture of width D, is given by

$$
\sigma_{\text{FAO}}^2 = -2(6.88)2^{-5/3}\pi^{-4}\left(\frac{D}{r_0}\right)^{5/3}\left(\frac{\lambda}{D}\right)^2\left[\int C_n^2(z)\,\mathrm{d}z\right]^{-1}
$$
$$
\times\left\{\int C_n^2(z)G\left(1-\frac{z}{H},\,1-\frac{z}{H},\,0\right)\mathrm{d}z\right.
$$
$$
\left.-\int C_n^2(z)G\left(1-\frac{z}{H},\,1-\frac{z}{H},\,\frac{z}{H}\frac{s}{D}\right)\mathrm{d}z\right\} \qquad (12.34)
$$

where the function $G(a,\,b,\,c)$ in (12.34) is given by

$$
G(a,\,b,\,c) = \pi\int_0^{2\pi}(2Ac)^{5/6}\cos tg(2c/A)\,\mathrm{d}t, \qquad (12.35)
$$

the function $g(x)$ in (12.34) is given by

$$
g(x) = |x|^{-5/6}\cdot{}_2F_1(-\tfrac{5}{6},\,-\tfrac{5}{6},\,1,\,x^2) \text{ for } |x| \leqslant 1
$$
$$
= |x|^{5/6}\cdot{}_2F_1(-\tfrac{5}{6},\,-\tfrac{5}{6},\,1,\,x^{-2}) \text{ for } |x| > 1, \qquad (12.36)
$$

and where

$$A = (a^2 + b^2 - 2ab \cos t)^{1/2}.$$

Numerical evaluation of Eqs. (12.34)–(12.36) shows that for most cases of interest for astronomical AO, the residual SW error for FAO/FAB is very small. For $s = D/5$, $\sigma_{FAO} \simeq 0.007\sigma_{SW}$ for $H = 90$ km; for Rayleigh spots at $H = 20$ km, $\sigma_{FAO} \simeq 0.02\sigma_{SW}$. These values correspond to $\sigma_{FAO}(90) \simeq 5.6$ milliarcseconds and $\sigma_{FAO}(20) \simeq 16$ milliarcseconds, respectively, for sodium and Rayleigh beacons.

Because the residual SW errors after FAO will be weakly correlated, and the wave-front reconstructor will have an error propagator ~ 1, we can safely take σ_{FAO}^2 as an upper bound to the integrated wave-front error for the entire beacon array. We conclude that SW, if not corrected for, will render multiple beacon methods useless in principle. However, if the capability to image the spot array at altitude is incorporated into the LGS AO design, FAO effectively reduces SW error to a negligible level.

Field experiments by MIT/Lincoln Laboratory confirming the efficacy of FAO for two Rayleigh beacons are reported in Murphy *et al.* (1991). Detailed laboratory experiments demonstrating the performance of FAO for 4–16 beacons are described in Sandler and Massey (1989).

12.3.2.4 Limiting performance for finite beacon arrays

We have seen that, to a good approximation, the performance of multiple beacon LGS AO is limited primarily by the r_0^* effect. The performance limitation is not large for multiple sodium LGS arrays, but is severe for attempts to obtain diffraction-limited performance using Rayleigh guide stars. These conclusions have been obtained from analytical arguments, based on a large number of spots. In this section, we illustrate the performance of finite beacon arrays.

Figure 12.6 shows results of a Monte Carlo simulation of multiple beacon AO, for $D = 4$ m and $\lambda_i = 0.7$ μm. The simulation includes the errors described above, and uses FAO to limit SW errors. The Strehl ratio is plotted as a function of the number of beacons, for the case of Rayleigh scattering at $H = 20$ km. (This plot is an updated version of results originally reported by Sandler in 1984 (see Sandler and Massey 1989), for correction with $D = 1$ m aperture at very short wavlength ($\lambda = 0.35$ μm).) The Strehl ratio clearly shows a leveling out, or saturation, with increasing numbers of spots. From the figure, it can be seen that most of the Strehl increase occurs for just 5 beacons, with modest increase for 15 beacons, and very little increase thereafter. Recall that

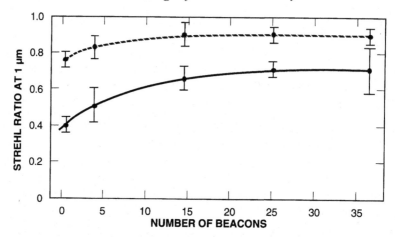

Fig. 12.6. Dependence of Strehl ratio on the number of Rayleigh spots across a 4-m telescope, for an imaging wavelength of 0.7 µm. Solid line, Rayleigh beacons at 20 km, dashed line, beacons in the mesospheric sodium layer.

the limiting Strehl for this case is $S_{r_0^*} = 0.4$, which is seen to be close to the saturation Strehl value. Thus, we see that nearly optimal performance can be obtained with a modest number of beacons.

Since the r_0^* error scales as $D^{5/3}\lambda^{-6/5}$, we can extend the above remarks to multiple sodium beacons. The same saturation behavior should be observed, and this is borne out by Monte Carlo simulations.

12.3.2.5 Advanced schemes using multiple beacons

At this time, there is strong consensus within the astronomical community that sodium beacons are preferable to Rayleigh beacons. Of course, this is partly because a single sodium beacon can provide diffraction-limited correction even for large telescopes, in a wavelength region H–K which is of immediate scientific interest. There is enough work and science to be done in this area alone for many years of research. In addition, many astronomers also believe it is better to advance sodium laser technology which may someday enable multiple sodium beacons, rather than consider the use of multiple Rayleigh beacons. This 'prejudice' is illustrated by the fact that only one astronomical LGS AO project has chosen a Rayleigh LGS approach (Chapter 13).

However, the high-power sodium lasers required for optical correction will certainly be expensive, so it is worth asking whether there is any way to overcome the fundamental limitation of Rayleigh beacons. Further, there will soon be off-the-shelf solid-state lasers which operate near $\lambda = 0.5$ µm which provide the required laser power. Much work has been done in the area of novel

Rayleigh LGS techniques, since this is the crux of the problem for extending defense applications of LGS AO to 4-m telescopes.

One interesting scheme combines the use of multiple Rayleigh beacons with a single sodium beacon. This scheme, called a hybrid method (Jelonek *et al.* 1994; Ellerbroek 1994), uses the Rayleigh spots to correct high spatial frequency wave-front distortion, and a low-power sodium beacon to correct for low order modes. It is motivated by the fact that the r_0^* error has large low frequency components (since it has a nearly Kolmogorov spectrum), and a the low order FA for a sodium beacon is considerably less than the full σ_{FA}^2 (Sandler *et al.* 1994a). The major advantage to the hybrid method is that the sodium beacon sensor can have large subapertures, and hence the required laser power can be kept low enough that existing laser designs can be used. Another potential advantage is that FAO of the Rayleigh spots may not be required, since the SW error is by its very nature a low frequency error. The drawback is that while performance can be obtained exceeding $S_{r_0^*}$ for Rayleigh beacons, one is still limited by the low order FA for sodium.

The most exotic, but potentially the most powerful, multiple beacon methods use Rayleigh spots at two altitudes, say $H_1 = 10-15$ km and $H_2 = 20$ km. The potential of this approach was first recognized over a decade ago independently by Hudgin and Spivey, and Sandler, and has recently been studied by Tyler (1994b).

The use of two beacon altitudes can, in theory, overcome the limitations of the r_0^* error, because it can estimate the unsampled turbulence by extrapolation, as follows. Let $\Delta\phi_1$ be the wave-front phase difference measured using a laser beacon at altitude H_1, at a point on the aperture within the laser beacon's section. Let $\Delta\phi_2$ be the corresponding phase difference measurement at the same point, but using photon return from a beacon directly above at H_2. Then we can form the phase difference estimate

$$\widehat{\Delta\phi} = \frac{H_2}{H_2 - H_1}\Delta\phi_2 - \frac{H_1}{H_2 - H_1}\Delta\phi_1. \qquad (12.37)$$

It can be shown (Smith and Sandler 1989) that the resulting r_0^* error now becomes

$$\sigma_{2-\text{alt}}^2 = \frac{2.17}{\lambda^2} D^{5/3} \int_{H_1}^{H_2} \left(\frac{z - H_1}{H_2 - H_1}\right)^2 C_n^2(z)\, dz \qquad (12.38)$$

Numerical evaluation of Eq. (12.38) shows that it significantly reduces the r_0^* error, and recovers the potential for diffraction-limited performance. For example, take $H_2 = 20$ km and the MHV turbulence model, and consider the cases of $D = 4$ m and $D = 8$ m telescopes imaging at $\lambda_i = 0.7$ μm. We find

$\sigma^2_{2-\mathrm{alt}}(D = 4) = 0.236$ rad^2 for the lower set of beacons at $H_1 = 10$ km, and $\sigma^2_{2-\mathrm{alt}}(D = 4) = 0.064$ rad^2 for $H_1 = 15$ km. For an 8-m telescope, $\sigma^2_{2-\mathrm{alt}}$ $(D = 8) = 0.836$ rad^2 for $H_1 = 10$ km, and $\sigma^2_{2-\mathrm{alt}}(D = 8) = 0.202$ rad^2 for $H_1 = 15$ km. Thus, we see that values of $S_{\mathrm{FA}} \lesssim 0.8$ should be possible at optical wavelengths for large telescopes, using two dense sets of Rayleigh spots.

Importantly, because the r_0^* effect has been overcome, the saturation of Strehl ratio versus the number of laser beacons is no longer observed. Simulations have shown (Smith and Sandler 1989) that very high Strehl ratios can be obtained in this manner using a dense array of spots, and an increase in the Strehl ratio with decreasing s (as predicted by Eq. (12.29)) is found to hold. To implement this scheme, one can envisage two arrays of 100 spots over an 8-m telescope. It is clear that the hardware implementation would have to be carefully designed for such a scheme to be viable. Detailed designs have in fact been made, which include use of a phase grating at a pupil of the outgoing laser beam to form the spots, and use of a shearing interferometer wave-front sensor to give the flexibility to combine signals from adjacent spots. This design has been implemented in field experiments by ThermoTrex in 1992, which projected 20 Rayleigh spots over a 1-m aperture. While data was obtained verifying the spot-signal addition concept, closed-loop Strehl measurements were not made, so that the question of significant Strehl improvement using this scheme remains open.

The above discussion on novel schemes has taken us far from our primary goal of designing a LGS AO using a single sodium beacon. However, within a few years, diffraction-limited images should be common using LGS AO on large telescopes, and then the question of pushing to optical wavelengths will resurface. We have shown that it is likely that either several high-power sodium lasers will be required to achieve high performance in the optical, or novel schemes will be required which use Rayleigh beacons and put the burden on innovative hardware designs and optimal processing of extensive wave-front data from many sources.

We conclude this section by mentioning that there has been preliminary work on using multiple laser beacons to extend the field of view of LGS AO. The basic idea is that if the turbulent atmosphere is composed of nearly discrete layers, then an array of beacons can be used to probe the phase profiles from different directions. The data can then be inverted to solve for the phase at each layer, and multi-conjugate AO can be applied to increase the isoplanatic angle. Tallon and Foy (1990) (see also Tallon *et al.* 1992) study an inversion method applied to model C_n^2 profiles, and find that use of even a few spots shows promise when the number of discrete layers is small.

Baharav, Ribak, and Shamir (1993) have analyzed the use of a fringe pattern projected at the sodium layer to probe the structure of multiple layers. (We note that the use of fringe patterns for Rayleigh backscatter was proposed and analyzed in the early 1980s by R. Hudgin.) The pattern is then imaged by a lenslet array with a FOV larger than 1 arcminute, onto a large CCD, and the data is analyzed to retrieve two discrete layers (at low and high altitudes). The method appears promising in simulations, reducing the rms phase at the edge of the field by a factor of two or more.

12.3.3 Wave-front sensing with laser beacons

12.3.3.1 Differences between NGS and LGS wave-front sensing

From an operational point of view, wave-front sensing of high order aberrations is the same for both LGS and NGS AO (see Section 3.2.2). Most commonly, for LGS AO a Shack–Hartmann sensor is used to measure the instantaneous centroids of laser beacon images over subapertures of width d. For astronomical imaging in H–K bands, typically $d \sim 0.5$ m will be used for the new large telescopes, since $r_0(2 \; \mu m) \sim 1$ m. The value of r_0 at the sodium laser wavelength $\widehat{r_0}(0.589 \; \mu m) \sim 20$ cm, so typically $d/\widehat{r_0} = 3$, which is near the optimum subaperture size for maximum resolution. For 3-m class telescopes developing LGS AO, the subaperture size is typically smaller to aim for the goal of diffraction-limited imaging of bright objects near $\lambda_i = 1 \; \mu m$, so $d/\widehat{r_0} \sim 3$ for these systems.

The major differences between NGS and LGS wave-front sensing are twofold. First, the Shack–Hartmann images will be larger, because the laser beacon propagates through the atmosphere and is then imaged through turbulence. For a site with 0.7″ seeing, the short exposure fwhm is ~ 0.5″ for one-way propagation, so spot widths $w \sim 0.7 – 1$″ can be expected. In addition, in some cases the sodium laser beam may not be diffraction-limited, so its finite laser beam quality can add to the beacon image width as recorded on the CCD. We will see below that the sodium image size is a crucial ingredient in determining the SNR of the slope measurements for a given laser power, and considerable savings in power requirement can be achieved through optimization of the sodium projector aperture size and optics.

Second, most NGS AO systems optimized for astronomy are designed for WFS integration times $\Delta t \leqslant 5 – 10$ ms to reach to faint field stars. However, to limit temporal delay errors most LGS AO systems take advantage of the bright laser beacon and operate with $\Delta t \leqslant 1$ ms. Thus, WFS CCD detectors are required which operate at low noise at high read-out rates. For diffraction-

limited correction with large telescopes, these detectors must be at least
64×64, with 128×128 CCDs required for correction near 1 μm.

In this section, error formulas are given for the accuracy of slope measurements for LGS AO, as a function of beacon brightness, integration time, and CCD read-out noise.

12.3.3.2 Photon noise errors

The estimate of instantaneous image motion obtained from a Shack–Hartmann WFS will have error $\epsilon_\theta^{\text{noise}}$ arising from the finite photon return per subaperture, and the finite beacon image size. From photon statistics it can be shown that the rms error in measuring the centroid of an image due to Poisson and detector read-out noise is proportional to the 1/SNR, where

$$\text{SNR} = \frac{N}{\sqrt{N + pn^2}}. \tag{12.39}$$

Here N is the number of photons in the image and p is the number of pixels used to compute the centroid. The parameter n is the the rms detector read-out noise (n_d) plus background noise (n_b) per pixel:

$$n = \sqrt{n_d^2 + n_b^2}. \tag{12.40}$$

For a quadrant detector ($p = 4$), Tyler and Fried (1982) showed that for an unresolved point source image

$$\sigma_{SH}^{\text{noise}} = \frac{c}{\text{SNR}} \frac{\lambda_b}{D} \tag{12.41}$$

where

$$c = \left[4 \int_0^1 H(u, 0) \, du \right]^{-1} \tag{12.42}$$

and λ_b is the mean tilt sensing wavelength at which the centroid is measured. In Eq. (12.42) $H(u_x, u_y)$ is the optical transfer function of the imaging system in terms of dimensionless frequency variables $u_x = f_x/f_{dl}$, $u_y = f_y/f_{dl}$, where $f_{dl} \equiv D/\lambda$. For centroiding the image of a natural star through uncompensated turbulence, we take the short-term optical transfer function for Kolmogorov turbulence and a circular aperture

$$H(u) = \frac{2}{\pi} \exp\left[-3.44 \left(\frac{d}{r_0} \right)^{5/3} u^{5/3} (1 - u^{1/3}) \right] \left[\cos^{-1}(u) - u(1 - u^2)^{1/2} \right]. \tag{12.43}$$

When c is plotted as a function of d/r_0, we see a linear increase in c, with a slope (to within 10%) of $\simeq 0.45$ over a wide range of d/r_0. Since (in units of λ/d) the short exposure image width w is approximately equal to $0.7d/r_0$, we can write $c = \alpha w$, where $\alpha = 0.65$. Thus, we see from Eq. (12.41) that the centroid noise error is proportional to the width of the image.

Equation (12.41) gives the instantaneous centroid error due to photon and read-out/background noise. For closed-loop control of image motion, there will be some low-pass filtering of noise by the servo-loop, resulting in reduction of σ_{SH}^{noise}. The reduction factor γ for a single-pole control system is given by

$$\gamma = \left[\frac{2}{\Omega} \tan^{-1}\left(\frac{\Omega}{2}\right)\right]^{1/2}. \tag{12.44}$$

The parameter Ω is defined by

$$\Omega = \frac{1}{\Delta t f_c} \tag{12.45}$$

where f_c is the closed-loop bandwidth of the AO control system. Ω is thus the number of AO samples/updates contained in the time interval $T = 1/f_c$. For example, if the closed-loop servo bandwidth is 100 Hz, for $\Omega = 10$ the update interval is $\Delta t = 1$ ms. For $\Omega = 10$, $\gamma = 0.52$.

The WFS photon-noise error is now

$$\sigma_{SH}^{noise} \simeq \frac{\alpha \gamma w}{\sqrt{N}}\left(1 + \frac{4n^2}{N}\right)^{1/2}\frac{\lambda_b}{d}. \tag{12.46}$$

We can convert this slope measurement noise to an equivalent phase-difference error σ_{pd}, by noting that the diffraction angle λ_b/d corresponds to 2π radians of phase error. We can also convert to radians of phase error at the imaging wavelength by scaling by the ratio of beacon wavelength to imaging wavelength. Combining these we have

$$\sigma_{pd} \simeq \frac{2\pi \alpha \gamma w}{\sqrt{N}}\left(1 + \frac{4n^2}{N}\right)^{1/2}\frac{\lambda_b}{\lambda_i}. \tag{12.47}$$

As an example, we consider the case of sodium beacon images and $d = 0.5$ m $(\lambda_b/d = 0.24'')$, and consider $N = 100$ detected photons/subaperture. The width factor is taken to be $w = 3$, corresponding to $0.72''$ sodium images on the CCD. We consider two cases for read-out noise: a CCD with $n = 3$ electrons/pixel, and one with $n = 10$. The corresponding phase-difference measurement error is plotted in Fig. 12.7. We see that for this photon return, to keep the measurement noise $< 1/10$ wave rms for imaging in H–K bands, it is necessary to have a CCD with $n = 3$.

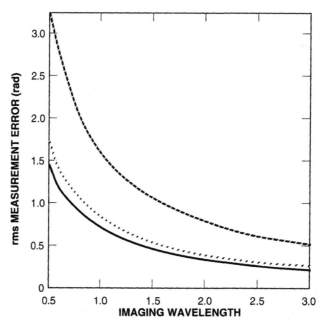

Fig. 12.7. Phase-difference measurement error as a function of imaging wavelength, for sodium beacon AO. Three values for the rms noise/pixel of the CCD are treated, $n = 0$ (solid line), $n = 3$ (dotted line), and $n = 10$ (dashed line).

Figure 12.8 plots σ_{pd} versus integration time for sodium return, for three equivalent stellar V magnitudes of the sodium beacon, at an imaging wavelength $\lambda_i = 2$ μm, for $n = 3$. The photometric assumptions are the same as those used in Section 11.1.2, except that a wavelength band of $\Delta\lambda = 100$ nm is now used. From the figure, we see that a $m_v = 10$ sodium beacon, and a 1 ms integration time give an rms phase-difference error of 0.4 rad for an $n = 3$ electrons read-out noise CCD.

Before discussing how the equivalent stellar magnitude of a sodium beacon translates into the required laser power, first we turn to a brief discussion of wave-front reconstructors, which determine how the slope errors σ_{pd} are converted into errors in wave-front commands sent to the DM.

12.3.3.3 Wave-front reconstructors for laser beacons

The expression for the mean square phase error resulting from reconstructing slopes with rms error σ_{pd} is given by

$$\sigma_{rec}^2 = G\sigma_{pd}^2 + \beta\left(\frac{D}{r_0}\right)^{5/3} \tag{12.48}$$

where G is the error (or noise) propagator for the given reconstruction

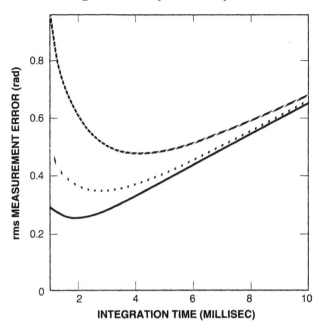

Fig. 12.8. Plot of the phase-difference measurement error as a function of time delay between servo updates, for three equivalent visual magnitudes of the sodium laser beacons. (*solid line*: magnitude 9 beacon; *dotted line*: magnitude 10 beacon; *dashed line*: magnitude 11 beacon.) The results are for an imaging wavelength of 2 μm, and $n = 3$ electrons/pixel rms noise for the CCD detector.

algorithm. The second term on the rhs is the reconstruction error in the limit of zero photon-noise. It depends on the geometrical arrangement and number of subapertures, and represents the error in converting slopes to phase values at discrete points (hence the dependence on (D/r_0)). The value of the coefficient β typically varies from 0.02 to 0.1, and is higher for modal control than for zonal control. For numerical examples, we take the nominal value $\beta = 0.05$.

For a single laser beacon, the wave-front reconstructor algorithm can be the same as for reconstructing the wave-front from slope measurements obtained using a natural guide star. To date, all existing LGS AO systems carry over NGS AO reconstructors directly, so the value of the error propagator is to first order independent of the type of beacon. As discussed in Chapter 5, wave-front reconstructors have been studied extensively for NGS AO. The standard least-squares reconstructors all yield error propagators in the range $G = 0.7–2$. For example, consider the standard Fried geometry, corresponding to slope measurements across the interior of four actuator points of a continuous DM. In this case, the error propagator is bounded by $G \leqslant 1.6$, increasing as $N_a \ln N_a$ (where N_a is the number of actuators) to the limiting value at $N_a \sim 30$. When the actuator response is folded in to the total control matrix, the result is that $G \simeq 1$, which is indeed the requirement that one usually imposes on the system noise propagation.

For multiple beacon or hybrid LGS AO, the reconstructors must be more complicated, incorporating stitching procedures, and possibly two altitude weighting as discussed above. However, it can be shown that these procedures do not greatly affect the propagation of noise in the control matrix, so that to a good approximation we can assume that even for multiple beacon LGS AO the error propagator is nearly unity.

The subject of optimal control algorithms refers generally to going beyond a simple, Fried-like least-squares reconstructor by including the effects of atmospheric statistics and the spectrum of measurement noise. In the simplest case, one can use a Wiener-like filter (see, for instance, Sandler *et al.* 1994b) constructed from Kolmogorov statistics of open-loop wave-front sensor measurements. Recently, Wild (1996) has produced a series of studies on optimal reconstructors, including the extension required for closed-loop AO operation. Reconstructors have been tested by Wild with C. Shelton at the Mt Wilson 100-inch NGS AO system, demonstrating significant improvement in both Strehl ratio performance and robustness for fainter operation.

Ellerbroek (1994) has derived a very general formalism for deriving closed-loop reconstructors for LGS AO. Recently, Rhodarmer and Ellerbroek (1995) have extended the theory to allow development of optimal closed-loop reconstructors from field data obtained at the telescope, for the case when a second, 'scoring' wave-front sensor is available. Experimental tests were performed at SOR, for the case of hybrid AO combining measurements from a bright NGS and a single Rayleigh LGS. The results of the tests were promising, showing an improvement over the single LGS Strehl ratio when low spatial-frequency NGS data were used to supplement the LGS slopes.

12.3.3.4 Laser power requirements

In Section 12.3.3.2, we derived formulas for the phase-difference measurement error as a function of the number of photons per measurement. In order to determine the laser power requirements for LGS AO, it is necessary to know how many watts of power are required to backscatter N photons into the telescope aperture.

Rayleigh beacon. To generate a Rayleigh beacon for LGS AO, a pulse of radiation is transmitted and the detector integrates during the short interval corresponding to a well-defined spot being formed at the chosen focus altitude,

H. The air molecules at every altitude contribute to Rayleigh scattering, and the return is proportional to atmospheric density whose e-folding distance is about 2.5 km. Thus the return is much stronger at low altitudes, where it is undesirable because of strong FA. To gather photons scattered in a narrow region around the focus altitude of the pulse is referred to as range-gating; for a total elongated spot of vertical width L, a shutter in front of the detector is timed so as to collect photons scattered from $H - L/2$ to $H + L/2$, where $L \ll H$. The range-gate length is typically $L = 1-2$ km for spot altitudes of $H = 10-20$ km.

The radiometry of Rayleigh scattering is well understood from many years of lidar applications. We refer the reader to Gardner *et al.* (1990) for explicit formulas for the backscatter flux, as a function of altitude, range gate distance around focus, and properties of the atmosphere. Note that the Rayleigh cross-section depends on wavelength as λ^{-4}, so that shorter wavelength lasers are more efficient in producing backscattered photons. Because the artificial beacon returns a spherical wave, the return flux falls off as H^{-2}, where H is the range of the laser beacon above the telescope.

Two examples from current systems serve to estimate power requirements for astronomical AO systems using Rayleigh beacons. The SOR 1.5-m system currently uses $P = 200$ W of average power, with equal power produced at laser wavelengths of 0.5106 and 0.5782 μm. The Copper vapor laser produces 5000 pulses/s, with a 50 ns pulse width. The radiometric results indicate that an average of five LGS pulses, each focused at a range of $H = 10$ km with $L = 2.4$ km long range gate around focus, produce 190 photons in a $d = 9.2$ cm subaperture. This corresponds to approximately 23 000 ph/m^2/ms. The University of Illinois LGS AO system (Chapter 13) will use a $P = 50$ W laser operating at $\lambda = 0.35$ μm with 200 Hz pulse repetition rate. It is estimated that this power produces the equivalent of a $m_v = 10$ laser beacon at an altitude of 18 km, or 11 500 ph/m^2/ms.

For LGS AO correction for imaging near 2 μm wavelength, the subaperture size is $d \sim 0.5$ m. Scaling the SOR results to this case and a Rayleigh beacon at $H = 20$ km, we find that approximately $P = 25$ W of power is required for $N = 200$ ph/ms backscattered into each subaperture. We see that lasers are in operation now for wave-front sensing using a Rayleigh beacon at about 20 km range, which generate enough backscattered photons for sufficient signal-to-noise for imaging at 2 μm wavelength or longer. At some sites, the FA for this case may be weak enough even for 3.5-m telescopes to produce reasonable, even good, correction at infrared wavelengths. However, the pulsed lasers in use now have very poor beam quality. The output beam must be projected through the full telescope aperture to limit the beam divergence, in order to produce a small enough spot to use for wave-front sensing. This means that high fluence levels are produced in the telescope optics, and scattered light could interfere with the tilt signal from faint field stars. In addition, at SOR there is some evidence that phosphorescence has occured, causing faint stray light internal to the beam-train at wavelengths shifted longward from the laser wavelength. Since these wavelengths are in the red region of the visible spectrum where avalanche photodiodes are often used for tilt sensing, this problem must be overcome if Rayleigh lasers are to be used for infrared astronomy with LGS AO. The University of Illinois system will be the first to use a Rayleigh beacon for infrared astronomy, so more information on the limitations for use with faint field stars should be forthcoming.

Sodium beacon. The mesospheric sodium layer is believed to be the result of meteoric ablation. The layer occurs at approximately 94 km altitude and its average thickness is about 10 km (fwhm ~ 4 km). The sodium column density is $n_{\text{sod}} \sim 10^{13}$ atoms/m^2, and the temperature is approximately 200 K. However, the density of the layer can change by a factor of two or more depending on the time of year, and smaller changes in density and mean altitude of the layer can occur on a time scale of hours. Because of the importance of characterizing these changes, the University of Illinois lidar group has performed the Giant Aperture Lidar Experiment with SOR at the 3.5-m telescope (Papen *et al.* 1996).

The use of resonant excitation of the sodium layer for LGS generation was first proposed by Happer in 1982 (Happer *et al.* 1994). The physics of absorption and radiation by sodium atoms in the mesospheric layer is quite complicated at the atomic physics level, and beyond the scope of this chapter. (See, for instance, Morris 1994). For a review of the important physics, see the review article by Jeys (1991). The transition of interest for LGS AO is the D_2 transition at $\lambda = 589$ nm, between the $3^2S_{1/2}$ and $3^2P_{3/2}$ states of the sodium atom. The upper level has several hyperfine states. The absorption profile for a single hyperfine transition for a single atom is Gaussian with a center frequency of 10 MHz, corresponding to a radiative lifetime $\tau = 16$ ns for spontaneous emission. In the sodium layer, Doppler broadening of the energy levels occurs because of the finite temperature of the layer, which produces a velocity distribution of the atoms. The net absorption profile for gound and hyperfine transitions is inhomogeneously broadened, with two broad peaks separated by 1.77 GHz extending over 5 GHz, with fwhm of $\Delta\nu_D = 3$ GHz. The lowest frequency peak is a factor of two stronger than the higher one. Experimental results for sodium return versus wavelength are shown in Fig. 12.9.

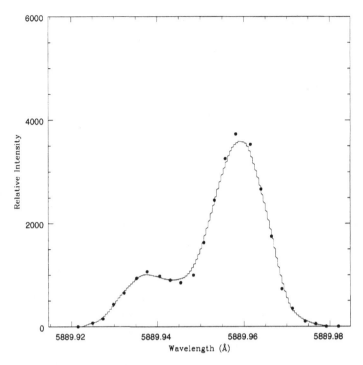

Fig. 12.9. Intensity of circularly polarized sodium laser resonant backscatter versus wavelength, obtained by the Arizona group at the MMT. The dots are measurements. The solid line is the best fit Gaussian profile with the fwhm of 0.140 nm, corresponding to a mesosphere temperature of 251 K if only thermal broadening contributes.

The column density per 10 MHz linewidth is about 1.7×10^{11} atoms/m^2, and the absorption cross-section is 1.7×10^{-13} m^2 for a single atom. This results in absorption by only a few percent of the resonant photons injected into the layer. Thus, the laser line is often broadened so as to cause absorption and radiation over a substantial portion of the inhomogeneously broadened profile. For optimal operation, this should be done in a way that efficiently takes into account the stronger absorption at the first Doppler broadened peak, while avoiding saturation. The term saturation refers to the fact that as the intensity of the laser is increased, stimulated emission can become comparable to resonant absorption and emission. This results in a leveling out of the return flux as the power is increased indefinitely. For the case of continuous wave (CW) laser radiation, the single-frequency (10 MHz) saturation intensity is 6.3 mW/cm^2. Over the full Doppler-broadened profile, the saturation intensity is 3 GHz/10 MHz larger, or $I_{sat} = 1.9$ W/cm^2.

For CW illumination, the probability of an excited state for incident intensity I is given by

$$p_e = \frac{1}{2}\left(1 + \frac{I_{sat}}{I}\right)^{-1}. \tag{12.49}$$

Thus, at $I = I_{sat}$, one-quarter of the atoms are available for fluorescense, and as the intensity is increased indefinitely to $I \gg I_{sat}$, the fraction of atoms available is limited to one-half. At the saturation intensity, it takes a single atom a time equal to $4\tau = 64$ ns to emit one photon on average. A patch of the layer of area $a = 1$ cm^2 will then radiate $n_{sod}/64$ ns $\sim 1.6 \times 10^{16}$ ph/sec/sr. For a $d = 1$ cm receiver on the ground, the solid angle is $(d/H_{sod})^2$, and the integrated signal is approximately 200 ph/s/cm^2.

Consider the case of a 1″ sodium spot. Then the sodium spot at 90 km will have fwhm ~ 45 cm, so the measured backscatter flux at CW saturation intensity $I_{sat} = 3.8$ kW should be $\sim 4 \times 10^5$ ph/s/cm^2, or 400 ph/ms/cm^2. Since the return is linear with intensity up to the saturation threshold, we conclude that the sodium return per watt of power for CW lasers should be about $F_{cw} = 0.1$ ph/ms/cm^2. W. Jacobsen *et al.* (1994) performed radiometry experiments at the MMT array using several different dye lasers tuned to the sodium D$_2$ line, and found for a commercial CW ring dye laser the return per watt of power was equivalent to the flux from a $m_v = 11$ natural star. More precisely, the Arizona group has measured values in the range F = 0.045−0.07 ph/ms/cm^2.

CW sodium dye lasers can produce up to 4 W, or the equivalent of about a $m_v = 9.5$ laser beacon. Later, we will see that this is the theoretical required power for LGS AO correction for K band imaging with large telescopes, for $d = 0.5$ m subapertures. (The backscatter in this case for $P = 4$ W is 450 ph/ms.) However, for imaging at shorter wavelengths, the power requirements increase significantly, as $\lambda^{-18/5}$. Even for imaging at near-infrared wavelengths, one would like 10 W of power, to accomodate poorer seeing, off-zenith science targets, and generally more robust operation.

Many different approaches to developing pulsed lasers which are affordable and practical for astronomical AO have been conceived and tested over the past 15 years; a summary is given in Fugate (1996). Three approaches are currently being pursued for use at astronomical observatories, and these are discussed briefly in Chapter 13. They are sum-frequency combination of YAG lines (ARC telescope, University of Chicago); Nd:YAG pumped dye laser (Lick 3.5-m telescope and Keck II Telescope, LLNL); and a new solid state Raman laser using a CaWO$_4$ crystal (MMT conversion, University of Arizona).

In order to develop efficient pulsed sodium lasers for LGS AO one has to come as close as possible to the theoretical CW flux, F_{cw}, while avoiding saturation. Pulsed lasers have a much lower duty cycle than CW radiation, where duty cycle is the temporal pulse width multiplied by the pulse repetition bandwidth. By definition, CW lasers have duty cycle of one. The probability of operating in a saturated mode increases for low duty cycle lasers, especially as the pulse energy increases.

Table 12.1 summarizes the properties of sodium lasers now in use or planned for operation at astronomical telescopes (following Martinez 1998). Included in the table are measured photon return, laser power, pulse format, Doppler broadened bandwidth, and measured spot size. The final column gives the power multiplied by the return and divided by the square of the spot size; from Eq. (12.47), we see that this quantity should be maximized to achieve optimal slope accuracy per watt of laser power.

From the table, we see that the CW dye laser is the most efficient, with a return close to the limiting estimated CW return given above. The spot sizes are only approximate, and of course will vary with atmospheric conditions. However, they are given as illustrative of the potential limitations due to laser beam quality or intentionally spoiling of the laser beam to avoid saturation. The Arizona group has measured spot sizes close to 1″, resulting in efficiency factors which are substantially higher than for pulsed lasers. In some cases, pulsed lasers must operate not only with broaded Doppler profiles but also with intentionally large spot sizes, in order to limit the power per unit area at the sodium layer. The last entry of Table 12.1 is the Raman-shifted solid-state laser (Murray *et al.* 1997) under development by LITE CYCLES in Tucson. Because it is truly quasi-CW, it promises to approach the limiting performance of a 10 W CW laser. Because the values for output energy and spot size have not yet been measured, they are listed as unknown in the table.

It is difficult to compare directly the values for photon return in Table 12.1, because of natural variations in the absolute sodium column density. Recently, Ge *et al.* (1997) have obtained measurements of sodium abundance spectro-scopically, as shown in Fig. 12.10, using the Advanced Fiber Optic Echelle spectrograph at the Center for Astrophysics 60-inch telescope on Mt Hopkins, near the MMT. This has allowed simultaneous measurement of sodium beacon return and absolute abundance, for the first time. The return value shown in Table 12.1 for a CW dye laser corresponds to a sodium density of $n_{sod} = 2 \times 10^9$ cm^{-2}.

To conclude, there exist lasers which meet the power requirements for diffraction-limited imaging at near-infrared wavelengths, and in the near term at least four different types of sodium laser will be in use for astronomical LGS AO. However, for imaging at wavelengths ≤ 1 μm, much higher power is required. Fugate estimates that 50–200 W will be required for imaging in I band at the SOR 3.5-m telescope. Proposals have been

Table 12.1. *Sodium laser operating parameters for various guide star lasers in operation or planned for the near future. The parameter in the last column is the efficiency parameter discussed in the text*

Type (Group)	Photon return ($cm^{-2}ms^{-1}W^{-1}$)	Power (W)	Pulse format (μs, kHz)	Bandwidth (GHz)	Spot size (arcsec)	Efficiency factor
Sum frequency solid-state (Chicago-SOR)	0.03	7–12	50, 0.84	3.5	2	0.07
Pulsed dye (LLNL–Keck)	0.01–0.03	15	0.1, 11	3	1.6–2	0.06–0.18
CW dye (Arizona, MPI)	0.07	3–4	CW	<0.1	1.1	0.17
Raman-shifted solid-state (Arizona)	unknown	10	8×10^{-4}, 10^5	0.8	unknown	

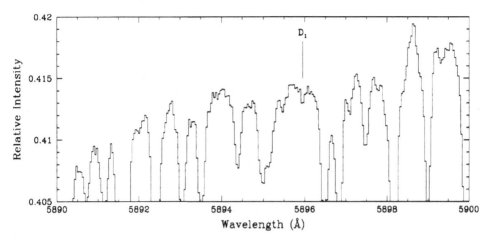

Fig. 12.10. Telluric sodium D_1 absorption in Alpha Leo ($V = 1.4$) spectrum with 120 min integrations. The data were obtained with the Advanced Fiber Optic Echelle at the 60-inch telescope on Mt Hopkins, and correspond to a sodium abundance of $(2.0 \pm 0.3) \times 10^9$ cm^{-2} (Data supplied by J. Ge).

made, including the coherent addition of 50 ns pulses at many thousand pulses/s, which has the potential to increase average power without increasing peak power (Hogan *et al.* 1996). At this time it is unclear whether lasers at this power level will be developed which are practical and affordable for astronomical telescopes. An open area is to investigate the use of resonant excitation of other naturally occuring species, such as other metals in the mesosphere (Papen *et al.* 1996).

12.3.4 Total high order LGS AO system error

We conclude this section by bringing together the major sources of high order wave-front errors required to calculate S_ϕ. In the Marechal approximation,

$$S_\phi = \exp(-\sigma_\phi^2), \tag{12.50}$$

where σ_ϕ is the total rms wave-front error for LGS AO correction. The major error sources included in σ_ϕ are the FA error (or the total multiple beacon LGS error for use of more than one LGS), the reconstruction error given by Eq. (12.48), and the additional dominant error sources normally included for NGS AO. (See Chapters 4–7.) The first of these errors is the fitting error, which for most DMs used in LGS AO is given by

$$\sigma_{\text{fit}}^2 \simeq 0.3 \left(\frac{r_\text{s}}{r_0} \right)^{5/3}, \tag{12.51}$$

where r_s is the interactuator spacing. The temporal decorrelation error for full correction can be written

$$\sigma_{\text{time}}^2 = \left(\frac{\tau}{\tau_0}\right)^{5/3}. \tag{12.52}$$

The temporal error is often written in terms of the closed-loop bandwidth f_c of the high order loop as

$$\sigma_{\text{BW}}^2 = \left(\frac{f_G}{f_c}\right)^{5/3}, \tag{12.53}$$

where f_G is the Greenwood frequency for high order wave-front correction. Thus, we may summarize the total mean square high order error as

$$\sigma_\phi^2 = \sigma_{\text{FA}}^2 + \sigma_{\text{rec}}^2 + \sigma_{\text{fit}}^2 + \sigma_{\text{time}}^2. \tag{12.54}$$

In general, an optimized LGS AO design is one which comes close to minimizing the total error with respect to the AO system parameters, which include the number of actuators, N_a; the number and size, d, of wave-front sensor subapertures; the closed-loop bandwidth, f_c; and the laser guide star power, P.

Later in this Chapter we will give examples of LGS AO performance. First, however, we turn to the important issue of estimating the Strehl ratio S_θ for image motion correction.

12.4 Image motion correction

12.4.1 Image motion Strehl ratio

The blurring of the long exposure PSF from image motion errors is a convolution operation of the Airy pattern with a Gaussian jitter function about the central axis corresponding to rms tilt errors $(\sigma_\theta)_x$ and $(\sigma_\theta)_y$, measured in arcseconds. For high image motion Strehl ratios, it is clear that the residual jitter must be a small fraction of λ/D (or equivalently, the residual tilt phase error must be a small fraction of 2π radians $= 1$ wave). If we approximate the core of the Airy pattern (within the first disk) as a Gaussian profile of fwhm $1.06\,\lambda/D$, then the reduction in the on-axis intensity can be computed as

$$S_\theta = \frac{1}{1 + \dfrac{\pi^2}{2}\left(\sigma_\theta \dfrac{\lambda}{D}\right)^2} \tag{12.55}$$

where $\sigma_\theta{}^2$ is an effective one-axis image motion given by

$$\sigma_\theta^2 = \tfrac{1}{2}[(\sigma_\theta)_x^2 + (\sigma_\theta)_y^2]. \tag{12.56}$$

There are several contributions to σ_θ, which we analyze below.

12.4.1.1 Tilt anisoplanatism

Tilt anisoplanatism (TA), $\epsilon_\theta^{\mathrm{TA}}$, is the error introduced in by estimating the true atmospheric tilt in the target direction using a measurement derived from photons from an off-axis field star. TA is an example of the angular decorrelation of Zernike modes treated in Section 3.1, and the formalism described there can be used to determine the magnitude of $\sigma_\theta^{\mathrm{TA}}$. We find:

$$(\sigma_\theta^{\mathrm{TA}})^2 = \frac{8}{\pi^2}(0.448)[1 - \Gamma_{\mathrm{tilt}}(\theta)]\left(\frac{D}{r_0}\right)^{5/3}\left(\frac{\lambda}{D}\right)^2 \tag{12.57}$$

where θ is the angular separation of science target and field star. The correlation $\Gamma_{\mathrm{tilt}} = (\Gamma_2 + \Gamma_3)/2$, where Γ_2 (Γ_3) is the normalized correlation parameter for the tilt Zernike mode parallel (perpendicular) to the axis joining science target and field star. The correlation parameters can be written as a weighted integral over the C_n^2 profile:

$$\Gamma_{2,3}(\theta) = \frac{\int \left(\gamma_0\left(\frac{z\theta}{D}\right) \pm \gamma_2\left(\frac{z\theta}{D}\right)\right) C_n^2(z)\,\mathrm{d}z}{\int C_n^2(z)\,\mathrm{d}z}, \tag{12.58}$$

where the $+$ sign refers to mode index $m = 2$, and the minus sign to $m = 3$. The functions $\gamma_{0,2}(x)$ are given by

$$\gamma_{0,2}(x) = \int y^{-14/3} J_{0,2}(2xy)[J_2(y)]^2\,\mathrm{d}y. \tag{12.59}$$

Numerical approximations to the integral in Eq. (12.59) are given in Olivier and Gavel (1994). In Sandler *et al.* (1994a) a useful formula is given which approximates the TA error in terms of D/r_0 and θ/θ_0:

$$\frac{(\sigma_\theta^{\mathrm{TA}})_x^2}{(\lambda/D)^2} = 0.0472\left(\frac{\theta}{\theta_0}\right)^2\left(\frac{D}{r_0}\right)^{-1/3} - 0.0107\left(\frac{\theta}{\theta_0}\right)^4\left(\frac{D}{r_0}\right)^{-7/3} + \cdots \tag{12.60}$$

$$\frac{(\sigma_\theta^{\mathrm{TA}})_y^2}{(\lambda/D)^2} = 0.0157\left(\frac{\theta}{\theta_0}\right)^2\left(\frac{D}{r_0}\right)^{-1/3} - 0.002\,14\left(\frac{\theta}{\theta_0}\right)^4\left(\frac{D}{r_0}\right)^{-7/3} + \cdots \tag{12.61}$$

Truncating Eqs. (12.60) and (12.61) after the first terms on the right-hand side results in a good approximation out to field angles $\theta \leqslant 0.5(D/r_0)\theta_0$; including both terms on the right-hand side is valid out to $\theta \leqslant (D/r_0)\theta_0$.

We now present numerical examples for TA. In Sandler *et al.* 1994a, the correlation functions given by Eq. (12.58) for longitudinal and transverse tilt modes are plotted for the M1 and M2 atmospheric models. It is shown that the tilt measured with an off-axis field star remains correlated with the on-axis tilt ($\Gamma_x \geqslant 0.5$) out to large field radius ($\theta = 5$ arcmin for strong M2 turbulence). However, to obtain high TA Strehl ratios S_θ^{TA} much smaller fields are required. The TA Strehl ratio for turbulence model MHV, calculated from Eq. (12.55) using Eq. (12.57) for a $D = 8$-m telescope is shown in Fig. 12.11, for imaging wavelengths of $\lambda = 0.5$ μm, $\lambda = 1$ μm, and $\lambda = 2$ μm.

From the plots, we see that TA can significantly limit the natural star field for high-Strehl ratio correction. In fact, if we specify a value of $(S_\theta)_{\mathrm{TA}} \geqslant 0.6$ (leaving a tilt Strehl contribution > 0.8 for other error sources to meet the goal of $S_\theta = 0.5$), this leads at $\lambda = 2$ μm to limiting radii at of $\theta_{\mathrm{lim}} = 45''$ for MHV turbulence. From Sandler *et al.* (1994a), the limiting radius for M1 is $57''$ and for M2 $\theta_{\mathrm{lim}} = 25''$. At $\lambda = 1$ μm, the angular fields for $(S_\theta)_{\mathrm{TA}} \geqslant 0.6$ are approximately $1/2$ as large, with another factor of two reduction for imaging at $\lambda = 0.5$ μm.

On average, then, we can conclude that natural star fields of 1–2 arcmin diameter can be used for high-Strehl image motion correction when imaging at 2 μm wavelength, with a reduction in field diameter proportional to imaging wavelength for shorter wavelengths.

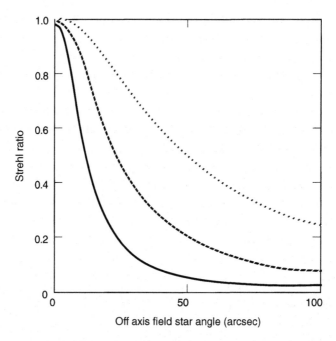

Fig. 12.11. Strehl ratio for TA as a function of off-axis tilt field star angle, for MHV turbulence. Results for three imaging wavelengths are shown: 0.5 (solid line), 1 (dashed line), and 2 (dotted line) μm.

12.4.1.2 Time delay error

Because of the finite time used for integrating field star photons, and the finite temporal response of the detector, computational electronics, and deformable (or tip/tilt) mirror, the tilt correction will contain an error ϵ_θ^{time} arising from the temporal evolution of turbulence. As discussed in Chapter 1, wave-front errors due to a finite time delay between sampling and correction can be analyzed on the same footing as anisoplanatism, since within the Taylor hypothesis both errors arise from a shift in phase screens relative to their positions in the science target direction. Thus, the rms time delay error for tilt sensing, σ_θ^{time}, can be calculated using Eq. (12.57), with the shift $z\theta$ replaced by $v(z)\Delta t$ in Eq. (12.58), where $v(z)$ is the wind speed at altitude z. Equivalently, the expansions Eqs. (12.60) and (12.61) can be used to calculate σ_θ^{time}, if θ is replaced by Δt and θ_0 by t_0, the atmospheric decorrelation time. Averaging Eqs. (12.60) and (12.61) to obtain an effective one-axis angular error, and keeping only the first terms on the right hand sides (valid out to time delays of $\Delta t = 5t_0$ for $D/r_0 = 10$), we obtain

$$\sigma_\theta^{time} \simeq 0.18 \frac{\Delta t}{t_0} \left(\frac{r_0}{D}\right)^{1/6} \frac{\lambda}{D}. \tag{12.62}$$

Thus, if we define an effective decorrelation time for tilt sensing for which the rms time error is λ/D, we find $t_0^{tilt} \simeq 5t_0(D/r_0)^{1/6}$. For $D/r_0 = 10$, we find $t_0^{tilt} \simeq 7t_0$. Importantly, note that for high-Strehl correction, which requires accuracy of a small fraction of a wave, a sampling time on the order of t_0 is required, even though the tilt aberration remains correlated over much longer time scales.

In real AO systems, image motion correction is implemented via a closed-loop servo. Consider a servo with closed-loop transfer function $H(f) = (1 + if/f_c)^{-1}$, where f_c is the 3-dB servo bandwidth. Tyler (1994c) has shown that the rms error due to finite servo bandwidth is given by

$$\sigma_\theta^{BW} = \frac{f_{tilt}}{f_c} \frac{\lambda}{D}. \tag{12.63}$$

The frequency f_{tilt} is a property of the atmospheric turbulence profile:

$$f_{tilt} = \frac{0.08}{r_0^{5/6} D^{1/6}} v_{eff}^{tilt}, \tag{12.64}$$

where v_{eff}^{tilt} is an effective wind speed for tilt correction given by

$$v_{eff}^{tilt} = \left[\frac{\int C_n^2(z)v^2(z)\,dz}{\int C_n^2(z)\,dz} \right]^{1/2}. \tag{12.65}$$

The wind profile Eq. (12.28) gives $v_{\mathrm{eff}}^{\mathrm{tilt}} = 23$ m/s. For $r_0 = 0.9$ m ($0.7''$ seeing) and $D = 8$ m, Eq. (12.64) gives $f_{\mathrm{tilt}} = 1.4$ Hz at $\lambda_i = 2$ μm. From Eq. (12.64), f_{tilt} is proportional to λ^{-1}, so that $f_{\mathrm{tilt}} = 5.6$ Hz at $\lambda_i = 0.5$ μm. Note that for diffraction-limited correction ($\sigma_\theta^{\mathrm{BW}} \ll \lambda/D$) bandwidths 5–10 times higher than f_{tilt} are required.

The errors $\sigma_\theta^{\mathrm{BW}}$ and $\sigma_\theta^{\mathrm{time}}$ can be related by recalling that closed-loop servo operation is equivalent to a filtering operation on successive updates of the error signal, separated by delay Δt. The ratio of update (or sampling) frequency $f_s = 1/\Delta t$ to the closed loop frequency

$$\Omega = \frac{f_s}{f_c} \tag{12.66}$$

is chosen to optimize servo operation. Typical servo loops for AO use $\Omega \simeq 10$, so that $\Delta t \simeq 1/(10\, f_c)$. In general, Eq. (12.63) becomes

$$\sigma_\theta^{\mathrm{BW}} = \Omega \Delta t f_{\mathrm{tilt}} \frac{\lambda}{D}, \tag{12.67}$$

and we see that for one-tenth wave accuracy and $\Omega = 10$, $\Delta t = 0.01/f_{\mathrm{tilt}}$, or about 10 ms for the above parameters. This is in accord with our earlier observation that for high-Strehl image motion correction delay times are limited to approximately the value of t_0.

Figure 12.12 shows the dependence of the image motion Strehl due to finite bandwidth, S_θ^{BW} (calculated from Eqs. (12.63)–(12.65)) as a function of the closed-loop bandwidth, for imaging wavelengths of $\lambda = 0.5$, 1, and 2 μm. Below the bandwidth-axis is shown the corresponding delay time Δt as set by Eq. (12.67). We see from the figure that to obtain tilt bandwidth Strehl ratios of $S_\theta^{\mathrm{BW}} \geqslant 0.9$, a servo bandwidth of $f_c \geqslant 9$ Hz are required for imaging at 2 μm, increasing to $f_c \geqslant 36$ Hz at $\lambda = 0.5$ μm.

12.4.1.3 Centroid measurement error

The estimate of instantaneous image motion obtained from the tilt sensor will have error $\epsilon_\theta^{\mathrm{noise}}$ due to the fact that the stellar image used to compute the centroid contains a finite number of photons. The accuracy with which the instantaneous atmospheric tilt can be measured using faint sources sets the limiting magnitude for field stars, and hence sky coverage for LGS applications.

The centroid error due to photon and detector/background noise, $\sigma_\theta^{\mathrm{noise}}$, can be calculated directly from Eq. (12.46) with the Shack–Hartmann subaperture width d replaced by the telescope diameter D. For numerical application, we will take $\alpha\gamma = 0.23$, appropriate for treating large tele-

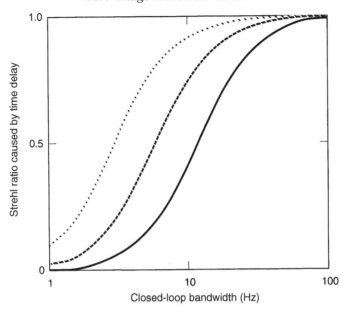

Fig. 12.12. Image motion Strehl ratio as a function of closed-loop bandwidth of the tilt correction loop, for wavelengths of 0.5 (solid line), 1 (dashed line), and 2 (dotted line) μm. Also shown on the horizontal axis are the corresponding delay times between updates for the case of a sampling frequency 10 times the closed-loop bandwidth.

scopes, moderate values of D/r_0, and servo-loops with sampling frequency equal to ten times the servo bandwidth ($\Omega = 10$). For smaller telescopes and/or very strong turbulence, or for different servo parameters, one has to evaluate the centroid error directly from Eq. (12.41) using Eqs. (12.42) and (12.43).

Consider the case of a tilt sensing wavelength $\lambda_t = 0.7$ μm, so that $w = D/r_0 \sim 30$. For the case of uncorrected stellar images, corresponding to a field star well outside the isoplanatic patch at λ_t, we find that $\sigma_\theta^{\mathrm{noise}} = 0.75\lambda_t/D$ for $N = 100$ detected photons and $n = 0$. For $n = 3$, the rms error increases to $\sigma_\theta^{\mathrm{noise}} = 0.87\lambda_t/D$, and for $n = 5$ the error increases to $> \lambda_t/D$. At an imaging wavelength of $\lambda_i = 2$ μm, from Eq. (12.46) the rms error in units of λ_i/D will be less by a factor of $\lambda_t/\lambda_i = 0.35$, which for $N = 100$ gives $\sigma_\theta^{\mathrm{noise}} = 0.26\lambda_i/D$, or a Strehl ratio of $S_\theta^{\mathrm{noise}} = 0.75$. For $S_\theta^{\mathrm{noise}} = 0.9$ at λ_i, $N = 300$ detected photons are required. For this case, $n = 3$ noise photons increases the centroid error only slightly, the effect of read noise becoming significant at $n = 5$. Thus, while for operation at very low photon levels it clearly is important to have nearly zero read noise for centroiding with uncorrected images, very high accuracy centroiding with $N \gg 100$ can tolerate up to $n = 3$.

12.4.1.4 Spurious contribution from higher order modes

An implicit assumption made in the previous sections is that the ideal tilt measurement would be furnished by a centroid determination using photons from a very bright field star in the direction of the science target. In reality this is not true in all cases due to a subtle difference between the tilt inferred from a centroid measurement and the coefficients of modes Z_2 and Z_3 in a Zernike decomposition of the wave front. It is the 'Zernike tilt' which is desired for usual AO operation, because the deformable (or separate steering) mirror corrects for the angular deviations of the plane normal to the propagation direction. The strengths of the Zernike modes give the orientation of the best-fit plane.

However, a centroid measurement is proportional to the average of the gradient of the wave front over the pupil. For field stars outside the isoplanatic patch for high order correction, the reference wave front entering the tilt sensor is composed of the atmospheric wedge (whose strength is desired) as well as the full spectrum of high-order aberrations. In particular, Zernike modes with azimuthal degree $m = 1$ and odd radial degree $n > 1$ will contribute to the integral of the phase gradient. These coma terms have nothing to do with the best-fit plane, and therefore contribute to an error $\epsilon_\theta^{\text{coma}}$ contributing spuriously to the value of the measured tilt. The rms value of this error is

$$\sigma_\theta^{\text{coma}} = 0.06 \left(\frac{D}{r_0}\right)^{5/6} \frac{\lambda}{D}. \tag{12.68}$$

Note that this error can indeed be large. For $D/r_0 = 10$, $\sigma_\theta^{\text{coma}} = 0.4\lambda/D$, which (using Eq. (12.55)) gives $S_\theta^{\text{coma}} = 0.56$. In many applications, this will be the dominant error limiting performance of the tilt correction system.

12.4.1.5 Total tilt error

The total tilt error is the sum in quadrature of the error sources discussed in Sections 5.2.4.2–5.2.4.5:

$$\sigma_\theta^2 = (\sigma_\theta^{\text{TA}})^2 + (\sigma_\theta^{\text{BW}})^2 + (\sigma_\theta^{\text{noise}})^2 + (\sigma_\theta^{\text{coma}})^2. \tag{12.69}$$

The total image motion Strehl can be computed by inserting Eq. (12.69) in Eq. (12.55).

In designing a tilt control system for LGS AO, the general goal is to minimize Eq. (12.69) while maximizing the sky coverage for tilt field stars, keeping in mind the performance and cost of available hardware. Global optimization is clearly ambitious, given the large number of free parameters. As a first step toward optimization, we note that a trade-off exists in balancing BW and noise errors. Let us treat the case where the delay time Δt is dominated by integration time, with the response time of tip/tilt mirror and electronics small in comparison. Longer integration times are favored to lower the noise error by increasing N for a given stellar magnitude m. But larger values of Δt imply smaller values of f_c (for a given value of Ω, which we assume is fixed at $\Omega = 10$), which increases the BW error. For $n = 0$ (appropriate to a very low-noise tilt sensor and low sky background) direct minimization of the sum of σ_θ^{BW} and σ_θ^{noise} yields for the optimal integration time

$$\Delta t_{opt} = (2F\eta)^{-1/3} \left(\frac{\alpha \gamma w}{f_{tilt} \Omega D} \right)^{2/3} \tag{12.70}$$

where η is the total quantum efficiency. Equation (12.70) is valid only when the number of photons collected during Δt_{opt},

$$N_{opt} = (2\Omega)^{1/3} (FD)^{4/3} \left(\frac{\alpha \gamma w \eta}{f_{tilt}} \right)^{2/3} \tag{12.71}$$

is large compared to $4n^2$. Denoting the rms of the sum of the BW and noise errors evaluated at $\Delta t = \Delta t_{opt}$ as σ'_θ, we then find

$$\frac{(\sigma'_\theta)^2}{(\lambda/D)^2} = 2^{2/3} \Omega^{10/3} \left(\frac{\alpha \gamma w}{D} \right)^{4/3} \left(\frac{f_{tilt}}{F\eta} \right)^{2/3}$$

$$+ 2^{-1/3} \Omega^{-2/3} \left(\frac{\alpha \gamma w}{D} \right)^{4/3} \left(\frac{f_{tilt}}{F\eta} \right)^{2/3} [1 + 2^{-1/3}\Omega^{-2/3} 4n^2 D^{-4/3} (F\eta\alpha\gamma w)^{-2/3}]. \tag{12.72}$$

To illustrate the behavior of the BW plus noise error, we now give two numerical examples. Both address a $D = 8$ m telescope with 0.7″ seeing. We take $f_{tilt} = 1.4$ Hz corresponding to the Bufton wind model, $\alpha w = 0.25$, $\eta = 0.4$, and again consider the case of negibible read and background noise.

1. The first example is the case of a field star very close to (or within) the science object, such that the field radius $\theta \leq \theta_0(\lambda_t)$. We take $\lambda_t = 0.8$ μm, with a spectral band $\Delta\lambda = 0.3$ μm. For this example, the laser beacon AO system will partially correct the wave front entering the tilt sensor imaging system. If we assume the LGS AO produces a high-order Strehl ratio $S_\phi = 0.6$ ($\sigma_\phi^2 = 0.6$ rad^2) at $\lambda_i = 2$ μm, the corresponding Strehl will be $\simeq 0.05$ at λ_t. As an estimate, we may thus take $w = (0.05)^{-1/2} = 4.5$, implying that the stellar image on the tilt sensor is about 1/6 of the width of an uncorrected image. Figure 12.13(a) plots for three wavelengths the optimal integration time Δt_{opt} versus stellar magnitude m, for this case. We next plot the BW plus noise Strehl ratios S'_θ (found by inserting Eq. (12.72) in Eq. (12.55)) as a function of m, in Fig. 12.14(a). The limiting magnitudes which yield $S'_\theta > 0.8$ is found to be $m_{on-axis} = 24$ ($\lambda = 2$ μm), 21 ($\lambda = 1$ μm), and 18 ($\lambda = 0.5$ μm).

To determine the total tilt Strehl ratio for this example, it remains to add TA

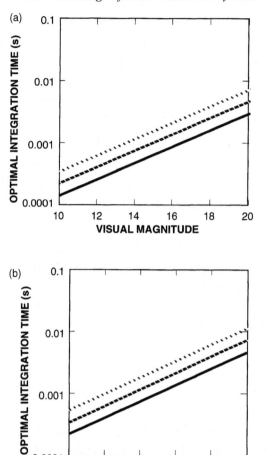

Fig. 12.13. Optimal integration times at imaging wavelengths of 0.5, 1, and 2 μm for tilt correction as a function of field star magnitude. (a): The case of a tilt field star within the isoplanatic patch at the tilt-sensing wavelength. (b): Field star outside the isoplanatic patch. Solid line, 0.5 μm; dashed line, 1.0 μm; dotted line, 2.0 μm.

and coma terms in Eq. (12.69). For the small field angle under consideration, we assume that TA is very small, and hence neglect its contribution. We take for the coma tilt error the uncorrected error given in Eq. (12.68) reduced by the ratio $w/(D/r_0)$. Results for the total tilt Strehl ratio are given in Fig. 12.15(a) as a function of stellar magnitude. It is seen that a total tilt Strehl $S_\theta > 0.5$ is obtained for limiting magnitudes $m_{\text{on-axis}} = 21$ ($\lambda = 2$ μm), 19 ($\lambda = 1$ μm), and 15 ($\lambda = 0.5$ μm).

2. The next example is for the same parameters above, except for off-axis field stars such that $\theta \gg \theta_0$ at $\lambda = \lambda_t$, so that $w = 30$. Figure 12.13(b) gives the

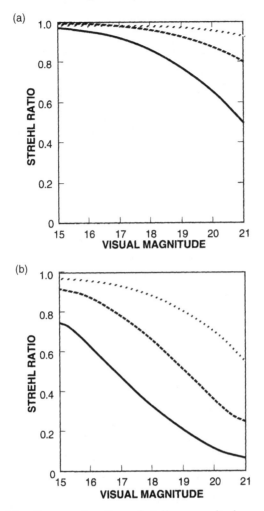

Fig. 12.14. Tilt Strehl ratio as a function of stellar magnitude, corresponding to the partial error consisting of the rms quadrature sum of tilt BW and noise errors, at imaging wavelengths of 0.5, 1, and 2 μm. (a): on-axis field star. (b): off-axis field star. solid line, 0.5 μm; dashed line, 1 μm; dotted line, 2 μm.

optimal integration time versus m, and Fig. 12.14(b) shows the corresponding Strehl S'_θ for BW plus noise errors. We see that optimal tilt operation giving $S'_\theta > 0.8$ is pushed toward smaller values of limiting magnitude, as expected. The total tilt error (now including the full coma tilt error given by Eq. (12.68)) is shown in Fig. 12.15 (b), for off-axis angular radius $\theta = 45''$. The TA Strehl ratios are taken from Fig. 12.11, for turbulence model MHV. The tilt Strehl ratios for a given m have decreased substantially compared to the values for on-axis field stars, and now the TA and coma-induced errors dominate at all

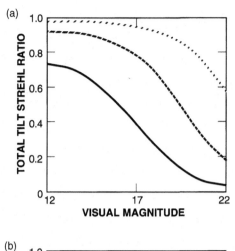

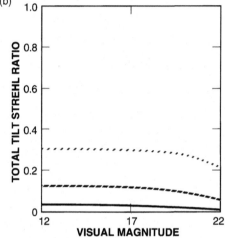

Fig. 12.15. Total tilt Strehl ratio for $D = 8$ m as a function of field star magnitude. (a): on-axis field star. (b): field angle of 45″. Solid line, 0.5 μm; dashed line, 1 μm; dotted line, 2.0 μm.

magnitudes. We see that for all values of m, the total tilt Strehl is $\leqslant 0.3$, thus falling short of our goal of $S_\theta > 0.5$ for diffraction-limited correction. Since the tilt Strehl consumes all of the system error budget, leaving no room for high-order wave-front errors, we must resort to implementation of advanced concepts to meet our design goals.

12.4.1.6 Methods for extending sky coverage

As illustrated by the above examples, the main factors limiting sky coverage for image motion sensing are the presence of coma-induced centroid error and the use of severely speckled images (of width $w \sim D/r_0$) for centroiding. The most obvious approach to overcoming coma-induced error is to increase the number of pixels in the tilt sensor to enable measurement of first order coma. For natural guide star systems, modification of the tilt loop to account for coma correction is efficacious, as discussed by Roddier (1992). However, for LGS AO this solution is self-defeating, since a brighter field star is required to measure coma with sufficient accuracy, so much of the potential gain is lost.

Several promising methods have been proposed for extending sky coverage. One is the use of 'dual AO', as proposed by Rigaut and Gendron (1992). Here a second LGS AO system operating from the same telescope is used to correct the field star, resulting in sharpened images on the tilt sensor, and also reducing coma-induced error since the coma aberrations are compensated. The potential rewards in terms of sky coverage are great. The limiting magnitudes for $S_\theta \geqslant 0.5$ at $\theta = 45''$ can be increased by dual AO to $m_{45} = 24$ ($\lambda = 2$ μm), $m_{45} = 22$ ($\lambda = 1$ μm), and $m_{45} = 18$ ($\lambda = 0.5$ μm).

As discussed in Chapter 2, the probability for finding at least one field star brighter than magnitude m within a field of radius θ is given by Poisson statistics

$$P(\theta, m) = 1 - \exp[-\pi\theta^2\Sigma(m, l)], \tag{12.73}$$

where $\Sigma(m, l)$ is the number of stars per arcsec^2 that are brighter than magnitude m at galactic latitude l. $P(\theta, m)$ can be interpreted as the fraction of the sky that is brighter than m within a field of radius θ. Data on $\Sigma(m, l)$ can be found in Allen (1973), and more recently in Bachall and Soniera (1981). At the Galactic equator, we find $P(45, 24) = 1$, $P(45, 22) = 1$, and $P(45, 18) = 1$; at the pole, the corresponding probabilities are 1, 1, and 0.1, respectively. Thus, using dual AO full sky coverage is obtained for infrared imaging, with very good coverage for imaging at optical wavlengths.

The basic drawbacks of dual AO are, of course, complexity and cost, and it is likely to be many years before a dual AO scheme is implemented on an astronomical telescope. The possible use of dithering a single LGS AO system between science object and field star has also been discussed, although this also appears to be complex and increases the bandwidth requirements of the AO system.

The simplest near-term solution has been analyzed by Sandler *et al.* (1994a). They proposed the use of infrared field star photons for tilt correction, using an infrared quad-cell tilt sensor. Because of the larger isoplanatic angle in the H–K bands, the image of the field star will share laser beacon correction with the science object, automatically resulting in sharp images on the infrared quad-cell, and significantly reducing the coma-induced error. Figure 12.16 shows

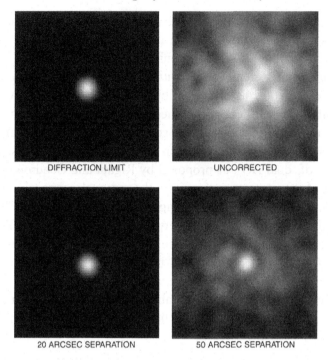

Fig. 12.16. Monte Carlo simulations of field star images at 1.5 μm wavelength. A sodium LGS has been used in the science direction, taken to be on axis. *Upper left:* diffraction-limited image; *Upper right:* seeing-limited image; *Lower left:* field radius of 20″; *Lower right:* field radius of 50″.

simulated images of infrared field stars at $\lambda = 1.5$ μm, corresponding to field radii of $\theta = 20''$ and $\theta = 50''$. A $D = 8$ m telescope equipped with an AO system with $d = 1$ m subapertures ($N_a \sim 50$) was treated via Monte Carlo simulations for M1 and M2 turbulence, using a sodium LGS in the target direction ($\theta = 0$). Also shown are the diffraction-limited image and the seeing-limited image, for comparison. From the figure, it can be seen that a well-defined image core exists even out to the largest radius, with $w = 1.2$ (3) for $\theta = 20$ (50)″.

The result is that a limiting magnitude $m_H = 17.5$ for imaging wavelength $\lambda_i = 2$ μm yields an optimum bandwidth plus noise tilt Strehl (for $n = 5$ electrons/pixel read noise) of $S'_\theta = 0.8$, with $m_H = 16.7$ for $\lambda_i = 1.6$ μm. Including the effects of TA, it was found that $S_\theta > 0.6$ for $\theta = 60''$ for K band imaging with M1 turbulence, and for $\theta = 40''$ radius for H band. These results illustrate that infrared tilt sensing using LSG AO goes a long way toward the theoretical dual AO limiting performance.

The probabilities for finding at least one field star are estimated by Sandler *et al.* (1994a) to be $P(60, 17.5) = 1$ and $P(40, 16.7) = 0.65$ at 30° Galactic elevation, and $P(60, 17.5) = 0.75$ and $P(40, 16.7) = 0.33$ at the pole. The upshot is that very close to full sky coverage for tilt stars can be obtained for LGS infrared imaging with an 8-m telescope, by using an infrared detector to sense image motion.

12.4.1.7 Advanced concepts for tilt sensing with LGS AO

In this section, we briefly summarize an important area of research which aims at eliminating the need for use of a natural tilt star for laser beacon AO. This is a new field of investigation, and it is too early to give a prognosis of the viability of these approaches. Nevertheless, if ways can be developed to use the laser beacon signal itself to sense tilt, it would be a major step forward for LGS AO.

As discussed in Chapter 11, the problem in using the LGS signal to measure overall tilt is caused by the fact that the laser beam suffers image motion on both the upward and downward paths. If the laser is projected through the full telescope aperture, its measured jitter using the full telescope will to first order be zero, because the two paths cancel. If the laser beacon is projected from the side of the telescope or from behind the secondary mirror obstruction, using optics of diameter $\ll D$, then the measured image motion using the full aperture will be closer to the desired atmospheric tilt, but still in error by the magnitude of the motion suffered during uplink. Belen'kii first analyzed this problem, and showed that the resulting error is substantial. In order to determine the overall tilt, additional measurements are needed which can be correlated with motion of the LGS from the full telescope, using either additional telescopes to view the motion or additional laser beams to provide more measurements.

Recently, both Belen'kii (1995) and Ragazzoni (1996a,b) proposed using crossed laser beams projected from telescopes separated by several kilometres from the main telescope. By viewing the configuration from the ground, an estimate can be made of the overall tilt in the direction of the science target. Ragazzoni *et al.* (1995) have analyzed other potential solutions, including the use of two or more outrigger telescopes straddling the observatory. These telescopes are used to measure the differential tilt between the elongated sodium beacon and surrounding natural stars, from which one can infer the one-way tilt experienced by the LGS beacon. An additional concept analyzed by Ragazzoni involves attempting to infer the motion of the LGS projected from behind the secondary mirror, by monitoring its motion over time through by imaging through the projection telescope, and estimating the tilt from the measured temporal drift.

These methods require experimental tests at facilities with LGS AO. Neyman (1996) has pointed out that even if tilt can be accurately extracted from differential motion of more than one laser beacon, the measurement is still not perfect because of the tilt component of FA. Because the laser beacon samples a cone on either upward or downward paths, the motion of the beacon does not correspond exactly to the desired tilt, because turbulence outside of the interior of the cone is not sampled. Neyman finds that the FA tilt error becomes large for imaging wavelengths $\lambda_i \geq 1$ μm. Since this is the domain where sky coverage for LGS AO becomes very low, there is some doubt as to the ultimate value of these methods.

Another potential solution to the tilt problem was proposed by Foy *et al.* (1995), and is undergoing field tests (Friedman *et al.* 1995a,b). The method uses two close laser wavelengths (569 nm and 589 nm) to optically pump the sodium atoms to a higher level, which results in a cascade of several lines. The various lines suffer different atmospheric dispersion, from which the atmospheric tilt can be extracted. In practice, the method uses a UV emission line at 330 nm and the sodium line at 590 nm.

The drawback of the technique is that it requires very high-power lasers, several hundred watts at least. This is because the relative dispersion is only a few percent. Image motion at the two wavelengths must be measured using filters, to within several milliarcseconds accuracy. Tests of the concept are currently being performed at LLNL, using a high-power laser developed for isotope separation. Initial results indicate that the necessary SNR may be achievable using a high-power laser, split between the two lines. Unfortunately, this method also suffers from the tilt FA analyzed by Neyman, and therefore may not be of much use for LGS AO at optical wavelengths.

12.5 Application of design analysis to LGS AO for the 6.5-m MMT conversion

12.5.1 The MMT AO project

In 1999, the Multiple Mirror Telescope on Mt Hopkins in Arizona will be converted to a 6.5-m telescope with a single primary mirror. Drawing on several years of experimentation with LGS AO at the current MMT array (see, for example, Lloyd-Hart *et al.* 1995), Steward Observatory at the University of Arizona is developing a sodium LGS AO system for the new MMT. The hardware design for this system is discussed in Chapter 13, which is devoted to a summary of existing and planned LGS AO systems. In this section, we briefly describe the parameters for the design of the system, and then describe the expected performance. This will serve to give a concrete example of the application of the concepts and formulas of this Chapter.

The MMT AO system is based on use of an adaptive secondary mirror, which directly feeds a compensated beam to an infrared imaging detector. The goal of LGS AO for the new MMT is to routinely produce diffraction-limited correction at $\lambda_i \geqslant 1.6$ μm, for high-resolution imaging and spectroscopy. Much of the important science will be performed in K band ($\lambda_i = 2.2$ μm) and in the 3−5 μm band (to take advantage of the low emissivity resulting from very few optical elements in the imaging arm resulting from use of an adaptive secondary). A dichroic beam splitter in front of the science instrument reflects light below 1 μm wavelength into the wave-front sensor, which will sense photons from a sodium laser beacon. Although a more powerful 10 W sodium laser is also planned, for the present discussion we assume a CW sodium laser of $P = 4$ W, which corresponds to a visual magnitude $m_v = 9.5$ laser beacon.

12.5.2 Tilt performance

As shown in Sandler *et al.* (1994a), and discussed in Section 12.4.1.6, even at infrared wavelengths in order to obtain full sky coverage for diffraction-limited correction of $D = 6.5−8$ m telescopes it is necessary to go beyond conventional tilt sensing using visible field star photons. Because the field star images are uncompensated at visible wavelengths, it is difficult to achieve the required centroid accuracy to give $S_\theta \geqslant 0.5$, when the effects of both tilt anisoplanatism and measurement/decorrelation errors are included. The simplest solution for infrared imaging is to measure image motion using infrared field star photons, which for field angles $\leqslant 1$ arcminute results in significant sharpening of the images (Fig. 12.16).

Figures 12.17 and 12.18 illustrate the performance of this scheme as applied to the 6.5-m MMT. The first figure shows the predicted tilt Strehl ratio for combined noise and time delay errors, corresponding to the error σ'_θ of Eq. (12.72). Results are given for the MK turbulence model, which has been shown to predict LGS AO parameters at the MMT site, and for three values of field star flux in a sensing band from 1.3 to 2 μm ($F = 400, 1000,$ and 2500 ph/m^2/s). One can see that even for the lowest flux considered, the optimal Strehl ratio remains above 0.8, as required. The optimum integration time is $\Delta t = 15$ ms for $F = 400$.

The total Strehl ratio is given by Eq. (12.69). We take the coma contribution to be negligible, since the low-order modes should be well compensated by sharing laser correction, although this strictly begins to break down for $\theta \geqslant 30''$. The TA Strehl ratios for the MK model are given in Sandler *et al.*

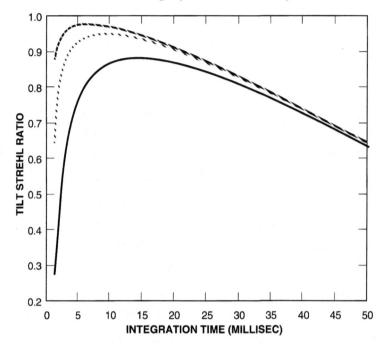

Fig. 12.17. Strehl ratio at three imaging wavelengths as a function of integration time for the sensing of infrared field star photons using the new 6.5-m MMT. Results are given for three values of field star flux: $F = 400$ (solid line), $F = 1000$ (dotted line), and $F = 2500$ (dashed line) $ph\,m^{-2}\,s^{-1}$. A constant sky background of 9000 $ph\,m^{-2}\,s^{-1}\,arcsec^{-2}$ is assumed. The infrared quad-cell size is $0.15''$ with a read-out noise of 5 electrons rms.

(1994a), and are somewhat higher at a given field angle than the results shown in Fig. 12.11. The results for the total tilt Strehl are shown in Fig. 12.18, which corresponds to an image width w for the infrared image of three times the diffraction-limited at a wavelength of 1.5 μm. We see that even for $\lambda_i = 1.6$ μm, the Strehl ratio goal of $S_\theta \geq 0.5$ is nearly met for the limiting flux value above.

Using arguments similar to those given in Section 12.4.1.6, we can estimate probabilities for finding field stars. An infrared flux $F = 400$ corresponds to H magnitude $m_H = 18.2$, and the number of stars/deg^2 is 6000/deg^2 at 30 degrees Galactic latitude and 2500/deg^2 at the pole. The outer radii for TA Strehl ratio $S_\theta^{TA} = 0.6$ for MK turbulence are $r_K = 44''$ and $r_H = 29''$. We then find that $P(44, 18.2) = 0.89$, and $P(29, 18.2) = 0.42$, for 30 degrees latitude. Thus, very close to full sky coverage obtains for diffraction-limited imaging in K band, with good probability approaching 50% in H band. At the pole, the numbers reduce to $P(44, 18.2) = 0.36$, and $P(29, 18.2) = 0.17$, still high enough to permit high-resolution imaging over much of the sky.

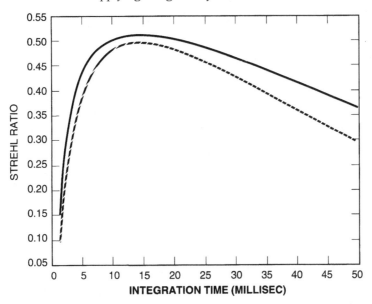

Fig. 12.18. Total tilt Strehl ratio in H and K bands as a function of integration time, for the 6.5-m MMT. Solid line, 2.2 μm; dotted line, 1.6 μm.

12.5.3 High order system performance

To meet the total Strehl goal of $S_{sys} \geqslant 0.3$, we must have S_ϕ given by Eq. (12.50) $\geqslant 0.6$.

Focus anisoplanatism. The magnitude of FA at Mt Hopkins has been measured to be (Lloyd-Hart *et al.* 1995) $d_0 = 3$ m at $\lambda = 0.5$ μm, which corresponds to $d_0 = 18$ m at $\lambda = 2.2$ μm. This gives $S_{FA} = 0.83$ (0.71) at $\lambda = 2.2(1.6)$ μm. In accordance with our general design philosophy, we wish to keep the remainder of the high-order errors of comparable magnitude to the FA. We note at $\lambda = 3$ μm the FA Strehl is 0.9, so for imaging and spectroscopy in a 3–5 μm band the FA becomes negligible.

Fitting error. For $N_a = 300$, the effective interactuator spacing is $r_s = 0.33$ m. For a mean value of $r_0 = 0.91$ m at 2.2 μm at the MMT site, this gives a fitting error Strehl ratio of $S_{fit} = 0.85$ (0.75) in K (H) band.

Temporal error. The temporal errors for the new MMT will be minimized by a fast reconstructor allowing cycle times for the AO to be limited only by CCD frame rate. For 1 ms read-out of the detector, the rms noise per pixel will be $n = 3$, which meets the read-out noise goal. For delay times $\Delta t \leqslant 2$ ms, we find $\sigma_{\text{time}} \geqslant 0.95$ even for both H and K bands. The upshot is that combined fitting plus temporal errors are kept small, with total Strehl ratio near that for FA alone.

Wave-front sensing and reconstruction. The remaining error is given by the reconstruction error, Eq. (12.48). The MMT system will use $d = 0.5$ subapertures, 13×13 across the pupil. Custom projector optics for the sodium LGS have been designed to produce laser beacons of width $w = 1''$ for $0.7''$ seeing. For a total system efficiency of $\eta = 0.4$, and a $P = 4$ W, we thus find optimum

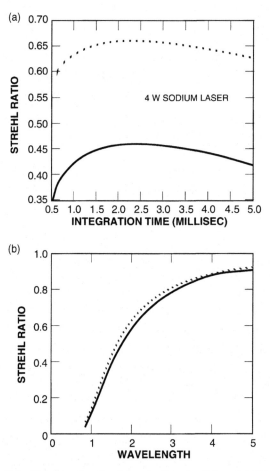

Fig. 12.19. (a): Predicted high-order Strehl ratio for the 6.5-m MMT LGS AO system, assuming a 4 W CW sodium laser. Results are shown for imaging in H and K bands (solid line: 1.6 μm, dotted line: 2.2 μm). (b): Strehl ratio versus imaging wavelength for 4 W and 10 W sodium lasers (solid and dotted lines respectively) and imaging at zenith. The advantage of a margin in laser power is more easily seen for off-axis imaging or situations in which the seeing becomes worse.

integration time of 2 ms for laser photons, and a resulting slope measurement error of $\sigma_{pd} = 0.13$ (0.18) rad at $\lambda = 2.2$ (1.6) μm. Note that these errors are very small, indicating the 4 W of laser power allows in principle very accurate measurements of the wave front. The resulting reconstruction Strehl ratio is given by 0.87 at $\lambda = 2.2$ μm. However, at $\lambda = 1.6$ μm, this Strehl contribution drops to 0.75, because of the static limit term in Eq. (12.48).

These results are summarized in Fig. 12.19(a) and 12.19(b), which plots the high-order Strehl ratio as a function of wave-front sensor integration time. It

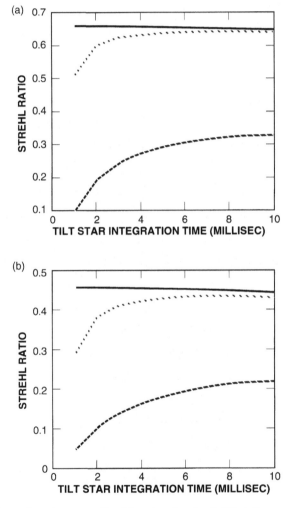

Fig. 12.20. Total predicted system Strehl ratios for the LGS AO system for the 6.5-m MMT conversion. Results are given for $\lambda = 2.2$ μm (a) and $\lambda = 1.6$ μm (b). Solid line, H = 15 tilt star, on axis; dotted line, H = 18 tilt star, on axis; dashed line, H = 18 tilt star, 44″ in (a) and 29″ in (b).

can be seen from the plot that through this analysis we have achieved an optimum design for diffraction-limited imaging in K band and longward. In Fig. 12.20(a) and 12.20(b), we plot the corresponding system Strehl ratios for three cases: bright on-axis tilt star; faint on-axis tilt star; and faint off-axis tilt star. We see that the total Strehl ratio drops to $S_{sys} \sim 0.23$ in H band. It can be shown that to increase the Strehl to 0.3 in H band, the number of actuators must be increased by 30% and the laser power by a factor of two.

The 300-actuator design, nonethless, is seen to have great power for astronomy at infrared wavelengths, the increased complexity required to increase the Strehl at shorter wavelengths is probably not worth the price, especially for a first-generation system. Figure 12.21 is a simulation of the imaging capability of the new 6.5-m MMT LGS AO system. Shown are three

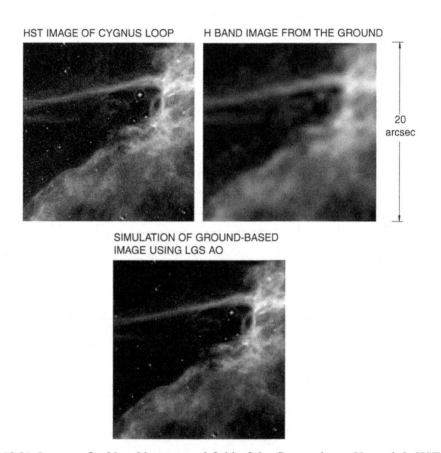

Fig. 12.21. Images of a 20×20 arcsecond field of the Cygnus loop. *Upper left*: HST image at 0.7 μm. *Upper right*: simulated image at 1.6 μm, taken from the ground without AO. *Bottom:* simulated image at 1.6 μm as obtained with LGS AO on the 6.5-m MMT.

images of a portion of the Cygnus loop. The first image was taken by the Hubble Space Telescope, showing a 20 × 20 arcsecond field at a wavelength of 0.7 μm. The second image is a simulation of an image taken from the ground at $\lambda = 1.6$ μm, without AO. The third image is a simulation of the expected image in average seeing, with the LGS AO loop closed. The natural star used for tilt sensing is in the upper right hand quadrant of the image. It can be seen from the figure that the resolution and contrast obtained from the ground are very comparable to the HST performance.

We note that while 4 W of laser power is very robust on paper, in reality it is extremely desirable to have more power. A factor of two margin will allow the LGS AO system to operate farther off zenith and for worse seeing. Common sense dictates that the more power the better. However, cost and reliability are also very important, and should be weighed in the balance. The goal for the LGS AO developers should be to produce enough power to make the system flexible and robust, with the constraint that astronomers using the system should have to pay no attention to the laser.

References

Allen, C. W. (1973) *Astrophysical Quantities* (3rd edn.), p. 243. The Athlone Press, London.

Bahcall, J. N. and Soneira, R. M. (1981) Predicted star counts in selected fields and photometric bands: applications to galactic structure, the disk luminosity function, and the detection of a massive halo. *Astrophys. J. Suppl. Ser.* **47**, 357–401.

Baharav, Y., Ribak, E. and Shamir, I. (1993) Multi-layer wavefront sensing using a projected finge pattern. In: *Adaptive Optics*, ESO/OSA Topical Meeting, ed. M. Cullum, pp. 335–61, Garching bei Munchen, Aug. 2–5, 1993.

Beland, R. R. (1992) A decade of balloon microthermal probe measurements of optical turbulence. In: *Adaptive Optics for Large Telescopes*, OSA Tech. Digest Series, pp. 14–16. Optical Society of America, Washington, DC.

Belen'kii, M. S. (1994) Fundamental limitation in adaptive optics: How to eliminate it? A full aperture tilt measurement technique with a laser guide star. In: *Adaptive Optics in Astronomy*, eds M. A. Ealey and F. Merkle Proc. SPIE 2201, pp. 321–3.

Bufton, J. L. (1973) Comparison of vertical profile turbulence structure with stellar observations. *Appl. Opt.* **12** (8), 1785–93.

Ellerbroek, B. L. (1984) Measuring the angular separation between two objects in the presence of atmospheric turbulence. tOSC Report No. TR-605, the Optical Sciences Company, Abuquerque, NM.

Ellerbroek, B. L. (1994) First-order performance evaluation of adaptive-optics systems for atmospheric turbulence compensation in extended-field-of-view astronomical telescopes. *J. Opt. Soc. Am.* A **11**, 783–805, 1994.

Foy, R., Migus, A., Biraben, F., Grynberg, G., McCullough, P. R. and Tallon, M.

(1995) The polychromatic artificial sodium star: A new concept for correcting the atmospheric tilt. *Astron. Astrophy. Suppl.* **111**, 569–78.

Fried, D. L. and Belsher, J. F. (1994) Analysis of fundamental limits to artificial-guide-star adaptive optics system performance for astronomical imaging. *J. Opt. Soc. Am.* A **11**, 277–87.

Friedman, H. W., Erbert, G. V., Kuklo, T. C., Malik, J. G., Salmon, J. T., Smauley, D. A., *et al.* (1995a) Sodium beacon laser systems for the Lick Observatory. In: *Adaptive Optical Systems and Applications*, eds. R. K. Tyson and R. Q. Fugate, Proc SPIE 2534, pp. 150–60.

Friedman, H. W., Erbert, G. V., Gavel, D. T., Kuklo, T. C., Malik, J. G., Salmon J. T., *et al.* (1995b) Sodium laser guide star results at the Lick Observatory. In: Proc. Topical Meeting, *Adaptive Optics* ESO Conference and Workshop Proceedings No. 54, ed. M. Cullum, Garching bei München, October 2–6.

Fugate, R. Q. (1996) Review of laser technology and status. In: *Adaptive Optics*, OSA Tech. Digest Series pp. 90–2. Optical Society of America, Washington DC.

Gardner, C. S., Welsh, B. M. and Thompson, L. A. (1990) Design and performance analysis of adaptive optical telescopes using laser guide stars, *Proc. IEEE* **78**, 1721–43.

Gavel, D. T., Morris, J. R. and Vernon, R. G. (1994) Systematic design and analysis of laser-guide-star adaptive optics systems for large telescopes, *J. Opt. Soc. Am.* A **11**, 914–24.

Ge, J., Angel, R., Jacobsen, B., Roberts, T., Martinez, T., Livingston, W., *et al.* (1997) Mesosphere sodium column density and the sodium laser guide star brightness. In: Proc. ESO Workshop, *Laser Technology and Laser Guide Star Adaptive Optics Astronomy*, ed. N. Hubin, Garching, June 23–26.

Happer, W., MacDonald, G. J., Max, C. E. and Dyson, F. J. (1994) Atmospheric-turbulence compensation by resonant optical backscattering from the sodium layer in the upper atmosphere, *J. Opt. Soc. Am.* A **11**, 263–76.

Hogan, G. P., Booth, H., Montandon-Varoda, L. and Webb, C. E. (1996) In: *Adaptive Optics*, 1996 OSA Tech. Digest Series, pp. 71–3. Optical Society of America, Washington DC.

Jacobsen, B. P., Martinez, T., Angel, R., Lloyd-Hart, M., Benda, S., Middleton, D., *et al.* (1994) Field Evaluation of two new continuous wave dye laser systems optimized for sodium beacon excitation. In: *Adaptive Optics in Astronomy*, eds. M. A. Ealey and F. Merkle, Proc. SPIE 2201, pp. 342–51.

Jelonek, M. P., Fugate, R. Q., Lange, W. J., Slavin, A. C., Ruane, R. E. and Cleis, R. A. (1994) Characterization of artificial guide stars generated in the mesospheric sodium layer with a sum-frequency laser, *J. Opt. Soc. Am.* A **11**, 806–12.

Jeys, T. H. (1991) Development of a mesospheric sodium laser beacon for atmospheric adaptive optics. *The Lincoln Laboratory Journal*, **4**, 133–50.

Kibblewhite, E. (1992) Laser beacons for astronomy. In: Proc. Workshop, *Laser Guide Star Adaptive Optics*, Phillips Laboratory, Albuquerque, N.M.

Laurent, S., Deron, R., Sechaud, M., Rousset, G. and Molodij, G. (1995) Laser and natural guide star measured turbulent wavefront correlation, In: *Adaptive Optics*, ed. M. Cullum, pp. 453–58, ESO/OSA Topical Meeting, Garching bei Munchen, October 2–6, 1996.

Lloyd-Hart, M., Angel, J. R. P., Jacobsen, B., Wittman, D., Dekany, R. and McCarthy, D. (1995) Adaptive optics experiments using sodium laser guide stars, *Astrophys. J.* **439**, 455–73.

Martinez, T. (1998) PhD. thesis, Univ. of Arizona.

Morris, J. R. (1994) Efficient excitation of a mesospheric sodium laser guide star by intermediate-duration pulses. *J. Opt. Soc. Am.* A **11**, 832–45.

Murphy, D. V., Primmerman, C. A., Page, D. A., Zollars, B. G. and Barclay, H. T. (1991) Experimental demonstration of atmospheric compensation using multiple synthetic beacons, *Opt. Lett.* **16**, 1797–9.

Murray, J., Powell, R., Austin, W., Roberts, T., Angel, R., Shelton, C., *et al.* (1997) Sodium guidestar lasers based on solid-state Raman lasers. In: *Laser Technology and Laser Guide Star Adaptive Optics Astronomy*, ed. N. Hubin, ESO Workshop, Garching, June 23–26.

Neyman, C. R. (1996) Focus anisoplanatism: a limit to the determination of tip-tilt with laser guide stars. *Opt. Lett.* **21**, 1806–8.

Olivier, S. S. and Gavel, D. T. (1994) Tip-tilt compensation for astronomical imaging. *J. Opt. Soc. Am.* A **11**, 368–78.

Papen, G. C., Gardner, C. S. and Yu, T. (1996) Characterization of the mesospheric sodium layer. In: *Adaptive Optics*, OSA Tech. Digest Series pp. 96–9. Optical Society of America, Washington DC.

Parenti, R. R. and Sasiela, R. J. (1994) Laser guide star systems for astronomical applications. *J. Opt. Soc. Am.* A **11**, 288–309.

Ragazzoni, R. (1996a) Absolute tip-tilt determination with laser beacons. *Astron. Astrophys.* **305**, L13–L16.

Ragazzoni, R. (1996b) Propagation delay of a laser beacon as a tool to retrieve absolute tilt measurements. *Astrophys. J. Lett.* **465**, L73–L75.

Ragazzoni, R., Esposito, S. and Marchetti, E. (1995) Auxiliary telescopes for the absolute tip-tilt determination of a laser guide star. *Mon. Not. R. Astron. Soc.* **276**, L76–L78.

Rhoadarmer, T. A. and Ellerbroek, B. L. (1995) A method for optimizing closed-loop adaptive optics wavefront reconstruction algorithms on the basis of experimentally measured performance data. In: *Adaptive Optical Systems and Applications*, eds. R. K. Tyson and R. Q. Fugate, Proc. SPIE 2534, pp. 213–25.

Rigaut, F. and Gendron, E. (1992). Laser guide star in adaptive optics: the tilt determination problem. *Astron. Astrophy.* **261**, 677–84.

Roddier, F., Cowie, L., Graves, J. E., Songaila, S., McKenna, D., Vernin, J., *et al.* (1990) Seeing at Mauna Kea: a joint UH-UN-NOAO-CFHT study, In: *Advanced Technology Optical Telescope IV*, ed. L.D. Barr, Proc. SPIE 1236, pp. 485–91.

Roddier, F. (1992) The University of Hawaii Adaptive Optics System. Canada-France-Hawaii Telescope Information Bulletin No. 27, pp. 7–8.

Sandler, D., Stahl, S., Angel, J. R. P., Lloyd-Hart, M. and McCarthy, D. (1994a) Adaptive optics for diffraction-limited infrared imaging with 8 m telescopes, *J. Opt. Soc. Am.* A **11**, 925–45.

Sandler, D., Cuellar, L., Lefebvre, M., Barrett, T., Arnold, R., Johnson, P., *et al.* (1994b) Shearing interferometry for laser-guide star atmospheric correction at large D/r_0. *J. Opt. Soc. Am.* A **11**, 858–73.

Sandler, D. G. and Massey, N. (1989) Final report for the WRC adaptive optics laboratory experiment. ThermoTrex Corporation, San Diego, CA.

Sasiela, R. J. (1994) Wave-front correction by one or more synthetic beacons. *J. Opt. Soc. Am.* A **11**, 379–93.

Smith, A. and Sandler, D. (1989) Single Pulse Excimer Ground Based Laser Concept Definition Study, Vol. II, D-Method Scaling Code. Report No. TTC-1588-R. ThermoTrex Corporation.

Tallon, M. and Foy, R. (1990) Adaptive telescope with laser probe: isoplanatism and cone effect. *Astron. Astrophy.* **235**, 549–57.

Tallon, M., Foy, R. and Vernin, J. (1992) Wide field adaptive optics using a array of laser guide stars. In: Proc. Workshop *Laser Guide Star Adaptive Optics*, pp. 555–66, Phillips Laboratory, Albuquerque, N.M.

Tyler, G. A. (1994a) Rapid evaluation of d_0: the effective diameter of a laser guide star adaptive optics system. *J. Opt. Soc. Am.* A **11**, 325–38.

Tyler, G. A. (1994b) Merging: a new method for tomography through random media. *J. Opt. Soc. Am.* A **11**, 409–24.

Tyler, G. A. (1994c) Bandwidth considerations for tracking through turbulence. *J. Opt. Soc. Am.* A **11**, 358–67.

Tyler, G. and Fried, D. L. (1982) Image position error associated with a quadrant detector. *J. Opt. Soc. Am.* **72**, 804–8.

Wild, W. J. (1996) Predictive optimal estimators for adaptive-optics systems. *Opt. Lett.* **21**, 1–3.

13

Laser beacon adaptive optics systems

DAVID G. SANDLER

ThermoTrex Corporation, San Diego, California, USA

13.1 Introduction

This chapter presents an overview of laser guide star (LGS) AO systems which are in operation now or are under development. We will see that the systems encompass a wide range of technologies and implementation approaches. Some are built from well-tested components. Others explore new territory in terms of concepts and hardware, and are aimed at optimizing LGS AO for astronomy. As discussed in Chapter 11, one system (at the SOR 1.5-m telescope) has been operational for several years. Although it was developed for defense applications, the SOR system has been made available for astronomers to gain experience using LGS AO.

LGS AO systems are listed in Table 13.1. Included are those which exist or are under development and planned to be in operation by the year 2000. The table gives the telescope and the organization developing AO, along with: the diameter of the telescope; the number of DM actuators over the telescope pupil; the type of LGS (Rayleigh or sodium) and laser power; and the date of AO first light.

The remainder of this chapter briefly summarizes the systems in Table 13.1, emphasizing special features of each. For more detailed discussions, the reader is referred to technical papers on the systems in the proceedings of the 1994 Kona SPIE conference on AO (SPIE 1994), the 1995 Garching ESO/OSA (Garching 1995) and SPIE (SPIE 1995) conferences on AO, the 1996 OSA (OSA 1996) conference, and the 1998 SPIE and ESO/OSA meetings.

We begin with a broad overview of LGS AO systems.

13.2 The state of LGS AO

As mentioned in Chapter 11, several experiments have demonstrated the efficacy of Rayleigh LGS correction, including those by MIT/Lincoln Labora-

Table 13.1. *LGS AO systems in operation or under development*

Telescope	Diameter	Actuators	LGS type and power	Laser loop closed
SOR	1.5	241	10 kHz Rayleigh (CuV), 200 W	1989
MMT	6×1.8	12	CW sodium, 3 W	1995
Lick	3	69	10 kHz sodium, 20 W	1996
ARC	3.5	97	840 Hz sodium, 8 W	1998 goal
Mt Wilson	2.5	241	300 Hz Rayleigh (excimer), 50 W	1998 goal
Calar Alto	3.5	97	CW sodium, 3 W	1998
Keck II	10	349	20 kHz sodium, 20 W	1998 goal
MMT conversion	6.5	300	CaWO4 sodium, 10 W	2000 goal

tory, ThermoTrex, and Phillips Laboratory (SOR). The SOR 1.5-m LGS AO system has been in operation since 1989, and has consistently yielded diffraction-limited correction ($S_{tot} > 0.3$) in I band ($\lambda = 0.88$ μm) for tracking with bright objects (Fig. 11.4). Importantly, the performance of the SOR 1.5-m system has shown: (1) that AO components exist capable of fast correction of more than 200 actuators across the pupil, and (2) that the analysis tools presented in Chapter 12 can be trusted to design and predict performance of LGS AO systems. (See Table III in Fugate *et al.* 1994.)

The state of performance for sodium correction is less mature. As of spring 1998 only three field experiments have achieved significant image sharpening with the AO loop closed around a sodium beacon: the Univ. of Arizona/MMT system, the LLNL/Shane telescope system at Lick, and the MPIE/ALFA system at Calar Alto. In all three cases, the lapsed time from start of project to closing the loop was less than 2 years, short compared to the many years of study and experimentation devoted to NGS AO. Thus, from the LGS AO workshop sponsored by Phillips Laboratory in Albuquerque in 1992, when the first sodium systems were seriously contemplated for astronomy, we may have expected a period of testing and learning. Since most projects listed in Table 13.1 have as their goal diffraction-limited correction by 1998, the next 2 years will be an important period, and by then we should have a wealth of data to assess the performance of LGS AO for astronomy.

The state of component development for LGS AO is more advanced, soon reaching the requirements laid out in Chapter 12 for optimal performance. Continuous facesheet deformable mirrors with 100 actuators are routinely purchased, and DMs with 349 actuators have become standard (Chapter 4). Phillips Laboratory has purchased two DMs with 30 actuators across, and one is now in use at the SOR 3.5-m telescope. Thus, conventional pupil-plane DMs for LGS AO are no longer a risk factor. Later in this chapter we will see that a new technology, allowing correction at the secondary mirror of large telescopes, will be ready for field testing early in 1999. This promises to be a breakthrough for infrared imaging and spectroscopy, since much simpler optical systems are made possible, with lower emissivity.

As discussed in Chapters 11 and 12, fast low-noise CCDs and wave-front computers are required for diffraction-limited correction of large telescopes, and the state of the art in both areas is now reaching the requirement for optimized performance. In order for 4 W of sodium power to suffice for diffraction-limited correction in H and K bands, $n \leqslant 3$ electrons/pixel rms readout noise is required at $\sim$ 1 MHz pixel rates, and new detectors by EEV Ltd, and other firms reaching this limit have been tested and are being integrated into LGS AO. Special processors for reconstructors have been developed by several groups, including 'brute force' combination of individual processing units, and use of a fast commercial parallel processor (Stahl *et al.* 1995) originally developed for highly parallel neural network processing. These designs have all reached the $\sim 10^9$ operations/s required to process several hundred slope measurements in $\ll$ 1 ms.

The hardware requirements for image motion sensing with faint natural stars are also just being reached, with the use of the new generation of APDs with high quantum efficiency. In addition, the use of faint infrared field stars, partially sharpened by sodium LGS correction, is under design by Steward Observatory, and will be implemented on the 6.5-m MMT conversion.

The sodium beacon laser is the key remaining component, and the astronomical community is far from reaching a consensus on the best approach. Low-power continuous wave (CW) dye lasers have been demonstrated at a few watts (Kibblewhite *et al.* 1994; Jacobsen *et al.* 1994), exist commercially, and have been chosen by several groups for astronomical AO. Sodium lasers which produce more than 5 W exist, and have been tested (Kibblewhite *et al.* 1994; Avicola *et al.* 1994, Olivier *et al.* 1995). However, there is still much controversy as to the best laser for astronomical AO, and about the higher power of 50–100 W required for correction at optical wavelengths. For an excellent overview of the types of sodium beacon laser, and technical issues, and trade-offs, see Fugate (1996).

13.3 Current LGS AO systems

13.3.1 SOR 1.5-m telescope

The LGS AO system for the SOR 1.5-m telescope is described in detail in Fugate *et al.* (1994). The second generation system now in place uses a continuous facesheet deformable mirror, with 241 actuators used to correct the 1.5-m aperture. The subaperture size is $d = 9.2$ cm, which is about one coherence length at $\lambda_i = 0.88$ μm (the average wavelength for scoring) for $2''$ average seeing. There is a total of 208 Shack–Hartmann subapertures across the pupil. The system is typically operated at closed-loop bandwidth $f_c \simeq$ 100 Hz.

The Rayleigh laser beacon is a copper vapor laser, with 200 W of average power. The beacon is projected off the telescope primary mirror, to limit divergence caused by poor laser beam quality to less than the atmospheric spot diameter. The CuV laser operates at two wavelengths, 0.5106 and 0.5872 μm, with equal power produced at both wavelengths. The laser produces a laser beacon at 10 km range, with a range gate of 2.4 m.

The wave-front sensor detector was orignally an intensified CCD, but was upgraded some time ago to a 64×64 CCD built by MIT/Lincoln Laboratory. Lincoln was the first organization to build a fast, low-noise CCD for AO. Current Lincoln chips operate with a readout noise $n < 10$ e-, at MHz pixel rates. Several of these CCDs have been obtained by astronomical groups for LGS AO.

The SOR system has limitations for astronomy, mainly because of its small aperture and internal beam train scattering from the high-power laser. Nonetheless, it has produced several astronomical images which show the power of LGS AO for astronomy (Fugate *et al.* 1994). Figure 13.1 shows an image of the Trapezium stars in Orion through an Hα filter, before and after closing the LGS loop. The CuV laser was pointed at the brightest star in this field ($m_v = 5.4$), which was used for tilt guiding. The image was made in poor seeing, but still provided sufficient resolution to allow detailed study and comparison with data on the radio continuum (McCullough *et al.* 1993).

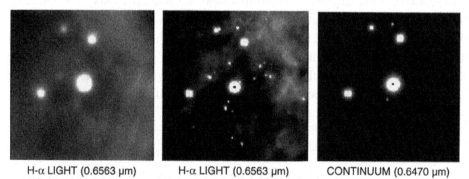

H-α LIGHT (0.6563 μm)   H-α LIGHT (0.6563 μm)   CONTINUUM (0.6470 μm)
NO COMPENSATION   IMAGES CORRECTED WITH LASER BEACON ADAPTIVE OPTICS

Fig. 13.1. Image of the Trapezium stars in Orion made through an Hα filter, using Rayleigh beacon AO at the SOR 1.5-m telescope. 4 min exposures, $40''$ field.

Typical PSF fwhm values for LGS AO at the SOR 1.5-m telescope in average seeing recover the 0.12 diffraction-limited core. In addition, optical transfer functions have been analyzed comparing NGS and LGS AO, that show a loss in power at intermediate spatial frequencies (0.2–0.4 D/λ) expected from a Rayleigh beacon, where the residual focus anisoplanatism (FA) error is quite high in I band ($\sigma_{FA}^2 \simeq 0.5$–1 rad^2) even for a 1.5-m telescope.

The SOR 3.5-m telescope has been equipped with NGS AO using a 756 actuator DM developed by XINETICS. Sodium LGS AO will be eventually implemented with plans to develop the required 50–100 W sodium laser.

13.3.2 Lick Observatory 3-m Shane telescope

The AO development for the Lick Observatory program is being performed by LLNL, and is part of a broader collaboration with the University of California and CALTECH culminating in LGS AO for the 10-m Keck II telescope on Mauna Kea (Wizinowich and Gleckler 1995). Led by Claire Max, LLNL entered the LGS AO field through its experience in very high-power lasers for isotope separation. In 1992, LLNL demonstrated the first very bright sodium beacon, created by the AVLIS pulsed dye laser (Avicola *et al.* 1994), pumped by CuV lasers. In this demonstration using a makeshift beam director, the laser produced over 1 kW of power. While the AVLIS laser is not suitable for astronomical AO, it did demonstrate, in principle, the possibilities of high-power sodium lasers and stimulated further work by LLNL in this field.

The system for the 3-m Shane telescope near San Jose is a testbed for sodium beacon AO. It is mounted at the $f/17$ Cassegrain focus, and uses a 127-actuator DM built by LLNL, with 61 actuators actively controlled to correct for atmospheric distortion. A tip/tilt unit built by PHYSIK INSTRU-MENTE corrects image motion, using data from a quad-cell APD sensor. There are 37 Shack–Hartmann subapertures ($d = 50$ cm) across the pupil to sense higher order aberrations. The WFS detector is a 64×64 Lincoln CCD, with readout noise $n = 13$ e-. The closed-loop bandwidth is $f_c = 30$ Hz.

The unique feature of the Lick system is its sodium laser (Friedman *et al.* 1995a,b), developed by LLNL. Flash-lamp-pumped frequency-doubled solid-state Nd:Yag lasers pump a dye laser, which produces light that can be tuned to the sodium resonance line. The pulse width is 100 ns, with a pulse repetition rate of 11 kHz. The pump lasers are in a room below the main telescope room and are fiber-optically coupled to the dye laser, mounted on the side of the 3-m telescope, along with a 30-cm launch telescope used to project the sodium beacon.

The laser has been routinely operated at 15 W of power, which has been

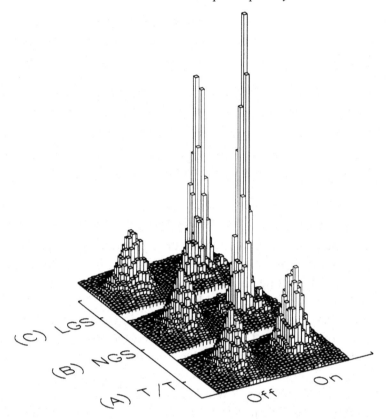

Fig. 13.2. PSFs before and after sodium LGS correction at the 3-m Shane telescope. These results, obtained by the LLNL AO group, are the first demonstration of closed-loop sodium beacon correction for a filled aperture telescope (Courtesy C. Max).

shown to correspond to $m_v = 9$. Spot sizes of 2″ fwhm have been measured. The large spot size is probably required for this laser to prevent saturation of the sodium layer.

Experimental results from the Lick system include the first closed-loop correction using a sodium LGS for a filled-aperture telescope, as shown in Fig. 13.2, which shows the measured PSF for a bright star before and after correction. The Strehl improvement for LGS correction is a factor of 2.4 over the Strehl for tilt correction alone, still far short of the expected performance, but nonetheless an important milestone for LGS AO for astronomy.

13.3.3 Apache Point 3.5-m telescope

The University of Chicago has developed a LGS AO system for the 3.5-m telescope on Apache Point in New Mexico. Chicago's program is led by Ed

Kibblewhite, who was the first astronomer to pursue AO technology transfer from the US DoD. Significant AO testing has also been carried out on the 1-m telescope at Yerke Observatory.

The Apache Point AO system currently uses a 97-actuator DM, developed in house, with plans to upgrade to a 201-actuator DM. A 16×16 Shack–Hartmann wave-front sensor is used to measure wave-front distortion. The WFS detector is a 64×64 Lincoln CCD, with $n = 13$ e- read-out noise.

The Apache Point AO system has a sodium laser of a type orginally developed by Lincoln for DoD applications (Jeys *et al.* 1991). It is often called a sum-frequency laser. Two Nd:YAG lines at 1.06 µm and 1.32 µm are summed in a LBO crystal, which produces sodium light at $\lambda = 0.589$ µm. SOR first took delivery of a flash-lamp pumped sum-frequency laser, and with it performed sodium return measurements (Jelonek *et al.* 1994). The laser developed for Chicago is laser-diode pumped, and has been tested at 8 W of power at the 3.5-m telescope.

The Chicago system has been tested with NGS operation, yielding a factor of ~ 10 improvement in I band. As of early 1998, closed-loop operation with a sodium beacon has not yet been accomplished.

13.3.4 Multiple mirror telescope

The first LGS AO program at a large astronomical telescope was initiated by the Steward Observatory AO group at the University of Arizona. The program to date has used the existing MMT array telescope, which consists of six 1.8-m primary mirrors covering a 6.9-m baseline, with an total area equivalent to a 4.5-m filled aperture. The major thrust of the Arizona program, however, is sodium LGS AO for the conversion of the MMT to 6.5-m primary mirror, to take place in 1999.

Roger Angel, who leads the Arizona program, was the first to realize the applicability of existing low-power CW dye lasers to LGS AO for large telescopes. Since relatively large subapertures, $d \geqslant 0.5$ m, suffice for diffraction-limited correction in H–K bands, it follows that a 3–4 W of CW sodium power can be used to measure the wave front to high accuracy. The Arizona group was also the first to propose projecting the sodium beacon from a $\sim 3r_0$ optical flat in back of the telescope primary mirror. This type of projection works to block the strongest sodium Rayleigh backscatter at low altitudes, with the higher altitude Rayleigh return scattered into a wide angle.

The MMT array has been used for a variety of sodium LGS experiments to test the above concepts. In Chapter 12 we described the measurements of FA and off-axis anisoplanatism at the MMT, using simultaneous images of natural and sodium beacons. These measurements were the first on a large telescope to confirm the major design assumptions of Chapter 12. Sodium beacon photometric measurements, first using Chicago's 1 W dye laser, and later comparing sodium standing wave and commercial dye lasers (Jacobsen *et al.* 1994; Lloyd-Hart *et al.* 1995), showed that the effective beacon brightness for CW lasers corresponds to a $m_v = 11$ star for a 1 W laser. Currently, a 3 W CW laser has been developed, which optimizes low-power operation for astronomical AO.

The 3 W sodium laser is used for closed-loop correction of the 12 tip/tilt modes for the six MMT mirrors. Note that control of relative piston modes between MMT mirrors is not possible using a sodium beacon, because the distances between the mirrors is larger than the coherence length for the extended sodium spot, thus ruling out direct interference measurements. A six-facet adaptive beam combiner built by ThermoTrex was used to correct image motion at 25 Hz closed-loop bandwidth, using slope measurements from a Shack–Hartmann sensor. A low-noise CCD is used for centroiding faint field stars, to correct combined image motion.

The MMT LGS adaptive system, called FASTTRAC II, was used for testing low-order sodium LGS AO, while the conversion of the MMT to a single 6.5-m mirror takes place. The adaptive beam combiner directly feeds an infrared imaging camera, with light below 1 μm sent to the WFS by a dichroic before the entrance window to the camera. In Chapter 11, we showed the first astronomical image (Fig. 11.5) obtained using a sodium beacon and a dim natural star for image motion correction. Since the diffraction limit of 0.3" is constrained by the size of the individual mirrors, the MMT telescope array could not reach the 50 milliarcsecond limit for full correction of the new large telescopes. However, the primary rationale for FASTTRAC II was to provide a development base and testbed for LSG AO concepts and technology for the the new 6.5-m MMT.

13.4 LGS AO systems under development

13.4.1 *Three-meter class telescopes*

13.4.1.1 *Mt Wilson 2.5-m telescope*

Laird Thompson at the University of Illinois is leading the UnISYS project to implement LGS AO on the Mt Wilson 2.5-m telescope. The system will have a 241 continuous facesheet DM, with a Shack–Hartmann wave-front sensor and MIT/Lincoln Laboratory CCD detector with $n = 12$ e- readout noise. A unique feature of the UnISYS system is that it will be the only AO system solely for astronomical use to use a Rayleigh beacon. There is evidence that the high-altitude seeing at Mt Wilson is often weak enough that the Rayleigh beacon should provide diffraction-limited correction in the near infrared, where a 256 × 256 NICMOS3 array will be used to cover a 20 × 20 arcsecond imaging field. In addition, this project includes future plans to experiment with multiple Rayleigh beacons, and may be the first testbed for investigation of the concepts in Section 12.3.2.

The Rayleigh LGS will be formed using an excimer laser operating at a wavelength of 351 nm (Thompson and Castle 1992). The XeF laser is commercially available, delivering 50 W of power at 300 Hz pulse repetition rate. With this laser, a Rayleigh LGS at 18 km should be about 10th magnitude.

Because the excimer laser has relatively poor raw beam quality, the Rayleigh beacon will be broadcast through the Coude path and project off the telescope's primary mirror. (Recall that the CuV Rayleigh beacon at SOR is also projected off the primary mirror to limit the beacon spot size at focus altitude.) Very accurate pulse timing and wave-front sensor range-gating are required to use this pulsed Rayleigh beacon.

The results of closed-loop operation of the UnISYS system should shed light on whether Rayleigh beacons have a future in astronomical AO. Because the value of d_0 can be five times smaller for Rayleigh correction (Section 12.3.1), the use of a single Rayleigh beacon is not currently favored by astronomers. In addition, the 300 Hz pulse repetition rate will limit the closed-loop bandwidth to $f_c = 30$ Hz, about a factor of three smaller than generally assumed necessary for diffraction-limited correction in J band and shortward. On the other hand, as discussed in Chapter 12, the use of multiple Rayleigh beacons may be very useful (perhaps even required, if practical high-power sodium lasers are not forthcoming) for correction of large telescopes at wavelengths below 1 μm.

13.4.1.2 Calar Alto 3.5-m telescope

The ALPHA LGS AO project for the 3.5-m telescope in Calar Alto, Spain is a joint project of the Max Planck Institutes for Astronomy and Extraterrestrial Physics. The project's goal is near-term use of sodium LGS AO for astronomical observations, and is conceived and designed to rapidly implement a high-performance system using the current state of LGS concepts and technology. The ALPHA instrument is mounted at the Cassegrain focus of the telescope.

The ALPHA system should produce diffraction-limited correction in H band and longward. It uses a 97-element continuous facesheet DM and a Shack–Hartmann wave-front sensor with an adjustable number of subapertures, from 32 to 120. The sodium laser beacon is produced using a commercial CW dye laser pumped by an Ar ion laser, the same type of device as tested by the Arizona group. The laser, located in the Coude room, will be expanded and focused by a telescope attached to the main body of the telescope. The laser produces 3–4 W of power routinely, corresponding to a 10th magnitude source. Tip/tilt is corrected by a separate steering mirror, using faint natural stars of 16–17th magnitude sensed in the optical spectrum.

It is worth noting that the MPIA and MPIE have taken a direct path to LGS AO. Without relying on detailed internal research and development, they realized the value of LGS AO, performed an assessment of existing technologies, and proceeded on a quick development path. The ALPHA program is thus the first illustration that LGS AO for astronomy has become mature enough that it has moved beyond a research field into a practical way to upgrade telescopes to produce diffraction-limited images and high-resolution spectroscopy. The ALPHA system produced modest closed-loop LGS AO results early in 1998, and current work concentrates on increasing the Strehl ratio performance.

13.4.2 *The new large telescopes*

In this section, we summarize the first LGS AO systems under development for the new generation of large telescopes. It is in this application that LGS AO will achieve its maximum reward and have the greatest impact on astronomy.

13.4.2.1 *Keck 10-m telescope*

The first of the new large telescopes are the twin Keck telescopes on Mauna Kea, which have already made a deep impact on ground-based astronomy. A program led by Peter Wizinowich to outfit the second Keck telescope, Keck II, with LGS AO has been in place since 1995, as a collaboration of the Keck Observatory and LLNL. The AO project for the 3-m Shane telescope (discussed in Section 13.3.2 above) has fed into this program as a field laboratory for LGS AO development and testing. Short informative papers on the optical bench design, user interface, wave-front controller, and system error budget for Keck AO are given in OSA (1996).

The Keck AO system will be mounted at the $f/15$ Nasmyth focus, and will feed both a 1024×1024 imaging detector and a near-infrared spectrograph. Because of its large telescope aperture, even for optimized imaging at 1.6–2.2 μm the Keck system requires the largest number of actuators currently under development for astronomical AO. A 349 continuous face-plate DM will be used with 241 Shack–Hartmann subapertures of size of $d = 56$ cm for wave-front sensing over the 10-m pupil. With the excellent seeing on Mauna Kea, this will allow diffraction-limited imaging down to H band routinely (32 milliarcsecond resolution), and often through J band. The wave-front sensor detector is a 64 × 64 CCD of the MIT/Lincoln Laboratory design, and should yield $n = 11$ electrons rms noise at 2 MHz frame rate. Tilt sensing with natural field stars should be possible at 19th magnitude, giving nearly full sky coverage for LGS AO in K band. Tilt control will be by a separate steering mirror, with centroids measured using 4 APD detectors.

The major new technology area for the Keck AO system is the 20 kHz pulsed sodium laser to produce a 20 W LGS, under development by LLNL. It will consist of a set of flash-lamp pumped frequency-doubled Nd:YAG lasers used to pump a three-stage dye laser. The pump lasers and dye master oscillator will be in a separate room on the dome floor, and the pump light will be fiber-optically coupled to the dye laser and 50-cm projection telescope mounted on the side of Keck II. The projection assembly will also contain elaborate diagnostics, capable of monitoring and correcting the laser beam quality.

The LLNL team has performed detailed design analysis, using the formalism presented in Chapter 12. They predict a high-order system Strehl of $S_{ho} = 0.6$ at $\lambda = 2.2$ μm, with roughly equal contributions of FA and fitting error dominating the error budget, in agreement with the general conclusions of Chapter 12 on optimal design of LGS AO. The total system Strehl at the limiting faintness for natural tilt stars should be $S \sim 0.3$ in K band, corresponding to diffraction-limited correction.

The Keck II AO system is expected to come on line in late 1998 for NGS correction, with LGS AO implemented soon after. The payoff for astronomy should be enormous. However, there are many new challenges and potential obstacles that the Keck project must overcome, as discussed in the Section 13.5.

13.4.2.2 6.5-m MMT conversion

The Multiple Mirror Telescope south of Tucson, Arizona will be replaced by a 6.5-m telescope in 1999, with a single borosilicate honeycomb primary mirror. The new 6.5-m MMT will be equipped with LGS AO, with the goal of routine diffraction-limited imaging and spectroscopy at wavelengths $\lambda \geqslant 1.6$ μm. As discussed in Chapter 11, Steward Observatory at the University of Arizona has had an active research program in LGS AO for several years. This work has included development of, and observing with the FASTTRAC II LGS AO system on the current MMT array. The AO group has tested several sodium lasers at the MMT, and has performed systematic experiments (e.g. see Lloyd-

Hart *et al.* 1995) to validate an optimized design for the new MMT, which has evolved from the earlier studies of Sandler *et al.* (1994). The expected performance of the system has already been discussed in Section 12.5.

The LGS AO system for the new MMT is shown schematically in Fig. 13.3. The design is distinguished by the use of a new scheme for astronomical AO, involving wave-front correction at the telescope's secondary mirror. A 300-actuator adaptive secondary mirror is being developed from first principles as part of the Arizona program, with close ties to Italian researchers at Arcetri Observatory (Salinari *et al.* 1993, Biasi *et al.* 1995). The design of the AO secondary consists of a thin 2-mm glass shell whose shape is controlled by 300 voice-coil actuators, about 2.5 cm apart. Capacitive position sensors located at each actuator measure the figure of the thin adaptive mirror relative to a stable aspheric aluminum reference surface, with a 100 μm gap between shell and reference surface. The actuators are inserted in holes drilled in the aluminum support structure. Force is applied to the thin adaptive mirror through the interaction of the current in the voice coils with small magnets

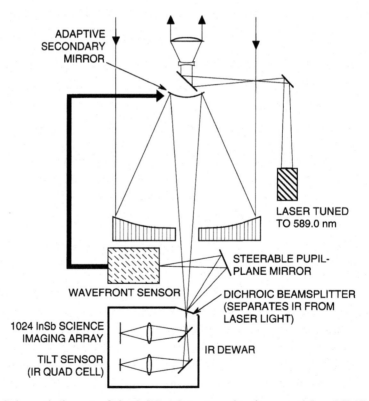

Fig. 13.3. Schematic layout of the LGS AO system for the new 6.5-m MMT conversion.

glued to the back of the thin shell; there is no direct contact between the shell and reference. The capacitive sensors measure the gap between the shell and reference surface at 10 kHz rate, effectively converting the force loop to very accurate position control.

A 5 × 5 actuator square prototype has been tested at ThermoTrex (Bruns *et al.* 1995), and with a 10 kHz fast figure sensor at SOR. The results show very good agreement with design predictions. A 53-cm spherical DM is shown in Fig. 13.4. It was configured as a prototype for a lightweight primary mirror for the Next Generation Space Telescope, the successor to Hubble to be launched in 2007. The mirror surface is a 2-mm glass shell, of the same type used for the adaptive secondary for the 6.5-m MMT conversion. The actuators are PZT, which are step and hold, whereas for the AO secondary voice coil actuators will provide rapid closed-loop updates of the mirror. For the AO secondary, the carbon fiber support will be replaced by an aluminum reference with holes to contain the actuators, and with cooling to remove the heat generated by electronics. After tests of this mirror in the spring and summer of 1998, the final 64-cm AO secondary will be built for installation at the new MMT.

The use of an adaptive secondary mirror greatly simplifies imaging at the $f/15$ Cassegrain focus. The corrected beam is relayed directly to the infrared science instrument, with a dichroic beam splitter passing light beyond 1 μm wavelength and reflecting optical light back into the wave-front sensing and acquisition cameras. Another advantage is very low emissivity, resulting in a design optimized for imaging and spectroscopy in the 3–5 μm band.

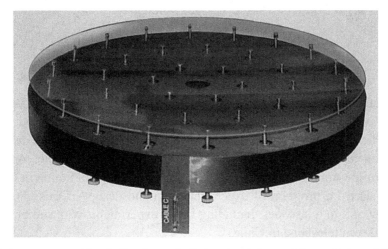

Fig. 13.4. Prototype thin shell mirror, built by the Arizona Mirror Lab and Thermo-Trex.

Two types of sodium laser will be available for use at the new MMT. The first is a CW dye laser of 3–4 W, based on a design similar to those tested at the MMT array (Jacobsen *et al.* 1994) but optimized for increased efficiency and wavelength stability. At 4 W of power, this laser will produce a magnitude 9.5 beacon, just sufficient for sensing with $d = 50$ cm Shack–Hartmann subapertures. The second is a new solid-state sodium laser under development by Lightwave for the University of Arizona, based on configurational tuned Nd:YAG laser intracavity Raman shifted by a $CaWO_4$ crystal, and then frequency doubled to 589 nm. The goal for the solid-state laser is 10 W of power for routine operation, with a pulse format which avoids saturation effects, potentially yielding a $m_v = 8.5$ sodium beacon.

Specialized 50-cm projection optics for the sodium LGS have been shown to give 1″ sodium images for 0.7″ seeing. In addition, the MMT AO system will use a new wave-front sensor CCD by EEV, which has been demonstrated at $n = 3$ electrons rms noise at 250 kHz pixel rate, allowing the AO system to be operated with 1 ms update, or $f = 100$ Hz closed loop bandwidth.

Another unique feature of the 6.5-m MMT LGS AO system is the use of infrared photons from field stars to sense overall tilt, as discussed in Section 12.4. Within a 0.5–1 arcminute radius from the center of an imaging field, infrared photons will share laser correction and be partially sharpened. Since the stellar images as recorded on the infrared quad-cell will have smaller width, the required number of photons for a given centroiding accuracy will be fewer, by up to a factor of 10. This implies that use of field stars of H magnitude $m_H \lesssim 17$–18 can be used, yielding nearly full sky coverage for diffraction-limited correction in K band. For centroiding in the visual band, one has to go as faint as $m_v = 20$ for full sky coverage, which gives a larger rms centroiding error.

13.4.2.3 *Plans for LGS AO on other large telescopes*

Several other large telescope projects have plans for LGS AO, and we should expect soon to find that this powerful new capability will be included from the outset, when plans are made for building large telescopes. Detailed design will begin soon for LGS AO at one of the new Gemini twin 8-m telescopes, with the goal of low emissivity infrared observing. The European Southern Observatory has started a program of experimentation and design of LGS AO for the VLT array of 8-m telescopes, including initial experiments with sodium LGS at the ESO 3.5-m telescope in Chile.

Planning has begun for LGS AO on the Large Binocular Telescope in Arizona, which will consist of two 8.4-m telescopes co-mounted on the same platform and separated by 12 m at their centers. The LGS AO systems will have adaptive secondaries, and capabilities for interferometric combination of two corrected beams. The Keck I telescope will also be equipped with AO, and both Keck I and Keck II AO systems will contain feeds from the Nasmyth platforms into common optics where the two corrected beams can be interfered.

13.5 Summary: key outstanding issues for LGS AO

In the next few years we should see numerous LGS AO systems in operation on 3-m telescopes, and the first systems yielding diffraction-limited infrared images on new large telescopes. There is an emerging concensus among astronomers involved in AO that LGS AO should eventually work as well as predicted. Numerous successful results from NGS AO systems, and many LGS experiments testing basic concepts and hardware, point toward general agreement with theory, and every year new hardware is tested which comes closer to meeting the requirements for optimized LGS AO systems.

However, the increased complexity of LGS AO leaves many new remaining challenges. It is worth reiterating that as of spring 1998, the predicted Strehl ratios for sodium laser correction have yet to be achieved, even for imaging a bright star. Although there is no evidence of unknown physical effects, routine demonstration of sodium LGS AO at the predicted performance level must remain the major milestone.

Although none of the challenges facing LGS AO seems insurmountable, each adds to long list of components and interfaces which must all come together at the telescope for successful operation. Thus, the real challenges will lie in integration and testing, and will require persistence and innovation to be met.

As an example, recent results (Papen *et al.* 1996) show that the sodium layer is variable in density at different times of the year, with the winter months often showing a factor of two higher column density than summer months. More bothersome is the fact that fluctuations in the mean height of the layer may occur on a time scale of less than an hour. Since the sodium wave-front sensor is self-referencing, it will not detect changes in overall system focus, so a separate sensor is required to monitor the system focus over relatively long time scales using photons from the tilt field star. In the Keck and MMT systems, the focus changes can be corrected by moving the wave-front sensor translation stage. In addition, many systems project the laser from the side of the telescope, so the sodium spot will appear slightly elongated when viewed from main telescope. It is not clear how this effect, or natural variations in beacon shape and size, will affect closed-loop operation.

Both the MMT and Keck systems have unique challenges. The Keck telescopes are composed of hexagonal 1.8-m segments, which will have residual piston and tilt errors due to imperfect phasing of the telescope. The continuous DM used in the Keck system will not be able to correct for these errors, and in addition errors between the segments may cause unwanted motion and distortion in the Shack–Hartmann images. Further, Keck has a rotating hexagonal pupil, and the registration between the actuators and wavefront measurements must be adjusted in real time, significantly complicating the reconstructor.

The MMT system is attempting many firsts, and the use of an adaptive secondary mirror has never been tried. The successful operation of the FASTTRAC II system, using an adaptive beam combiner, lends confidence to the overall strategy, but the secondary mirror itself will require flushing out at the telescope using bright starlight. In addition, the infrared tip/tilt sensor will be in the infrared dewar, and routine operation with faint field stars is required before the LGS AO system can meet its goals.

In summary, the next few years will be exciting for LGS AO. New results should start to come rapidly during the second half of 1997, and by the year 2000 we should see the promise of a new technology fulfilled for astronomy.

13.6 Acknowledgments

Chapters 11–13 were written with partial support from Air Force Office of Scientific Research grant no. F49620-94-100437 at the University of Arizona. The author is grateful for extended collaboration in LGS AO with many individuals at ThermoTrex, including Brett Spivey, Todd Barrett, Don Bruns, Steve Stahl, and Tim Brinkley. Steve Stahl helped greatly in generating artwork, and during many enjoyable in-depth discussions. My first introduction to AO was by Bob Stagat and Bill White at Mission Research Corporation. At the University of Arizona, I have had many stimulating conversations with Michael Lloyd-Hart and Nick Woolf, and I want to thank Roger Angel especially for stimulating my interest in astronomical applications of LGS AO nearly 10 years ago now. I have benefited enormously from working with him. I thank graduate students Jian Ge and Ty Martinez for reporting results on calibration and comparative returns for sodium laser beacons. I have had many very valuable informal discussions with David Fried, Peter Wizinowich, Chris Shelton, Bob Fugate, Brent Ellerbroek, Walter Wild, and with many other scientists and engineers too numerous to mention.

References

Avicola, K., Brase, J. M., Morris, J. R., Bissinger, H. D., Friedman, H. W., Gavel, D. T. *et al.* (1994) Sodium laser guide star system at Lawrence Livermore National Laboratory: system description and experimental results. In: *Adaptive Optics in Astronomy*, eds. M. Ealey and F. Merkle, Proc. SPIE 2201, pp. 326–41.

Biasi, R., Gallieni, D. and Mantegazza, P. (1995) Control law design for electromagnetic actuators at the secondary mirror. In: *Adaptive Optics*, ESO Conference and Workshop Proceedings No. 54, ed. M. Cullum, pp. 221–7, Garching bei München, October 2–6.

Bruns, D. G., Barrett, T. K., Sandler, D. G., Martin, H. M., Brusa, G., Angel, J. R. P. *et al.* (1995) Force-actuated adaptive secondary mirror prototype. In: *Adaptive Optics*, ESO Conference and Workshop Proceedings No. 54, ed. M. Cullum, pp. 251–6. Garching bei München, October 2–6.

Friedman, H. W., Erbert, G. V., Kuklo, T. C., Malik, J. G., Salmon, J. T., Smauley, D. A., *et al.* (1995a) Sodium beacon laser system for the Lick Observatory. In: *Adaptive Optical Systems and Applications*, eds. R. K. Tyson and R. Q. Fugate, Proc. SPIE 2534, pp. 150–60.

Friedman, H. W., Erbert, G. V., Gavel, D. T., Kuklo, T. C., Malik, J. G., Salmon, J. T., *et al.* (1995b) Sodium laser guide star results at the Lick Observatory. In: *Adaptive Optics*, ESO Conference and Workshop Proceedings No. 54, ed. M. Cullum, pp. 207–11, Garching bei München, October 2–6.

Fugate, R. Q., Ellerbroek, B. L., Higgins, C. H., Jelonek, M. P., Lange, W. J., Slavin, A. C. *et al.* (1994) Two generations of laser-guide-star adaptive optics experiments at the Starfire Range. *J. Opt. Soc. Am. A* **11**, 310–24.

Fugate, R. Q. (1996) Review of laser technology and status. In: *Adaptive Optics*, 13, OSA Tech. Digest Series, pp. 90–2. Optical Society of America, Washington, DC.

Garching (1995) *Adaptive Optics*, ESO Conference and Workshop Proceedings No. 54, ed. M. Cullum, Garching bei Munchen, October 2–6, 1995.

Jacobsen, B., Martinez, T., Angel, J. R. P., Lloyd-Hart, M., Benda, S., Middleton, D., *et al.* (1994) Field evaluation of two new continuous-wave dye laser systems optimized for sodium beacon excitation. In: *Adaptive Optics in Astronomy*, eds. M. Ealey and F. Merkle, Proc. SPIE 2201, pp. 342–51.

Jelonek, M. P., Fugate, R. Q., Lange, W. J., Slavin, A. C., Ruane, R. E. and Cleis, R. A. (1994) Characterization of artificial guide stars generated in the mesospheric sodium layer with a sum-frequency laser, *J. Opt. Soc. Am. A* **11**, 806–12.

Jeys, T. H., Brailove, A. A. and Mooradian, A. (1991) Sum frequency generation of sodium resonance radiation, *App. Opt.* **28**, 2588–91.

Kibblewhite, E. J., Vuilleumier, R., Carter, B., Wild, W. J. and Jeys, T. H. (1994) Implementation of cw and pulsed laser beacons for astronomical adaptive optics systems. In: *Adaptive Optics in Astronomy*, eds. M. Ealey and F. Merkle, Proc. SPIE 2201, pp. 272–83.

Lloyd-Hart, M., Angel, J. R. P., Jacobsen, B., Wittman, D., DeKany, R., McCarthy, D., *et al.* (1995) Adaptive optics experiments using sodium laser guide stars. *Astrophys. J.* **439**, 455–73.

McCullough, P. R., Fugate, R. Q., Ellerbroek, B. L., Higgins, C. H., Christou, J. C., Spinhirne, J. M., *et al.* (1993) Photoevaporating stellar envelopes observed with Rayleigh beacon adaptive optics. *Bull. Am. Astr. Soc.* **25**, 1341.

Olivier, S. S., An, K., Avicola, K., Bissinger, H. D., Brase, J. M., Friedman, H. W., *et al.* (1995) Initial results from the Lick Observatory laser guide star adaptive optics system. In: *Adaptive Optics*, ESO Conference and Workshop Proceeding No. 54, ed. M. Cullum, pp. 75–9, Garching bei München, October 2–6.

OSA (1996) *Adaptive Optics*, 1996 OSA Tech. Digest Series, Vol. 13, Optical Society of America, Washington, DC, July 8–12.

Papen, G. C., Gardner, C. S. and Yu, T. (1996) Characterization of the mesospheric sodium layer. In: *Adaptive Optics*, OSA Tech. Digest series, pp. 96–9. Optical Society of America, Washington DC, July 8–12.

Salinari, P., del Vecchio, C. and Biliotti, V. (1993) A Study of an adaptive secondary mirror. In: ICO-16 Conference on: *Active and Adaptive Optics*, ESO Conference and Workshop Proceedings No. 48, ed. F. Merkle, pp. 247–53, Garching bei München, August 2–5.

Sandler, D., Stahl, S., Angel, J. R. P., Lloyd-Hart, M., and McCarthy, D. (1994) Adaptive optics for diffraction-limited infrared imaging with 8 m telescopes. *J. Opt. Soc. Am. A* **11**, 925–45.

SPIE (1994) *Adaptive Optics in Astronomy*, eds. M. Ealey and F. Merkle, Proc. SPIE 2201, March 17–18.

SPIE (1995) *Adaptive Optics in Astronomy*, eds. R. K. Tyson and R. Q. Fugate. Proc. SPIE 2534, July 10–11.

Stahl, S. M., Barrett, T. K. and Sandler, D. G. (1995) Real-time reconstructor for adaptive secondary control of the 6.5 m single-mirror MMT. In: *Adaptive Optical Systems and Applications*, eds. R. K. Tyson and R. Q. Fugate, Proc. SPIE 2534, pp. 206–12.

Thompson, L. A. and Castle, R. A. (1992) Experimental demonstration of a Rayleigh-scattered laser guide star at 351 nm. *Opt. Lett.* **17**, 1485–7.

Wizinowich, P. L. and Gleckler, A. D. (1995) Keck Observatory adaptive optics program, *Proc. ESO/OSA Topical Meeting Adaptive Optics*, ed. M. Cullum, pp. 31–4, Garching bei München, October 2–6.

Part five

The impact of adaptive optics in astronomy

14

Observing with adaptive optics

PIERRE LÉNA

Université Paris VII & Observatoire de Paris, Meudon

OLIVIER LAI

Observatoire de Paris, Meudon

In this chapter, we consider how the astronomer may use an adaptive optics system, what kind of performance can be expected in a particular program, and how observations should be prepared. We also discuss what precautions must be taken during data acquisition, with special emphasis on how to keep proper track of the overall impulse response including the atmosphere. We give some advice, and discuss specific data reduction procedures.

14.1 Estimating performance

In addition to the art of imaging, well known to astronomers, the AO methodology adds the necessity of considering a constantly changing atmosphere and atmospheric seeing. Such a changeable state of affairs, departing from stationarity, is well illustrated by Fig. 14.1. Non-stationarity precludes a complete *a priori* knowledge of the actual performance a given system will reach at a given time: from the choice of pixel size or slit width to the selection of the operating loop frequency of the AO system or the adequacy of a given offset reference star, many observational parameters cannot be entirely pre-determined and will require real time decisions.

An *a priori* knowledge of the seeing at an astronomical site is therefore of importance for forecasting the atmospheric coherence time τ_0 and the coherence diameter r_0. At many modern observatories, programs are envisaged to deduce these values from meteorological observations, such as vertical thermal gradient and wind speed, local vorticity, etc.

As AO systems do not have the capability to entirely restore the wave front and give fully diffraction-limited images, an exact knowledge of the point-spread function (PSF) will be necessary. For optimal data reduction, one must account for the likelihood of the PSF changing with time, object, field-of-view, position in the sky, and observing conditions. There is a need for a

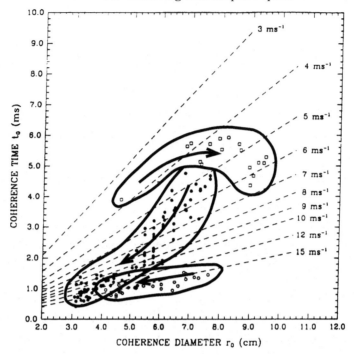

Fig. 14.1. The changing state of the atmospheric seeing. Each point in this graph corresponds to a measurement of the coherence diameter $r_0(\lambda = 0.5\ \mu m)$ (Fried's parameter) and the atmospheric coherence time $\tau_0(\lambda = 0.5\ \mu m)$ made at different instants in time, with the University of Sydney (Australia) optical interferometer SUSI, on its site in Narrabri. These are empirical measurements, without any particular assumption on the turbulence spectrum. In all cases the values plotted are the observed values – no corrections for zenith angle have been applied. *Filled circles*: the 112 data points were obtained at roughly 2.5 min intervals on the star α Eridani between hour angles 20:42–03:50 with some gaps in the records where the τ_0 values were not recorded. *Open circles* (clustered around 15 m/s): the 25 data points were obtained at roughly 18 min intervals on α Eridani between hour angles 21:50 and 05:14. The intervals varied a bit. *Open squares*: the 18 data points were obtained on ε and α Canis Majoris. The hour angle range covering both stars was 02:21–03:30 so the zenith angle range was limited to approx 30–45°. The observations were, on average, just under 4 min apart. Dotted lines are average windspeeds $\bar{v}$, related by the expression $\tau_0 = 0.314 r_0/\bar{v}$ (Eq. (2.25)). The arrows indicate the time evolution and show the well defined but varying state of the atmosphere in this astronomical site (Davis 1996).

strategy for obtaining the best possible PSF, and for estimating errors its use may lead to.

We give here the list of the parameters which affect the observations and their final result. We assume that only natural reference objects are available. The use of laser stars, still in its infancy, is discussed in Chapters 11–13.

- The astronomical source (A), namely the object to be imaged: satellite, comet, planet, star cluster, galaxy, etc.
  - wavelength of observation λ_{obs}: e.g. a photometric band such as V, R, I, J, H, K, L, M, N, ..., or a narrower spectral range in a spectroscopic observation,
  - zenith distance b,
  - magnitude at λ_{obs} (to determine integration time).
- The reference source (R), identical to or distinct from the astronomical source. The 'reference' is understood as the object used by the WFS during the observation of (A):
  - wavelength λ_{ref} at which the WFS operates,
  - magnitude m_{ref} at λ_{ref},
  - colour index at λ_{ref},
  - angular size: unresolved, or resolved by the telescope when larger than λ_{ref}/D, (D being the telescope diameter), or resolved by the subapertures of the WFS,
  - angular distance α from (A), and position angle if the isoplanatic field departs from circular symetry.
- The unresolved source (P), used for determination of the point-spread function (PSF):
  - angular distance from (A),
  - magnitude at λ_{obs},
  - colour at λ_{obs},
- Atmospheric properties:
  - Fried's parameter $r_0(\lambda, t)$,
  - characteristic (or coherence) time $\tau_0(\lambda, t)$,
  - isoplanatic angle $\theta_0(\lambda, t)$,
  - outer-scale of turbulence $L_0(t)$,
  - departure from Kolmogorov turbulence (if measurable).

It is convenient to refer to a standard Kolmogorov turbulence. In some circumstances, the atmosphere may depart from it, but this assumption, with its consequences and predictions, is generally made for the sake of simplicity. Finer analysis of the data collected by the WFS (e.g. sets of wavefronts, the variance of successive spatial modes, etc.) may help to understand the true nature of the atmospheric perturbations at a given time.

Given these specific parameters, the performance of a particular AO system may be predicted. Several metrics can be used: the Strehl ratio S of the AO corrected image, its full width at half-maximum (fwhm), the radii for various amounts of encircled energy. Ultimately, only the complete PSF or its Fourier transform, the optical transfer function (OTF), will give information which may be crucial for data reduction, such as the shape or extension of the residual halo. Figure 14.2 is a typical graph which helps assess the performance an astronomer can hope for. Based on theoretical estimates and/or validated by a

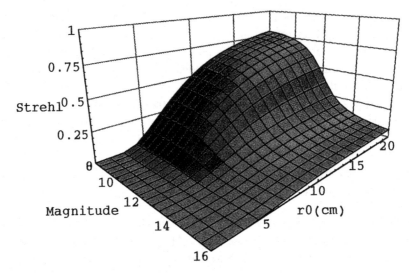

Fig. 14.2. Performance of an AO system. Characteristic of a given system, this family of curves gives the Strehl ratio S as a function of m_{ref} and $r_0(\lambda_{\text{ref}})$, for given values of λ_{obs} and $\tau_0(\lambda_{\text{ref}})$. There is such a set of curves per value of τ_0 and λ_{obs}. For this particular set of curves, $\lambda_{\text{ref}} = 0.55\ \mu\text{m}$ (V band), $\lambda_{\text{obs}} = 2.2\ \mu\text{m}$ (K band), $\tau_0 = 120$ ms. The curves given here are typical of the COME-ON/ADONIS system, allowing for the low (≈ 0.08) quantum efficiency of its EBCCD wave-front sensor (Gendron 1995). The r_0 dependance is derived from observations made with the PUEO system. The performance of PUEO is better (translation of the magnitude scale, not represented) due to the use of avalanche photodiodes in the WFS.

good knowledge of the system, it gives the Strehl ratio on the corrected image of a point source (A), observed at $\lambda = \lambda_{\text{obs}}$, versus the magnitude m_{ref} of the reference (R) located at a distance α from (A) and observed with a WFS working at λ_{ref}, the colour of (R) being assumed to be identical or close to the one of (A). Indeed there is a whole family of such curves $S = f(m_{\text{ref}}, \lambda_{\text{obs}}, r_0, \tau_0)$, for a given system but variable values of τ_0. One sometimes finds it convenient to label such a Strehl curve with values of the number $N(m_V)$ of equivalent fully corrected spatial modes of the wave front, understood in the following sense: for a given m_V, the resultant S is the one which would result from a system with infinite band pass, perfectly correcting $N(m_V)$ modes. It is obvious that for the faintest objects, $N = 2$ (tip/tilt correction only). Nevertheless, this may be misleading, as $N(m_V)$ will depend on the actual turbulence spectrum (Kolmogorov or not) and will not reflect the details of the modal optimization carried by the system and described in Chapter 8.

The astrophysical problem dictates the source, or the family of sources such as stars in a cluster, and the wavelength λ_{obs}. The performance of the camera, spectrograph, detectors, and the expected background, if any, set the required exposure time(s). The possible non-stationarity of the atmosphere should be taken into account in such calculation. The astronomer has to evaluate the minimal Strehl value he considers necessary, and decide if a proper reference is available (object, nearby star) to reach this performance for some range of expected atmospheric conditions. The choice of the reference is so crucial that we discuss it separately. Let us point out that the colour of the reference matters, as the WFS accepts a broad spectral band.

How can one define the limiting magnitude of a given system? This is of considerable interest, but quite difficult to do properly (Rigaut *et al.* 1991, 1992), as it is not a simple unique value for a given system. One could define it as the magnitude $m(\lambda_{ref})$ of the reference which gives a corrected image of a given Strehl ratio, at a given wavelength λ_{obs}, for given values of $r_0(\lambda_{ref})$ and $\tau_0(\lambda_{ref})$. Unfortunately this method has the drawback that it is seeing-dependent and therefore not directly related to the general performance of the instrument. Another 'rule of thumb', often used, is to define the limiting magnitude as the one for which the loop still closes, at the limit of improvement of the image quality by AO (tip/tilt correction only).

Performance graphs show how quickly the Strehl ratio degrades when, for instance, m_{ref} becomes too faint, everything else remaining the same: an excursion of less than two magnitudes at the transition will cover 80% of the total Strehl ratio excursion. AO correction becomes very sensitive at this point, before vanishing completely.

There are two ways to make sure that observing time is not wasted because of poor seeing. Back-up programs are necessary, since a brighter reference source or less stringent requirements on Strehl ratio can accomodate for smaller r_0. One may consider that the most efficient use of large telescopes such as the VLT or the Keck would result from queue observing, possibly helped by an available seeing forecast. Such a forecast could include anisoplanatism deduced from the height and velocity of the turbulent layers. Programs would be ordered in such a way that the ones having the highest need for excellent Strehl ratio or having the faintest references are selected when the seeing is best. The astronomer would not participate in the observing runs, and receive his data in much the same way as with the Hubble Space Telescope.

14.2 The observational procedures

14.2.1 *The choice of the reference*

This choice is crucial for the quality of the observations. It can be distinct from the choice of the star which will provide the PSF (see next subsection).

Whenever possible, the source itself will be used as a reference, but when it is too faint, or too extended, a better final Strehl may be obtained with some offset reference. The trade-off depends on the isoplanatic field $\theta_0(\lambda, t)$ at the time of observation. An example of the variation of S in the imaging field is given in Chapter 15, Fig. 15.12.

In peculiar cases, the exact time of observation may determine the choice of the reference. This happens when natural satellites are used to image the surface of a planet, its rings or other less bright satellites. This may also happen

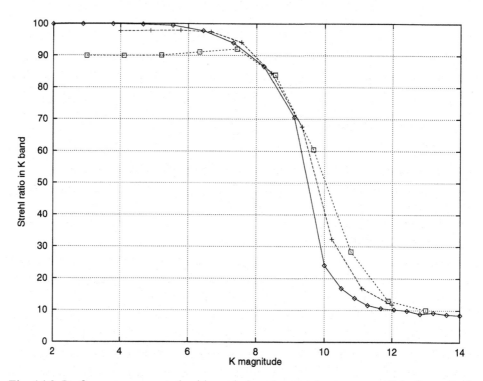

Fig. 14.3. Performance expected with an infrared wave-front sensor. This theoretical graph presents the Strehl ratio which can be reached by an AO system imaging at $\lambda_{obs} = 2.2\ \mu m$ (K band) while using a wave-front sensor also working in the infrared, with broad band coverage $1.6 \leqslant \lambda_{ref} \leqslant 2.3\ \mu m$ but observing a source of magnitude m_K assumed to emit a negligible flux below $\lambda = 2.0\ \mu m$, with a $1''$ seeing ($r_0(\lambda_{obs}) = 61$ cm). The three curves respectively refer to the following number of pixels per sub-pupil of a Shack–Hartmann sensor: ($\Diamond$) 6×6; ($+$) 4×4; ($\Box$) 3×3. The assumptions are the following: the SH sensor has 7×7 sub-pupils, its detector has 40 electrons rms read-out noise (a lower noise simply scales the curve by the noise improvement) and a quantum efficiency of 0.6. It is sampled at 100 Hz (for bright K magnitudes) or 20 Hz (for the fainter ones), and the coherence time τ_0 is assumed to be large. Overall system transmission is 0.6 and the simulation is made for a 3.6-m telescope, to be scaled for the European Very Large Telescope (Gendron, private communication).

when asteroids are used for reference. There exists a large number of asteroids brighter than magnitude $m_V = 14$, hence suitable for referencing: 10^4 to 10^5 of them scan the sky at low ecliptic latitudes. They provide movable reference objects which can be used at specified times to observe faint galactic or extragalactic sources (Ribak and Rigaut 1994).

Infrared wave-front sensing, as demonstrated by Rigaut *et al.* (1992), may find numerous applications in highly obscured regions (e.g. molecular clouds, the Galactic center) where no suitable visible reference may be present, but an adequate infrared one may be. Figure 14.3 gives the example of expected performance of an infrared wave-front analyzer.

14.2.2 The point-spread function

A proper understanding of as well as the reduction and/or deconvolution of data require knowledge of the overall impulse response $P(t)$ (point-spread function or PSF). The PSF includes contributions from the residual effects of the atmosphere and the system (AO, spectrograph if any, detector, electronics, etc.) and is unfortunately more or less time dependent because of the varying seeing. Great care must be taken to obtain a proper estimate of $P(t)$. If the PSF is better than the actual response of the system during the observation of the source, this will amplify the error in the deconvolution process. If it is worse, the effect in deconvolution will also be adverse by producing holes and rings around sharp structures.

Although the source (A), or some part of it, can be used as a reference (R) if adequately bright, its image itself cannot be used for determining $P(t)$, as (A) is supposed to be resolved. If an unresolved source in the field near the object is available for reference (R) within the isoplanatic field, it provides the ideal PSF, as the correction is simultaneous and almost identical to the one applied to the object (A). The isoplanatism condition is usually met for fields smaller than $10''$, $\lambda_{obs} \geqslant 1$ μm and a reasonable seeing. It is nevertheless wise to check the assumptions made on anisoplanatism, or angular decorrelation, against available theoretical models (Chassat 1989) but ideally, one would need a real or quasi-real time assessment of the isoplanatic angle, a quantity which is still difficult to obtain with current AO systems. This may currently represent a limitation to the ultimate performance that AO can reach on offset objects. Fried (1995) has proposed an interesting method to extend the isoplanatic field θ_0 by observing several stars surrounding (A), improving the tip/tilt determination.

Conversely, the PSF may be obtained by a separate exposure on a star (P), chosen as close as possible (a degree or so) to (R), in order to be affected by a similar slice of atmosphere, and as similar in magnitude and color, in order to obtain similar correction. When not available, one may resort to neutral density filters to match the WFS signal-to-noise ratios on (R) and (P), but this modifies the ratio of the sky background to the reference flux, hence the performance of the WFS which is background sensitive.

As stationarity of the atmosphere is never guaranteed, a minimum amount of time must separate the observations of (A) and (P), and bracketing is recommended, with alternate sequences A → P → A → P, etc. It is however important to get a good signal-to-noise ratio on the PSF, in order to avoid the introduction of artifacts in the deconvolution, therefore the integration time on (P) should not be excessively reduced.

An alternate and remarkable method to get a PSF is to use the data provided by the WFS itself, as proposed by Véran *et al.* (1997) and illustrated on real astronomical AO data in Figs. 15.22 and 15.23 in the next chapter. Even when a resolved object used for (R) (e.g. a Galilean satellite of Jupiter, size $\simeq 1''$, or a compact multiple star as shown on Fig. 15.10), the reconstructed wave front contains all the sampled spatial frequencies and this is the basic reason why the AO system can use its information to provide an improved correction. When the AO system runs, the residual phase errors of the wave front are known at any instant, as sums of the WFS measurements and the commands sent to the control mirror. These are indeed responsible for the final PSF. Hence the system can provide an estimate of the PSF, $P_{WFS}(t)$, directly from recorded WFS data and mirror commands without resorting to a separate measurement made on (R) or (P). It is remarkable that such a result can be obtained even when the signal-to-noise ratio per exposure per sub-pupil is not larger than unity!

Figure 14.4 shows the accuracy of such a PSF restitution. This apparently ideal method will not account for aberrations due to any optics located down-stream of the beam splitter which feeds either the WFS or the camera. In order to characterize these aberrations, a PSF taken from an internal artificial star has to be recorded. This can be done at the beginning of the night, but one has to ensure that it remains stable throughout, no matter how inclined the instrument. This so-called 'static' PSF is then convolved with $P_{WFS}(t)$.

The quantity P_{WFS} does not yet perfectly represent the actual PSF affecting the image of the source (A): this image includes the effects of the processing of the WFS data by the control loop, the deformable mirror(s), etc. Effects such as aliasing are also present, caused by the finite sampling of the WFS and leaving high spatial frequencies in the image. $P_{WFS}(t)$ has to be modified accordingly. High spatial frequencies, not corrected by the deformable mirror, produce a wide halo, also to be added to the PSF. This is done by computing

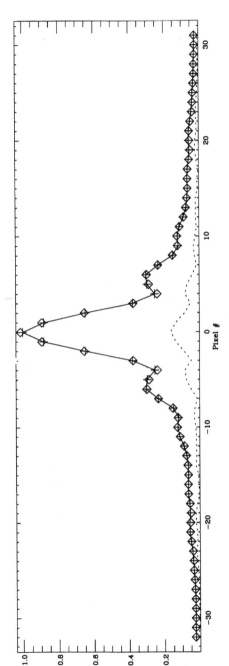

Fig. 14.4. The direct determination of the point-spread function. This simulation presents the result of estimating the PSF from the WFS instantaneous measurements in closed-loop and the commands sent to the AO mirror. The $(+)$ designate the initial unresolved object feeding the WFS. The $(\diamond)$ represent the reconstructed image, while the dotted curve gives the difference between the two, i.e. the residual error introduced by the method. Abcissae are pixel number, the pixel size is 0.034″. The curves are circular averages, the initial object being centro-symmetric. The initial Strehl value is $S = 0.622$, while the one on the reconstructed PSF is $S = 0.636$. The simulation conditions are as follows: star magnitude $m_V = 10.4$, $r_0(0.5 \ \mu\text{m}) = 18.7$ cm, AO system comparable to PUEO, as described in Chapter 9 (Véran *et al.* 1995, 1997).

the quantity $D/r_0(\lambda_{\mathrm{obs}})$ during the observations, D being the telescope diameter. Assuming at first order a Kolmogorov turbulence, one can compute an equivalent halo due to these uncorrected high spatial frequencies. Finally, the WFS noise can be limiting, in the case of objects so faint that the signal-to-noise ratio per sub-pupil becomes smaller than unity.

All these methods meet the requirement of obtaining a valid PSF, to be used in the data reduction and final interpretation. In any case, the strategy for PSF determination must be established before starting the observations.

14.2.3 At the telescope

Most AO systems are or will be user-friendly in that their presence is transparent for the user. Through expert systems, or artificial intelligence, they reduce the number of options or choices to be made in real time, and could almost be ignored once the proper parameters have been selected. They will automatically alternate sequences on object and reference, when needed, or perform other necessary tasks, in such a way that the overhead dead times for controls, system initialization, or reconfiguration be made very short, minutes or even seconds. It is nevertheless wise for the observer to understand as exactly as possible what happens. Not only will this improve the quality of the data and their reduction, especially when the system is pushed to its limits in sensitivity, dynamic range, or resolution, but it will also improve the organization and sequences of observations during the night.

The modal optimization procedures optimally configure the system for a given state of the atmosphere: they are described in Chapter 8 and are transparent to the user.

Keeping a sharp focus is a task simplified by AO: the small shifts of focus, encountered during the night, are usually taken care of by the AO correction and one focus at the beginning of the night may be sufficient, as long as mechanical stability is ensured and the WFS does not move along its z-axis with reference to the camera. One should account for chromatic effects, as the WFS takes a measurement at a wavelength λ_{ref} which usually differs from λ_{obs}.

The observing sequences must account for the actual state of the atmosphere and record it: an AO system can continuously provide values of r_0 and τ_0. The covariance matrices of WFS measurements and mirror(s) commands contain $r_0(t)$ and should be recorded. Computing in real time wave-fronts' cross-correlations provides $\tau_0(t)$. If in addition the average number of photons on the WFS is recorded, almost everything is available for later control and analysis. The direct PSF determination, discussed in the previous subsection, requires more data acquisition.

It is more difficult to obtain the isoplanatic angle θ_0, as it cannot be directly deduced and requires special measurements. Its knowledge may nevertheless be essential: suppose one observes Jupiter's surface or Saturn's rings while using, as reference (R), one of their satellites located at an angular distance α. A PSF can be deduced, despite the fact that (R) may be resolved, as the simple assumption of a uniform brightness disk may be made for (R). How good is this PSF to evaluate the resolution obtained on the surface or the rings? The answer depends on the value of $\theta_0(t)$ at the moment and in the direction of the observation. The field-of-view of AO corrected images is usually small, typically $10''$ or less, given the resolution and detector format. Hence $\theta_0(t)$ may not be determined by *a posteriori* examination of the PSF variations in the field. One could then imagine an optical scheme allowing one to superimpose

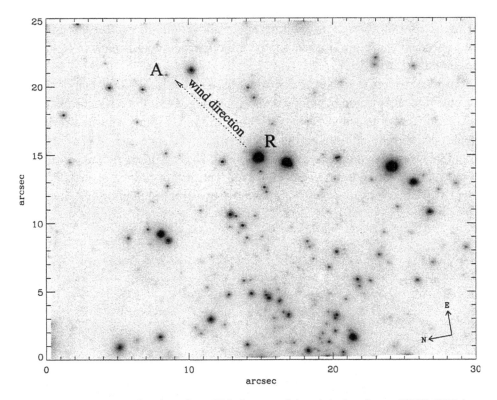

Fig. 14.5. A strange anisoplanatism. This image of the globular cluster NGC 6093 is a $25 \times 30''$ mosaic of individual $8 \times 8''$ images, made with the CFH 3.6-m telescope and the PUEO AO system during its first installation, at $\lambda_{obs} = 2.2$ μm. The reference is (R), the brightest star in the center of the cluster, but the highest Strehl ratio is found on the star (A), on the same individual image, $10''$ to the north-east of it, suggesting the wind direction and, with a loop delay of the order of 1 ms, a wind angular velocity of 0.05 rad/s (Lai 1996).

on the camera two fields, typically 0.5 to 1 arc-minute apart, both fields containing stars, as to provide for an estimate of $\theta_0(t)$ by alternate sequences of observation of the object (A) and of such a composite field.

One should add a word of caution on anisoplanatism, which may depart from the simple stationary model based on Kolmogorov turbulence and be site or telescope dependent. For instance, internal dome seeing, or mirror seeing, are known to exist with peculiar effects, such as convection or thermal bubbles over the primary mirror. Moving atmospheric layers may produce interesting effects: some observations show a better value of S off-axis than on-axis with the reference (R). Figure 14.5 is the image of a globular cluster where the Strehl ratio is higher off-axis than on the guide star used on-axis for (R): this can easily be explained if the phase corrugation is dominated by a single layer moving with high altitude wind. The time delay of the AO loop produces the best correction in the direction where the previously measured layer is located at the instant of correction: such an effect allows one to deduce the direction and speed of the high altitude winds, although one could think of a simpler anemometer!

Another demonstration of the 'frozen phase screen' formed by one or several turbulent layers pushed by winds is shown on Fig. 14.6: the inter-correlation of wave fronts, as measured by the WFS at different instants, shows clearly two peaks of correlation moving independently, each one being caused by a separate layer with a given altitude and wind. No one precisely knows yet how often such a configuration of frozen wave front may occur at a given site, but it clearly offers interesting opportunities for wave-front prediction, improvement

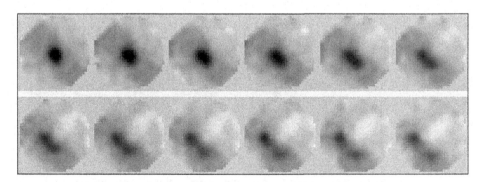

Fig. 14.6. A 'frozen' wave front. This sequence of images represents the temporal evolution of the intercorrelation function (*levels of gray*) $C_\phi(x, y, \tau)$, where $\phi(x, y, t)$ is the instantaneous wave-front phase, measured at $\lambda_{ref} = 0.55$ μm over the circular pupil of the 3.6-m ESO telescope. The views are separated by increments of 40 ms in τ, with $\tau = 0$ for the first one. Two correlation peaks are clearly distinguished, moving at different velocities in different directions (Gendron and Léna 1996).

of control loop performance and even use of an asynchronous reference (R), located up-stream of the object (A) with respect to the wind and providing an extended isoplanatic field (Gendron and Léna 1996).

It goes without saying that the standard procedures used in astronomical imaging such as flat fields, dark counts, background subtraction, photometric standards are needed also in AO observing. In addition, when images are made in the thermal infrared, longward of 2.5 µm, care must be exercized to control possible background modulations by the constantly changing deformable mirror(s) (Roddier and Eisenhardt 1986): this may introduce biases and spurious noise which must be carefully checked on an independent field.

14.3 Reducing the data

The images obtained with adaptive optics may be processed with the well-known techniques developed for high resolution imaging, speckle interfero-metry, shift-and-add, and deconvolution. These are now well mastered and dedicated software is available. The software usually uses pipelines, in which the data undergo clearly defined processes, such as dead pixel removal, de-biasing or flat fielding, sky or background emission removal. More advanced tasks, such as deconvolution or photometric calibration and analysis, require human intervention.

There are no rigid rules for data processing, as it depends mostly on experience: what is considered insignificant by some is crucial for others. For instance, dead pixel removal can be achieved by simple median (or sigma) filtering, but it can also be done in a more precise way using wavelet trans-forms. The same wavelet transforms may also be used in deconvolution to control or prevent an amplification of the noise in the image.

Once the basic operations have been done (sky subtraction, flat field, dead pixels removal, suppression of correlated noise, etc.), the data may be enhanced in terms of contrast and resolution by deconvolution. The result will depend of the exact goal assigned to the operation: it may be at the expense of accurate photometry. Conversely, the photometry may be improved, e.g. by spatial filtering, but then at the expense of contrast and resolution. As an example, it is easier to perform accurate photometric measurements on a globular cluster image, with a CLEAN procedure, than on very extended objects where faint structures are barely above the noise level. A discussion on the latter case can be found in Roddier, C. *et al.* 1996 (for an example see Fig. 15.16). There also exist deconvolution algorithms which preserve photometry on star-like objects. A good example of accurate photometric reduction can be found in the next chapter on Fig. 15.11, while the different task of finding with the utmost astrometric precision the location of an object, or of a set of stars in the image, is illustrated in Fig. 15.14 again in the next chapter.

In the future, larger fields will progressively become available with larger format detectors. With such fields, becoming larger than the isoplanatic field θ_0, the variations of the PSF within the field are no longer negligible, neither in width (or fwhm), nor in shape: the respective fractions of energy in the diffraction-limited core and in the halo may vary, creating additional difficulties for a precise photometry across the whole field.

14.4 The benefits of adaptive optics: an overview

Although the astronomical results presented in the next chapter speak for themselves, we summarize here some of the impacts of AO. Beating the seeing and restoring the resolution is not its only benefit: the reduction of the point-spread function étendue (or throughput) Ω from $(\alpha_{\text{seeing}}D)^2$ to λ_{obs}^2 improves the sensitivity by reducing the instrumental or atmospheric background reaching the detector, enhances the contrasts, helps to feed the light in spectrographs or optical fibers, and restores the coherence of the light in multi-telescope interferometry. Here is a list of properties affected by adaptive optics.

- *Angular resolution.* A perfect AO system would lead to perfect diffraction-limited images, if the aberrations after the beam separation between WFS and camera could be also corrected. For large telescopes (4–10 m) the resolution gain over the seeing (0.3″ at the very best, reaching 1 to 2″ in many sites) varies from 10 to 100, depending on the wavelength of observation λ_{obs}. This huge step reduces confusion in crowded fields (star clusters, central areas of galaxies) and allows high astrometric accuracy (of the order of a few milliarcseconds) on objects brighter than $m_V \approx 14$.
- *Sensitivity.* The sensitivity gain in AO depends on the spatial structure of the object, resolved or not in the image, and of the sources of noise. In Table 14.1 we present various cases of the evolution in signal-to-noise ratio per resolution element as a function of the resolution improvement. For a Strehl ratio S and considering the central core of the AO image, we define the resolution gain g as

$$g = \frac{\text{fwhm}_{\text{AO PSF}}}{\text{fwhm}_{\text{seeing PSF}}} = \frac{S\alpha_{\text{seeing}}D}{\lambda}. \tag{14.1}$$

The background noise is dominant for faint objects observed in the near infrared (OH airglow from 1 to 2 μm) and in the thermal infrared (longward of $\lambda = 2.5$ μm). With a one arcsecond seeing, it equals the signal for unresolved faint sources of typical magnitudes $m_J \geqslant 19$, $m_H \geqslant 17$, $m_K \geqslant 16.5$, $m_L \geqslant 16$, $m_M \geqslant 12.5$, assuming a 8.2-m telescope, negligible read-noise detectors and a low sky/instrument

Table 14.1. *Effect of AO on signal-to-noise ratio. Evolution of the signal-to-noise ratio, per resolution element, comparing AO imaging and seeing imaging. g is defined in the text. Loss appears when the gain in resolution is at the cost of a dilution of the resolved object over g^2 pixels*

Noise source	Signal[a]	Read-out[b]	Background[b]
Unresolved	even	even	gain g
Resolved, $\alpha \geqslant \alpha_{seeing}$	even	loss/g^2	loss/g

[a]Quantum detector in photon-counting mode, negligible background.
[b]The pixel size is matched to the resolution (seeing or diffraction).

emissivity (0.2). The background noise per pixel is reduced as the square root of the étendue. As values of g can reach 20 or more for large telescopes, AO leads to high sensitivity gains, reaching three or more magnitudes, for these unresolved faint sources.

- *Dynamic range*. A seeing limited PSF has large and time-fluctuating wings extending over several arcseconds. They drastically reduce the image dynamic range, and bright sources create halos hiding fainter sources. This limitation is severe when observing circumstellar environments, stellar clusters with repartition of stars over more than ten magnitudes, exo-planets, or companions. At a distance of $\alpha = 3\alpha_{seeing}$ from a bright source, the intensity drops at best only by $\exp(-3^{5/3})$, hence a dynamic range of 2000. In practice, the residual speckle noise reduces this value to a few hundred at best. On the other hand, the asymptotic form of the Airy function, normalized to unity at the origin $\alpha = 0$, is $(\pi\alpha D/\lambda)^{-2}$: at the same distance of 3″, for $\lambda = 1$ μm and $D = 10$ m, the intensity falls down to a level of 5×10^{-5}. Moreover, this may be improved by subtraction of the independently measured PSF: examples given in the next chapter show dynamic ranges of $10^4 - 10^5$ obtained with 3.6-m telescopes as close as 1″ from a bright object. In addition, a coronograph may be used to reduce saturation of pixels (Beuzit *et al.* 1997, Mouillet *et al.* 1997).

  The residual halo contains time-varying speckles, which still prevent detection below the levels just discussed. Labeyrie (1995) has proposed the use of their statistical distribution to extract faint companions of the central object, at levels as low as 10^{-9}, through the rare but real chance coincidence of dark speckles with such an object. A different approach to reach the same goal is proposed by Angel (1994) and illustrated in Fig. 14.7: a second stage of AO corrects the wave front to better than $\lambda/100$ in phase and has also to correct for the scintillation. This lowers the speckle noise to less than 10^{-7} per exposure, then averaging brings it below 10^{-9}.

- *Coherence*. The light in the image core (étendue λ_{obs}^2) of the corrected PSF, even with $S < 1$, is coherent and contains a fraction S of the PSF total energy. Its phase is that of the incoming wave front, or more precisely the average value of the phase over the entrance pupil of the telescope. This restored coherence has important

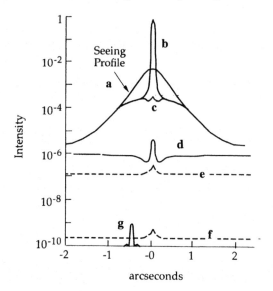

Fig. 14.7. Very high dynamic range with adaptive optics. This simulation shows the intensity profile around a bright, unresolved source (normalized to unity) for successive steps of correction, versus the angular distance to it. The seeing blurred profile (*curve a*) is reduced by a normal AO system (*curve b*), and the central light obscured by a coronograph, leaving the AO PSF halo unaffected (*curve c*). A next stage of high modes AO correction (optical path variance of the wave front reduced to less than $(\lambda/100)^2$ and scintillation corrected for) lowers the halo to a mean level of 10^{-6} (*curve d*), containing speckle noise in individual exposures of 500 μs at the indicated level (*curve e*). Averaging this noise over 5 h ideally lowers it to a level (*curve f*) where a companion, 10^{-9} fainter (*curve g*), could be detected at 5σ above the residual speckle noise (Angel 1994; graph provided by D. Sandler).

applications, discussed in the last section of this chapter. First, this spatial single mode signal propagating in free space may be efficiently coupled to a single mode waveguide (optical fiber), which can then transport the energy and phase. Second, the signal may be interfered with a similar one coming from another telescope. Demonstration of the efficiency of coherent beam transport and coupling in astronomical optical interferometry has been made, reaching accuracies better than 10^{-3} on the measurement of fringe visibility (Perrin *et al.* 1996; Coudé du Foresto *et al.* 1997).

- *Field-of-view.* The gain in resolution forces the use of smaller pixels, g^2 times more than for seeing limited observations. Current technology limits CCDs to a 2048 × 2048 format, juxtaposed to give at most $\approx 10^4 \times 10^4$ format. Visible AO correction of 10-m class telescopes requires 5 milliarcseconds pixels, hence a maximum field of 10″: this is very small, but of the order of the expected isoplanatic field in the visible. In the next chapter, examples are given of large mosaics made of independently corrected fields, each using a different reference (Fig. 15.13).

Table 14.2. *Sensitivity of the Very Large Telescope Interferometer with AO*

	V	J	K
2 × 1.8-m	9	12	10
2 × 8.2-m	13	16	14

The sensitivity is computed at the wavelengths of $\lambda = 0.55$ μm (V), $\lambda = 1.2$ μm (J), and $\lambda = 2.22$ μm (K), in two cases: making (*upper line*) two 1.8-m telescopes interfere or (*lower line*) two 8.2-m. Telescopes are assumed to be AO phased (with $r_0(0.5 \text{ μm}) = 20$ cm and $S \approx 0.4$ depending on wavelength), the object unresolved (visibility = 1); the spectral bandwidth is 20 nm, providing spectral resolution. Signal-to-noise ratio of 10 per spectral element is reached after 15 min integration, assuming overall throughput of 0.1 and an emissivity of 0.5 at 283 K in one coherence étendue. (von der Luehe *et al.* 1996).

- *Throughput and emissivity.* The current AO systems require a number of additional mirrors and beam splitters, which induce a loss in overall efficiency: 10% is considered as a good design goal for image and WFS channels, but improvements are possible. No cooled optics has yet been considered for the thermal infrared, but this may prove necessary to reduce the background, as in cooled spectrographs. Although 10-m class telescopes may not need AO for imaging at $\lambda_{\text{obs}} = 10$ μm ($r_0(10 \text{ μm}) \approx D$), special applications at this wavelength may require it, as the search for zodiacal thermal emission around solar type stars: then, the background modulation by the deformable mirror may introduce extra noise.

14.5 Multi-telescope optical interferometry with adaptive optics

Optical interferometry is not the subject of this book. Yet, as soon as the plan of the European Very Large Telescope was devised, the issue of coherently coupling large telescopes was raised. It was then demonstrated (Léna 1984, Roddier and Léna 1984) that, at a wavelength λ, a sensitivity gain would result only when the coherence of the light is preserved on each pupil, i.e. when $D \leqslant r_0(\lambda)$, a condition hardly met with large telescopes except possibly at $\lambda = 10$ μm. To satisfy this condition requires each of the individual pupils to be phased by AO. Once AO feasibility emerged, plans for such coupling could proceed, either for the VLT (von der Luehe *et al.* 1996) or for the Keck I and II telescopes. In most cases, only the coherent core (fraction S of the total energy) interferes efficiently. We show in Table 14.2 the expected

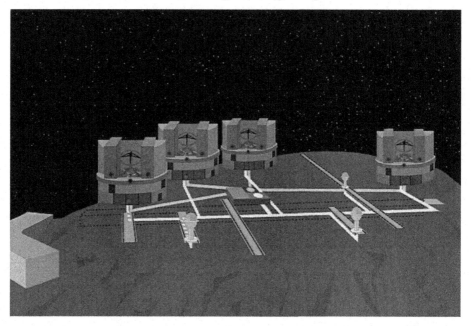

Fig. 14.8. The European Very Large Telescope Interferometer. This artist's view shows the four 8.2-m telescopes and three of the movable auxiliary ones, 1.8-m in diameter, located at three of the 30 possible fixed stations on the summit of Cerro Paranal, Chile. Beams are received from any of these telescopes in the central laboratory, where they are recombined and made to interfere after suitable optical delays are introduced. Adaptive optics will phase the pupil of each telescope. The first use of this VLT interferometric mode is expected in the year 2000. (Drawing courtesy of ESO.)

performance of the VLT interferometer, with telescopes equipped with AO (Fig. 14.8).

After providing the maximum spatial resolution that diffraction allows to large ground-based telescopes at infrared and progressively at visible wavelengths, adaptive optics will play a key role in optical interferometry to obtain the utmost sensitivity of ground-based optical interferometers.

References

Angel, R. (1994) Ground based imaging of extra-solar planets using adaptive optics. *Nature* **368**, 203–7.

Beuzit, J.-L., Mouillet, D., Lagrange, A.-M. and Paufique, J. (1997) A stellar coronograph for the COME-ON-PLUS adaptive optics system. I. Description and performance. *Astron. Astrosphys. Suppl. Ser.* **125**, 175–82.

Chassat, F. (1989) Calcul du domaine d'isoplanétisme d'un système d'optique adaptative fonctionnant à travers la turbulence atmosphérique. *J. Optics (Paris)* **20**, 13.

Coudé du Foresto, V., Ridgway, S. and Mariotti, J. M. (1997) Deriving object visibilities from interferometric observations with a fiber optical interferometer. *Astron. Astrophys. Suppl. Ser.* **121**, 379–92.

Davis, J. (1996) Observing with optical/infrared long baseline interferometers. In: *High angular resolution imaging in astronomy.* Les Houches Winter School, April 1996, eds. A. M. Lagrange, D. Mourard, and P. Léna, pp. 49–80 NATO ASP series, Kluwer (Dordrecht).

Fried, D. L. (1995) Artificial guide star tilt anisoplanatism: its magnitude and amelioration. In: *Adaptive Optics*, joint ESO-OSA meeting, ed. M. Cullum, pp. 363–9 ESO, Garching.

Gendron, E. (1995) Optimisation de la commande modale en optique adaptative: applications à l'astronomie. Thèse de doctorat, Université Paris VII.

Gendron, E. and Léna, P. (1996) Single layer atmospheric turbulence demonstrated by adaptive optics observations. *Astrophys. Sp. Sc.* **239**, 221–8.

Labeyrie, A. (1995) Images of exo-planets obtainable from dark speckles in adaptive telescopes. *Astron. Astrophys.* **298**, 544–6.

Lai, O. (1996) L'optique adaptative du télescope Canada-France-Hawaii et son utilisation pour l'étude des coeurs de galaxies à flambée d'étoiles. Thèse de doctorat, Université Paris VII.

Léna, P. (1984) Interferometry with large telescopes. In: Proc. IAU Coll. No. 79 *Very large telescopes, their instrumentation and programs*, eds. M. H. Ulrich and K. Kjär, pp. 245–55. ESO, Garching.

Luehe (von der), O. (1996) An introduction to interferometry with the ESO Very Large Telescope. In: *Science with the VLTI*, European Southern Observatory Workshop, eds. F. Paresce and O. von der Luehe, pp. 13–34, ESO, Garching.

Mouillet, D., Lagrange, A. M., Beuzit, J. L. and Renaud, N. (1997) A stellar coronograph for the COME-ON-PLUS adaptive optics system. II. First astronomical results, *Astron. Astrophys.* **324**, 1083–90.

Perrin, G., Coudé du Foresto, V., Mariotti, J. M., Ridgway, S. T., Carleton, N. P. and Traub, W. A. (1996) Observing stellar surfaces with a high precision infrared interferometer. In: *Science with the VLTI*, European Southern Observatory Workshop, eds F. Paresce and O. von der Luehe, pp. 318–25, ESO, Garching.

Ribak, E. and Rigaut, F. (1994) Asteroids as reference stars for high resolution astronomy. *Astron. Astrophys.* **289**, L47.

Rigaut, F., Rousset, G., Kern, P., Fontanella, J. C., Gaffard, J. P., Merkle, F. *et al.* (1991) Adaptive optics on a 3.6-m telescope: I. Results and performances. *Astron. Astrophys.* **261**, 280–90.

Rigaut, F., Cuby, J. G., Caes, M., Monin, J. L., Vittot, M., Richard, J. C. *et al.* (1992) Visible and infrared wavefront sensing for astronomical adaptive optics. *Astron. Astrophys.* **259**, L57–60.

Roddier, C., Roddier, F., Northcott, M. J., Graves, J. E. and Jim, K. (1996) Adaptive optics imaging of GG Tau: optical detection of the circumbinary ring. *Astrophys. J.* **463**, 326–35.

Roddier, F. and Léna, P. (1984) Long baseline Michelson interferometry with large ground-based telescopes operating at optical wavelengths. I & II. *J. Optics (Paris)* **4**, 171–82 and 363–74.

Roddier, F. and Eisenhardt, P. (1986) National Optical Astronomical Observatories (NOAO) infrared adaptive optics program IV: IR background speckle noise induced by adaptive optics in astronomical telescopes. *Advanced Technology Optical Telescopes III*, ed. L. Barr, Proc SPIE 628, pp. 314–22.

Véran, J. P., Rigaut, F. and Maître, H. (1995) Adaptive optics long exposure point spread function retrieval from wavefront sensor measurement. In: *Adaptive Optics*, Joint ESO-OSA meeting, ed. M. Cullum, pp. 497–502. ESO, Garching.

Véran, J. P., Rigaut, F., Maître, H. and Rouan, D. (1997) Estimation of the adaptive optics long exposure point spread function using control loop data. *J. Opt. Soc. Am.* **14**, 3057–69.

15

Astronomical results

PIERRE LÉNA

Université Paris VII & Observatoire de Paris, Meudon

OLIVIER LAI

Observatoire de Paris, Meudon

In this chapter we present astronomical observations obtained, at unprecedented high resolution, with the first adaptive optics systems installed on large telescopes and producing images of scientific value on a regular basis. These images cover many of the objects of interest to astronomers, from planets to quasars, and involve a number of approaches: straight imaging, spectro-imaging, polaro-imaging, coronography, all being done in wide or narrow fields. These results have been obtained less than seven years after the very first astronomical AO image was taken in 1989, we hope they will convince the reader of the great future adaptive optics should have in astronomy.

15.1 Scientific programs with adaptive optics

Astronomers rightly insist upon covering various spectral ranges with similar performance in terms of sensitivity and angular resolution. On Fig. 15.1 the performance of AO is compared with that of existing or planned space-borne telescopes and ground-based telescope arrays from near ultraviolet to millimetric wavelengths. It is worth noting the interesting match between the Hubble Space Telescope (HST) diffraction limit and the current AO systems on large telescopes, as will be demonstrated in several examples in this chapter. It is also apparent that AO observations provide the intermediate and necessary step between seeing limited images and multi-telescope interferometric observations. Optical interferometers provide the next step in improved angular resolution by one to two orders of magnitude, but require 'identification maps' of intermediate resolution, that AO observations can produce.

The benefits of AO have been discussed in the previous chapter. Its limitations are also known: the need for a suitable reference and the limited isoplanatic field. The reference magnitude, the seeing conditions and the

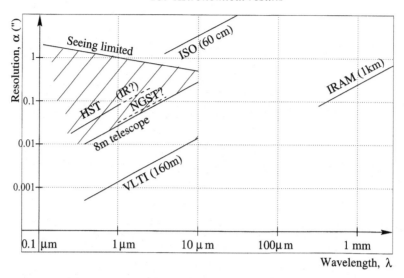

Fig. 15.1. *Performance in angular resolution*. Angular resolution obtained from near ultraviolet to millimetric wavelengths (HST = Hubble Space Telescope, possibly equipped with infrared detectors, IRAM = Institut de Radioastronomie Millimétrique, Plateau de Bure, France), by imaging and interferometric methods. The potential gain of adaptive optics over seeing limited imaging with a single telescope is indicated by the shaded area. AO is also necessary for optical interferometry as soon as the telescope diameter exceeds $r_0(\lambda)$.

system technical capabilities determine the ultimate achievable quality in the image. Artificial laser stars should relax the constraints imposed by a natural reference source. All the fields of astronomy can benefit from AO, and its use with even a moderate size telescope (e.g. 1 m diameter) and a low order correction can improve results. The emergence of adaptive optics has been rapid: Fig. 15.2 shows an exponential growth in the number of papers presenting scientific results obtained from adaptive optics observations since the first AO system of general use was opened in 1993.

The results presented in this chapter were obtained mainly with three AO systems: the COME-ON/ADONIS system located on the European Southern Observatory 3.6-m telescope at La Silla, Chile and described in Chapter 8, the University of Hawaii (UH) system operated on the Canada–France–Hawaii 3.6-m telescope (CFH) on Mauna Kea, and the CFH PUEO system, the last two being described in Chapter 9. A result obtained at Mt Wilson is also presented: this system, having 230 actuators, operates on the 100-inch telescope and can observe in the visible.

The examples given here are illustrative and they do not necessarily represent the very best achievable performance, as adaptive optics has continu-

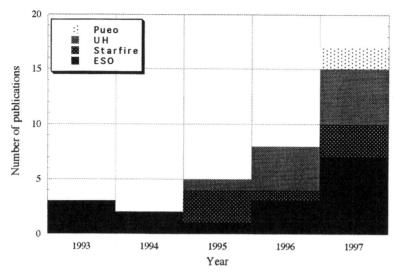

Fig. 15.2. *Growth of scientific results from adaptive optics observations.* The graph gives, per year, the number of papers published in refereed astronomy journals. The 1997 column shows the number of papers published, in press, or accepted as of September 30, 1997. All results come from four systems: the Airforce Starfire Optical Range (SOR), the CFHT PUEO system, the ESO COME-ON/ADONIS and the University of Hawaii instrument. This is omitting papers using 'tilt-only' correction. (Courtesy S. Ridgway and P. Wizinowich.)

ously progressed since the first images were obtained in 1989. The quality of the final image depends not only on system performance as discussed in Chapters 6 and 7 (number of corrected modes, bandpass, sensitivity of the wave-front sensors), but also on the state of the atmosphere (r_0 and τ_0, stationarity, height of the turbulence), on various calibrations (flat field and especially the point-spread function PSF) and post-processing (PSF estimate, deconvolution), as discussed in detail in the previous chapter. Whenever available, these various parameters are indicated, to allow for future comparisons. For this reason it is advisable that AO results always be published with the values of the observing and processing conditions.

15.2 From planets to quasars

15.2.1 Solar system objects

The case of the Sun itself, an extended and bright source affected by the day-time seeing, has been treated separately (Chapter 10) as solar adaptive optics requires a specific treatment. The Moon is obviously lacking a suitable ref-

Table 15.1. Solar system parameters

Object	d or s (")		Resol. (km)	Reference	m_V	Distance (")
1 Ceres	0.8	(d)	115	Ceres	7.4	–
2 Pallas	0.65	(d)	85	Pallas	7.3	–
4 Vesta	0.6	(d)	95	Vesta	5.9	–
216 Kleopatra	0.2?	(s)	150	Kleopatra	10.2	–
624 Hektor	0.1	(s)	300	Hektor	14.2	–
Venus	$\simeq 55$	(d)	25	Hot spots	$(m_K \simeq 4)$	
Mars	26	(d)	25	I (Phobos)	11.3	< 25
				II (Deimos)	12.4	< 65
Jupiter	49.5	(d)	300	I, II, III, IV	4.5–5.6	< 140 (I)
Io	1.2	(d)	300	Io or other sat.	5	–
Saturn	20.5	(d)	600	I to VI & VIII	8.3–12.9	< 30 (I)
Titan	0.85	(d)	600	Titan	8.3	–
Uranus		4.3	1200	Uranus	5.9	–
				III (Titania)	13.7	< 17
Neptune	2.9	(d)	1700	Neptune	7.7	–
				I (Triton)	13.5	< 17
Pluto/Charon	0.7	(s)	2800	Pluto	$\simeq 14$	–

First column is the object name, second column refers to angular size (d) or separation (s) in arcseconds. The third column shows the linear resolution achievable on the object with a 4-m class telescope in the near infrared ($\lambda \approx 2\ \mu m$). The fourth, fifth, and sixth columns show the name, the magnitude, and the distance of an adequate reference object other than a star (after Saint-Pé, private communication).

erence, except near the limb during stellar occultations, unless future space missions land such a source on the lunar soil. The resolution on the Moon would then reach 30 meters. Solar system objects range from the planets and their satellites to the asteroids and the comets. The vast majority of them is bright enough to provide adequate reference for AO (Table 15.1).

Most of the planetary discs, with the exception of Uranus and Neptune, are too extended, but their satellites can provide the reference, except indeed for Mercury and Venus: at shorter wavelengths, the size of the isoplanatic angle may not exceed 10″, causing the PSF to be non-uniform over the planetary surface. Meteorological studies of atmospheric motions become possible with high resolution and extended time coverage, and phenomena like the polar haze of Jupiter, the volcanism on Io (see Fig. 15.3) or collisions such as the impact of SL-9 on Jupiter in 1995 (all observed with AO) could be followed in time. Figure 15.4 is an image of Neptune, showing

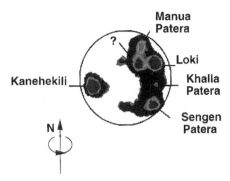

Fig. 15.3. *Volcanoes on Io*. Io is observed here during an eclipse in the shadow of Jupiter. Only the thermal radiation from the volcanoes is visible. Data were taken with the 3.6-m CFH telescope and the UH AO system through a narrow band filter centered at a wavelength of 2.3 μm where Jupiter's reflected sunlight is strongly attenuated by methane absorption. *Top*: This image is the sum of 40 exposures of 4 s each. The integration time of each exposure was limited because of the relative motion of Io and Europa, the latter being used as a reference source for the wavefront measurements. *Bottom*: The top image was deconvolved and slightly rotated so that the North pole of Io points straight up. The result is displayed here with false gray levels which enhance the faintest sources. On this image, hot spots can be precisely located and identified. The two brightest spots are volcanoes Loki (right) and Kanehekili (left). (Courtesy C. Dumas.)

its satellite Proteus observed here for the first time in the infrared. Figure 15.5(a) and (b) show observations made as Earth was crossing Saturn's ring plane in August 1995.

The same considerations apply for determination of rotation rates, or differential rotation between atmospheres and the ground, or imaging the atmosphere and/or the surfaces of satellites, such as the Galilean ones. When imaging is done in narrow spectral bands, information on velocity fields and local composition will be obtained. All these can be complementary to space

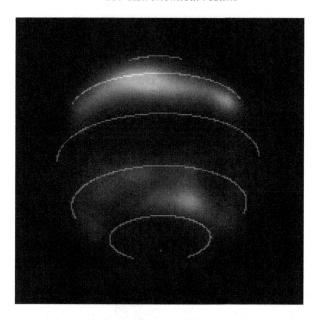

Fig. 15.4. *The planet Neptune.* Image is made at $\lambda = 2.2$ μm (K band) using the UH AO system on the CFH 3.6-m telescope, on August 12, 1995. Field-of-view is $4.5'' \times 5.6''$. North is up, east is left. Latitudes of $0°$, $\pm30°$ and $\pm60°$ are indicated as well as the location of the South pole. Pixel size is $0.035''$, or 740 km at Neptune's distance. Resolution is $0.13''$. The satellite Triton, $14''$ away and $0.13''$ diameter, was used as reference, and also for deconvolution, in addition to other stars. Neptune's atmosphere shows bright features, presumably clouds extending above the level where methane absorption becomes important at this wavelength. (Roddier F. *et al.* 1997a).

observations or prepare space probes' encounters. Images of Titan, satellite of Saturn, are given Fig. 15.6 and their quality compares well with the resolution obtained with the HST near $\lambda = 1$ μm. Figure 15.7(a) and (b) illustrates the power of mineralogical analysis of a solid surface with spectroscopic observations of the asteroid 4 Vesta, on which differences in soil composition indicate processing by meteorite collisions. As another example, Fig. 15.8 shows evidence for differences in surface composition between Pluto and its satellite Charon.

Accurate astrometric studies of satellites or asteroids may be done to improve their ephemerides. Comets can also be easily observed, with detailed follow-up of kilometer size details and spectrometric capabilities for chemical analysis, as was done on comets Hyukatake and Hale-Bopp in 1996 and illustrated on Fig. 15.9.

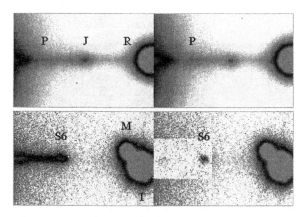

Fig. 15.5(a). *The rings and satellites of Saturn.* Images taken at $\lambda = 2.2$ μm with the ESO 3.6-m telescope and the ADONIS AO system, hours before the Earth crossed the plane of the rings on August 10, 1995 at 21:00 UT. Pixel size is 0.05″, i.e. 350 km on the rings. The resolution is 0.3″. Integration time is 60 s. Rhea is used as the reference. *Upper left*: Western ansa of Saturn's rings, one day and a half before crossing, at 06:41:20 UT. The halo on the left is caused by residual scattered light from the planet itself. Pandora (P), Janus (J), and Rhea (R) moving westward to the right are clearly visible. *Upper right*: Same image taken three minutes later, at 06:44:20 UT. *Lower left*: image of the western ansa taken twelve hours before crossing, at 08:52:10 UT on August 10, 1995, showing the brighter B ring and the Cassini division, the fainter A ring, plus a brighter clump at the tip of the ansa, corresponding to the F ring. *Lower right*: Subtraction, applied to the lower left image, of a symmetric image of the eastern ansa. The presence of an unresolved object is obvious (S6): it is identified with the new satellite 1995-S6 tracked by the HST a few hours after this image was taken. (Poulet and Sicardy 1996).

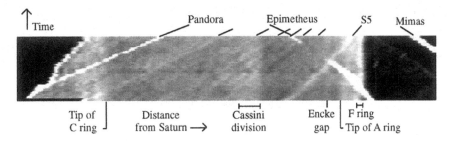

Fig. 15.5(b). *The rings and satellites of Saturn.* A plot of the illumination in the rings as a function of the distance from Saturn (horizontal axis) and as a function of time (vertical axis). Data were taken in the near infrared with the 3.6-m CFH telescope and the UH AO system. This 42-min long time-sequence was obtained on August 10, 1995 at a mid-time of 12:34 (UT), that is about 8 h before the Earth crossed the plane of the rings. One still sees the dark side of the rings and scattered sunlight can be seen transmitted through the C ring, the Cassini division, and the Encke gap as well as beyond the A ring up to the F ring. Objects in orbit produce inclined streaks. The brightest of them have been identified as indicated. S5 is one of the three objects discovered by the HST. Unlabelled marks indicate new objects discovered with AO (IAU circular No. 6515). This is an example where AO outperformed HST because of a larger telescope aperture and a more sensitive detector. Note the temporary disappearance of Epimetheus, in the shadow of the A ring. (C. Roddier 1996).

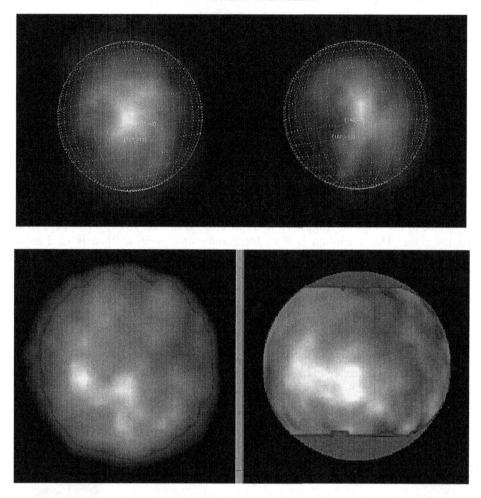

Fig. 15.6. *Titan, satellite of Saturn. Upper images*: Titan observed at its great eastern elongation with the ADONIS AO system on the ESO 3.6-m telescope near $\lambda = 2$ μm on September 17 and 18, 1994. North is up, east is left. A latitude–longitude grid has been added. Pixel size is 0.05″. Resolution is 0.13″ and Strehl ratios range from 0.2 to 0.6 before deconvolution. Titan itself is the reference. Nearby stars provide the PSF for deconvolution. The image contrast reaches 30%. The maps are obtained by subtracting images taken at $\lambda = 2.1$ μm and $\lambda = 2.2$ μm in order to remove most of the light scattered by the upper atmosphere. The residual emission shows the rotation rate of the surface and may possibly be interpreted as surface icy features of ethane or methane (Combes *et al.* 1997). *Lower images*: comparison of ADONIS-ESO and HST images, taken at identical times. HST images (Smith *et al.* 1996) are at $\lambda = 1.08$ μm.

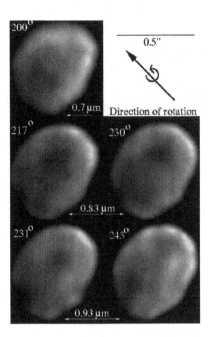

Fig. 15.7(a). *Mineralogical analysis of Vesta*. These images of Vesta were obtained on June 1996 (one month after its opposition), using the Shack–Hartmann AO system of the 2.5-m telescope at Mt Wilson Observatory. Vesta's dimensions (560 × 450 km) correspond here to angular sizes of $0.55'' \times 0.44''$. Observations were made through three narrow-band filters centered (from top to bottom) outside (0.7 μm), at mid-depth (0.83 μm), and at the bottom of a pyroxene absorption band. Impact craters are clearly visible. (Courtesy C. Dumas.)

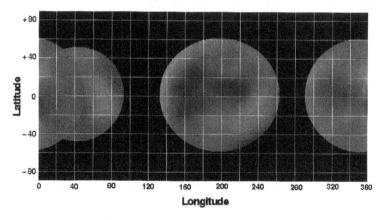

Fig. 15.7(b). *Mineralogical analysis of Vesta*. Map of Vesta obtained from the 0.8 μm images shown in Fig. 15.7(a). The longitude coverage is incomplete. Regions where data are missing are shown as black areas. Four large impact craters appear as dark areas near longitude 180°, between latitude −20° and +30°. In these areas the feldspar crust has apparently been removed by the impacts leaving the darker pyroxene rich mantle exposed (Dumas and Hainault 1996; Dumas 1997). As the sole intact differentiated asteroid, Vesta is a unique body in the solar system.

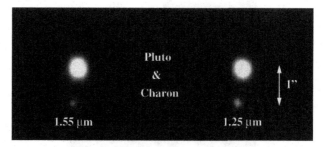

Fig. 15.8. *The Pluto–Charon system.* Images taken on June 14, 1997 (UT) with the UH AO system at the 3.6-m CFHT. The angular separation was 0.9″. These 20 min narrow-band images have a fwhm of 0.15″ (not deconvolved) which allows the first accurate photometry of Charon in the infrared. In particular, it was found that Charon is clearly enriched in frozen H_2O compared to Pluto (as is evident by the drop in brightness in the narrow-band 1.55 μm filter compared to the continuum 1.25 μm image). After monitoring the Pluto–Charon system throughout half an orbital period (3 days), it was found that Pluto has not only less H_2O but also a more heterogeneous surface than Charon (Roddier *et al.* 1997b). The visual magnitude of this object (used as a reference source) is V = 14.

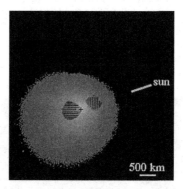

Fig. 15.9. *First AO image of a comet.* Comet Hyakutake (C1996B2) is imaged with the ESO 3.6-m telescope and the ADONIS AO system, in this composite [J–H] colour map (at respective wavelengths $\lambda_J = 1.2$ μm and $\lambda_H = 1.68$ μm), on March 8, 1996 at 14:54 UT. Pixel size is 0.05″, and the resolution in the coma reaches 0.2″ (90 km at the comet's distance of 0.59 AU) . Strehl ratios ranging from 0.05 to 0.13 were obtained; images are not deconvolved, but a control PSF was obtained on a nearby star. The nucleus ($m_V = 14.7$) is used as reference, sufficiently contrasted over the bright extended coma (integrated magnitude $m_V = 5.8$). Tracking the comet's motion was entirely taken care of by the AO system (tip/tilt control), allowing long exposures (approximately 1 min) without blurring. The [J–H] colour index values are respectively: 0.37 ± 0.05 (*gray*); 0.30 ± 0.02 (*vertical lines*); 0.50 ± 0.03 (*horizontal lines*). The cross indicates the nucleus. The 0.37 value is in good agreement with the one expected from silicate dust; 0.30 agrees with dirty ice grains spectral properties. The redder value (0.50) could be due to larger grains, the smaller ones being blown out by the gas outflow. (Marco *et al.* 1997).

15.2.2 Systems of stars

Systems of stars are easy targets for adaptive optics imaging: multiple stars or compact clusters are likely to contain at least one star bright enough to be used as reference and to determine the PSF. As stars are *a priori* known not to be resolved by AO, the deconvolution process can be more certain and extensive than for extended objects.

Astrometric determination of orbits in binary systems become more accurate, to within 10 milliarcseconds or better, allowing the determination of masses. Accuracy of positioning may evidence the presence of low mass companions such as red or brown dwarfs. Figures 15.10 and 15.11 illustrate the imaging capability of AO applied to star clusters. In the first case, a multiple star, initially supposed to be very massive, is separated into at least 12 components within less than 2″, the most massive one not exceeding a present mass of 35 $M_\odot$. In the second case, a cluster containing hundreds of stars in the Large Magellanic Cloud is imaged by AO: the dynamic range of more than 11 magnitudes and an accurate photometry allows coverage of a broad range of

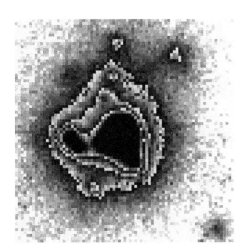 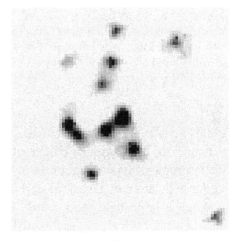

Fig. 15.10. *Resolution of a multiple star.* Images of the multiple star Sanduleak −66°41 in the Large Magellanic Cloud, before (*left*) and after (*right*) deconvolution, observed at $\lambda = 2.2$ μm with the ESO 3.6-m telescope and the ADONIS AO system in November 1993. North is up, east is left. Pixel size is 0.05″ and deconvolution leads to a resolution of 0.11″. Integration time is 20 min. The source itself is used as reference ($m_V = 11.7$) and a nearby star gives the PSF. Previous estimate of the 'star' mass was 120 $M_\odot$, then 90 $M_\odot$ after a partial resolution in six objects, while this image shows at least twelve components in the tight cluster and lowers the 'supermassive star' mass to the reasonable present value of 35 $M_\odot$. Establishing the upper limit of massive stars is fundamental for theories of birth and evolution of stars, as well as for the determination of the cosmic scale. (Heydari-Malayeri and Beuzit 1994).

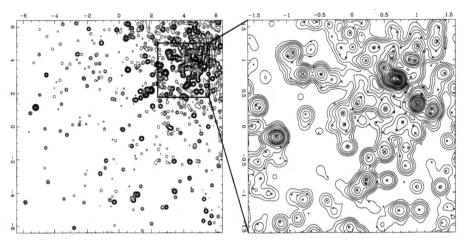

Fig. 15.11. *The R136 star cluster in 30 Doradus (Large Magellanic Cloud). Left*:
image of R136 obtained with the ESO 3.6-m telescope and the ADONIS AO system
at $\lambda = 1.68$ μm (H band) in December 1994 and March 1995. Field-of-view is
$12.8'' \times 12.8''$, contours extend fom 0.1% to 100% of the maximum intensity,
covering a magnitude range of 11.2. *Right*: enlargement of a $3.2'' \times 3.2''$ area.
Crosses indicate sources detected by the HST at $\lambda = 0.5$ μm, with $0.06''$ fwhm.
North is up, east is left. Pixel size is $0.05''$. With a seeing of $\approx 1''$, the Strehl ratio
reaches 0.24 (resolution of $0.18''$ before deconvolution) and remains uniform over
the field. Over 500 stars are detected, all but five of the HST stars with magnitude
$V < 16$ being detected in the H band. The initial mass function, the mass segrega-
tion and dynamical evolution of this young cluster can then be accurately derived.
Deconvolution is done using the bright star at the left of the left image. (Brandl
et al. 1996).

stellar masses. This image can be complemented by a similar HST field and
matched to it, providing colors of stars and a detailed description in the
Hertzsprung–Russel diagram. It gives the most detailed study to date of the
stellar population in a massive starburst.

When a large field (above $10''$) is imaged, the problem of anisoplanetism
may become critical, as the PSF will slowly degrade away from the reference,
chosen within the field. A demonstration of this effect, and of its magnitude, is
given in Fig. 15.12, where the progressive but moderate decrease of the image
sharpness in the field is quantitatively determined.

AO permits the mapping of even larger fields whenever several references
are available to allow building of a mosaic. A demonstrative illustration is
given on Fig. 15.13, where a large field of $\simeq 6$ square arcminutes is mapped
around the Trapezium star cluster in Orion using five different guide stars and
providing a final composite and extended map at a resolution of $0.25''$.

Figure 15.14 illustrates an astrometric application of sharp AO imaging: it

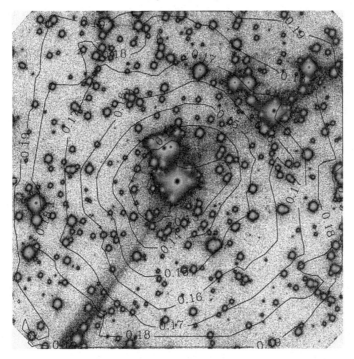

Fig. 15.12. *Anisoplanatism in a stellar field.* This stellar cluster (Messier 71) is imaged at the CFH 3.6-m telescope with the PUEO AO system in the visible ($\lambda = 0.5\ \mu m$). The field-of-view is $40'' \times 40''$ and over 100 stars are detected. North is up, east is left. Exposure time is 1200 s. The pixel size is $0.021''$ and the resolution is $0.12''$ at image center. The contours are iso-fwhm, with values ranging from $0.12''$ on the reference ($m_R = 11.7$) at center to $0.2''$ at a distance of $25''$. The relatively large value of the isoplanatic field at this wavelength is due to an exceptional seeing ($0.3''$ at $0.5\ \mu m$) and most likely a low altitude turbulence. (Lai 1996).

shows the area near the center of the Galaxy, with positions accurate to a few milli-arcseconds. With several such images separated by a few months, large proper motions are inferred and related to the mass distribution near the position of SgrA⋆. The Galactic Center is one area where an infrared wave-front sensor can be beneficial, as the bright IR source IRS7 ($m_K = 7$) is well within the isoplanatic field of the central area above $\lambda = 1\ \mu m$.

Another potential application is the imaging of globular clusters. It again allows resolution of crowded fields, to derive luminosity functions and spectral types, to analyze proper motions in their central area, and measure velocities possibly associated with the presence of black holes. Numerous stars are available to provide adequate references, the difficulty being to isolate a suitable one, or a compact arrangement, on the wave-front sensor.

Fig. 15.13. *The Trapezium star cluster and the Orion nebula.* This composite image is made at $\lambda = 2.2$ μm (K' band) with the University of Hawaii 2.2-m telescope and the UH AO system. The mosaic is made of seven different $62'' \times 62''$ frames, each of a 5 min exposure. The resolution is 0.25″. The images have not been deconvolved. Variable PSF fitting photometry found 293 stellar sources (down to K = 18.1). The binary frequency was found to be $14 \pm 2\%$ in this young cluster, which is the same fraction of binaries as in the solar neighborhood (Simon and Close 1998). The camera is the 1024×1024 UH IR Nicmos camera.

Fig. 15.14. (next page) *Proper motions in the center of the Galaxy. Top*: Image of the central area of the Galaxy, $13'' \times 13''$ (0.5×0.5 pc), imaged at $\lambda = 2.2$ μm by the CFH 3.6-m telescope and the PUEO AO system on June 21, 1996. North is up, east is left. Pixel size is 0.0344″ and resolution is 0.13″ (Airy rings show clearly on individual images, up to five of them on the brightest stars). The reference star has a magnitude $m_V = 14$ and is located 23″ away from image center: this leads to some anisoplanatism in the field, noticeable on slightly elongated star images. Deconvolution is made with the brightest star and is straightforward for a field of unresolved sources. The gray scale is proportional to the power 0.4 of intensity and the brightest objects are chopped off to improve the display. *Bottom*: Contour plot of the central area ($3'' \times 3''$) surrounding SgrA*. The position of SgrA* is marked with a box, whose sides are $\pm 1\sigma$ in length, at position (0, 0). After fitting the PSF on each object, the position is marked with a cross and identified with a number. This image was then compared with the positions of the most conspicuous objects previously obtained at the Keck telescope by speckle imaging, in order to determine proper motions as shown by the arrows. The length of each arrow is proportional to the amplitude of the proper motion, the largest one being 2500 km s^{-1}. The three

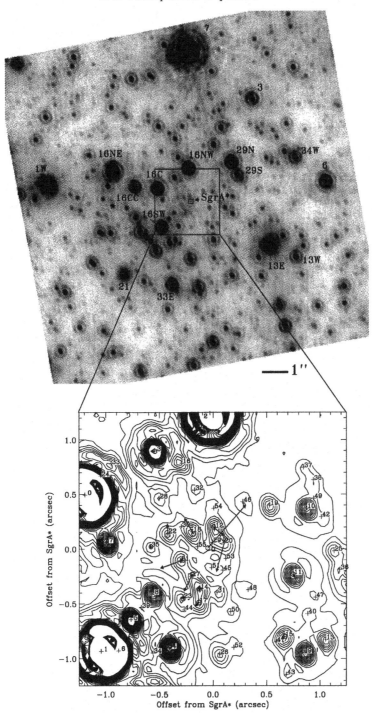

bright objects in the field are, from bottom to top, IRS–16SW, –16C, and –16NW. The astrometric accuracy reaches here 2 milliarcseconds for the brightest stars in the field (Rigaut *et al.* 1998).

15.2.3 Circumstellar and interstellar matter

Circumstellar environment is an ideal object for AO observations. The presence
of a star generally provides a suitable reference. At galactic distances of
100 pc, a 10-m telescope observing at $\lambda = 1$ μm will resolve 2 AU! The gain
in dynamic range, the reduction of stray light from the star are favorable for
imaging the faint structures, close to the star. Jets and flows in bipolar
structures, or discs, may present very low contrast with the star, as little as
10^{-5} or less of the stellar peak intensity for a pixel size of typically 0.05″ on a
3.6-m class telescope at $\lambda \geqslant 1$ μm. The dynamic range of AO was also
discussed in Chapter 14. Bright star environments will often be imaged, and the
full gain of AO (high number of corrected modes and high dynamic range) can
then be obtained: these interesting cases justify the construction of AO systems
with a large number of actuators [up to $(D/r_0)^2$]. Infrared wave-front sensing
(see Chapter 14) may be preferred in some cases where the central star is
heavily obscured in the visible.

One of the difficulties is the registration of maps obtained at different
wavelengths, first in a relative sense with respect to one another, then in an
absolute and astrometric sense. This requirement is often essential for a proper
physical interpretation of the observations: it may lead to inclusion in the AO
system of an additional imaging channel in the visible, exactly positioned (i.e.
within milliarcseconds) with respect to the infrared imaging channel and
operating simultaneously.

The environment of young stars represents a rich field of investigation.
Figure 15.15 shows the circumstellar disc of β Pictoris, seen edge on and much
closer to the star than previously observed with conventional imaging. Figure
15.16 shows the first optical image of the disc around the young star GG Tauri,
in an almost face-on configuration. A still younger object HL Tau is shown on
the book cover (bottom right image). It is a color composite made of AO and
HST images. The comparison with HST images at a similar spatial resolution is
often valuable, as shown on Fig. 15.17 which precisely locates young stars with
respect to a spectral emission of sulphur mapped by the HST on the star forming
region M16. The direct imaging of planets in such discs requires extremely high
dynamic ranges, an issue which has been discussed in Chapter 14.

Evolved stars, with their ejection of material, represent another favorable
opportunity. Quite compelling is the case of the proto-planetary nebula called
Frosty Leo, where the dynamics of the ejection could be observed with AO
at a sufficient accuracy to predict the existence of a central binary, which
indeed was discovered shortly later, again using AO (see top right illustration
on book cover).

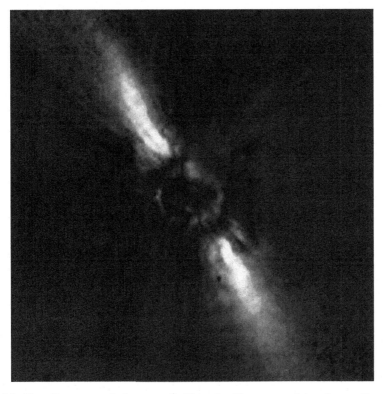

Fig. 15.15. *The disc around the star β Pictoris.* Coronographic observations of the proto-planetary disk around the star β Pictoris, on January 6 1996, with ESO 3.6-m telescope and the ADONIS AO system, equipped with a coronograph (diameter of the coronographic mask $= 1.5''$), in the J band ($\lambda = 1.28$ μm). North is up, east is left. Pixel size is $0.05''$. Total integration time is 6 min. Deconvolution is achieved with nearby stars ($m_V = 3$ to 4). Final accuracy on the positioning is $0.012''$. The disc is detected between 100 AU ($6''$ from the star) and as close as 24 AU ($1.5''$). A warp of the disc is clearly detected, with an amplitude of 2 AU, indicating the possible effect of a yet hidden planet. The dynamic range in the image of the disc is only 10, but the dynamic range extending from the star peak intensity to the faintest disk details reaches 10^5 at $1.5''$ from the star (Mouillet *et al.* 1997).

The study of the interstellar medium (molecular clouds, HII regions) at high angular resolution is possible whenever a suitable reference is available, either visible or infrared. A spectacular example is shown on Fig. 15.18: a reflection nebula is created by a bright star, close to it. Molecular hydrogen is excited by the ultraviolet radiation of the star and fluoresces in an infrared line at $\lambda = 2.12$ μm. Adaptive optics coupled to a narrow-band filter (resolution of 60) provides a map of the fine structure emission at scales as small as a few tenths of AU. No molecular cloud structure was

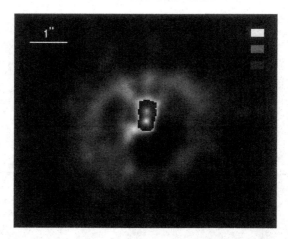

Fig. 15.16. *The disc around the young star GG Tau.* Image of the binary star GG Tau obtained in the J band ($\lambda = 1.2$ μm), with the CFH telescope and the UH AO system on Dec. 23, 1994. North is up, east is left. The image is a composite of several frames with exposure times ranging from 2 to 10 s. Pixel size is 0.035″. Deconvolution is achieved with a PSF measured on nearby stars. Dynamic range is indicated by the upper right corner, with respective intensity levels of 10^{-3}, 10^{-4}, and 10^{-5}. Maximum intensity has been normalized to unity and all intensities larger than 10^{-3} have been divided by 10^3. The close binary separation is 0.25″. The circumbinary ring is nearly circular, with apparent (i.e. projected onto the sky) semiminor and semimajor axes of 180 and 220 AU, leading to an inclination angle of 35°. These are the first images of a young binary clearing the inner part of its circumstellar disk, hence producing a circumbinary disk. This is thought to be a common and important part of the process in which a young star dissipates the circumstellar material around it. These images illustrate the high dynamic range obtained with AO (Roddier C. *et al.* 1996).

ever mapped at this resolution, and the physical hypothesis of a cascading turbulent energy can be traced down to scales where proto-planetary systems may form.

Polarization studies (polaro-imaging) of circumstellar environments is a natural extension, the difficulty here being the complex optical scheme of an AO system, which inevitably will introduce some instrumental polarization that needs to be carefully calibrated. Figure 15.19 illustrates this with a polarization image of a reflection nebula illuminated by the variable star R Monocerotis.

15.2.4 Galaxies and quasars

Observing extragalactic objects with AO is a challenge, as finding a suitable reference becomes more difficult. There are nevertheless a large number

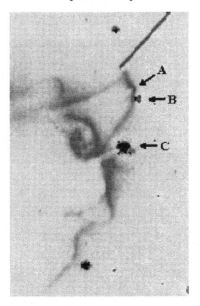

Fig. 15.17. *The star forming region Messier 16 ('The Eagle Nest')*. This composite picture superimposes: (i) star images obtained with the ESO 3.6-m telescope and the ADONIS AO system at $\lambda = 2.2$ μm; (ii) extended emission imaged by the HST in the sulphur II emission line at $\lambda = 673$ nm. This image is $17'' \times 27''$ in size and is a fraction of the complete AO image. The spatial registration between the two images is obtained by aligning about 15 stars, mostly outside this limited field-of-view. North is up, east is left. The resolution is $0.13''$ after deconvolution, using for reference and PSF determination a star ($m_V = 11.3$), $27''$ away from field center. This figure illustrates the importance of registration at high spatial resolution (Currie *et al.* 1996).

(hundreds) of favorable cases where the nucleus is sufficiently bright ($m_V \leqslant 16$ and contrasted (see Table 15.2 and Fig. 15.20). Then, imaging very faint structures in the vicinity of the nucleus becomes possible (Sol, private communication). Figures 15.21, 15.22, and 15.23 give examples of such cases and demonstrate the power of AO for imaging the central area of active galaxies. There, the accretion disc merges into the more classical old stellar component. In this region, cold molecular gas and star formation bursts occur. Jets ejected by nuclei may also be imaged, as for example in the quasars 3C273 (nucleus of $m_V = 12.8$ and jet at $10-20''$), or in PKS 0521-36 (nucleus of $m_V = 16$, jet at $10''$). Nodules showing supraluminal velocities as in 3C120 (nucleus of $m_V = 15$) require the high angular resolution of AO, as being as close as $0.1''$ from the nucleus.

Another approach is statistical: one could search for the coincidence of a sufficiently bright star and a galaxy, or area of a galaxy, belonging to a class

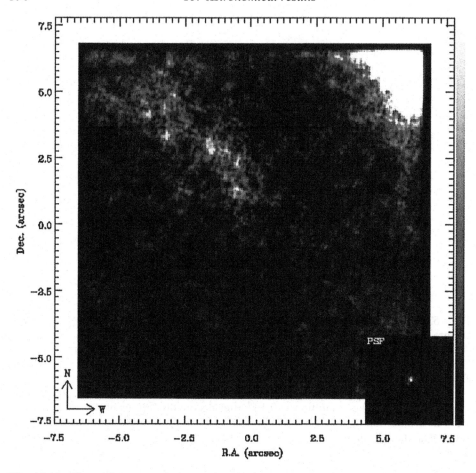

Fig. 15.18. *The reflection nebula NGC 2023 in Orion.* Monochromatic image obtained with the ESO 3.6-m telescope and the ADONIS AO system, at $\lambda = 2.12$ μm, in the $v = 1 - 0$, S(1) emission line of molecular hydrogen. North is up, east is left, the field-of-view is $12.8'' \times 12.8''$. Pixel size is $0.05''$. Resolution is $0.15''$ (i.e. 50 AU at this distance). The bright exciting star HD37903 ($m_V = 7.9$), $13''$ away from field center, is used as reference and for PSF determination (inset, showing a Strehl ratio of 0.42); it produces the scattered light seen in the NW corner of the image. The mottled structure is attributed to fine-scale inhomogeneities in the H_2 excitation. This structure, examined with a wavelet decomposition, shows a two-dimensional mean fractal dimension $D_0 = 1.32$, remaining constant over 5 or 6 orders of magnitude in size. This observation illustrates the power of AO spectral imaging. As the line is narrow, a higher spectral resolution would not decrease significantly the flux and could lead to the determination of the radial velocity field in the cloud (Rouan *et al.* 1997).

Table 15.2. *Number of bright galactic nuclei (Sol, private communication)*

Type	$V \leqslant 12$	$12 \leqslant V \leqslant 13$	$13 \leqslant V \leqslant 14$
Seyfert I	7	15	≈ 40
Seyfert II	2	11	≈ 50
Seyfert III	2	4	8
BL Lac	0	1	5
Quasars	0	3	10

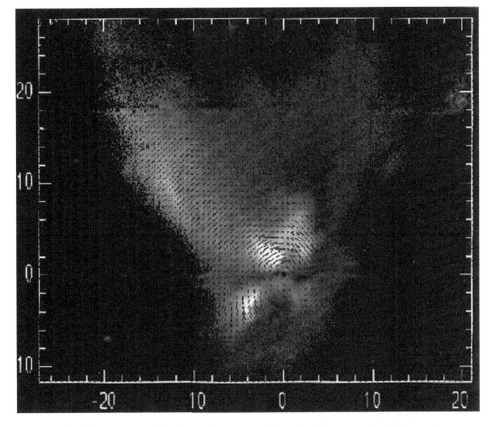

Fig. 15.19. *Polarization of the reflection nebula R Monocerotis.* Polarimetry map obtained with the CFH 3.6-m telescope and the UH AO system, at $\lambda = 1.6\,\mu m$. The pixel size is $0.035''$, the polarization vectors are deduced every $0.63''$ and the coordinates are in arcseconds. The polarization scale is given by levels of gray: black = 0%, white = 50%. The image is made as a composite out of four different linear polarization angles (in steps of 22.5° of a half-wave plate) with a total integration time of 24 min, using R Mon ($V = 13.1$) as reference. The centrosymmetric pattern demonstrates that the nebula around the star R Mon (at location $0.0''$, $0.0''$) is a pure reflection nebula illuminated by this star. A source, ≈ 100 times fainter and $0.69''$ away from R Mon, is detected as a thermal source and identified as a classical T Tauri star. This image is among the highest resolution ($0.2''$ fwhm) near-infrared polarimetry maps made to date (Close *et al.* 1997).

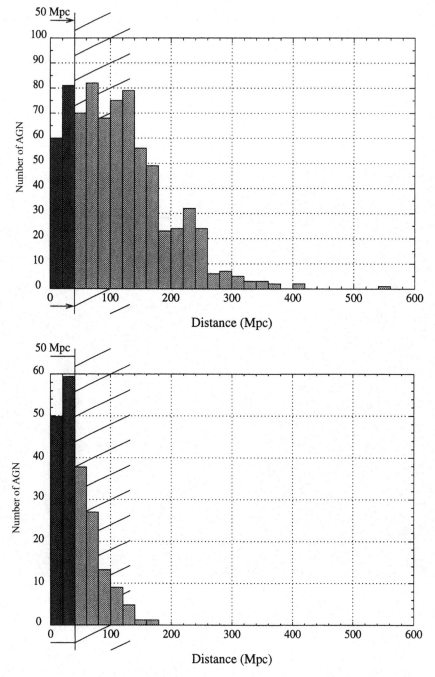

Fig. 15.20. *Magnitude histogram of active galaxies and quasars. Top*: Number of objects brighter than $m_V = 16$ versus their distance. *Bottom*: Number of objects brighter than $m_V = 14$ versus their distance. Data are from the catalogue of Véron-Cetty & Véron. A 10-m telescope observing at $\lambda = 1.0$ µm, AO corrected, will resolve 10 pc at a distance of 100 Mpc (Lai 1996).

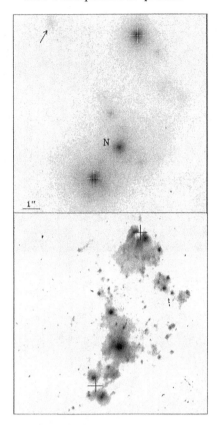

Fig. 15.21. *The nuclear region of NGC 3690, in the interacting galaxy Arp 299.*
Upper: This K band ($\lambda = 2.2$ μm) image was obtained at the CFH 3.6-m telescope
with the PUEO AO system in May 1996. The field-of-view of $11.8'' \times 11.8''$ is part of
a larger mosaic. North is up, east is left. Pixel size is $0.035''$ and the resolution is $0.2''$.
On this raw image, the Strehl ratio reaches 0.17 with a short (60 s), integration time.
The nucleus (N) is used for reference with $m_V \approx 15$. The true nucleus (N) is stellar.
The two regions (*white crosses*) are sites of intense star formation as demonstrated by
their color-index deduced from J,H,K photometry. The arrow points to the infrared
counterpart of a faint radio source. The sensitivity may be inferred from the
photometry of a faint source, SW of the bright upper source, of magnitude $m_K = 16.8$
in one resolution element, showing with a signal-to-noise ratio of 5 in a 1 min
exposure. *Bottom*: The same area observed at $\lambda = 0.5$ μm with the HST for comparison. It is interesting to note that the starburst regions (crosses), which are bright in the
K band, are completely enshrouded in dust and hence not visible on this image (Lai
et al. 1998b).

of interest and to be imaged. This is favorable to finding remote galaxies
($z \geqslant 0.1$) and determine by imaging their morphological type: their surface
density becomes high, of the order of 10^4 galaxies/deg² per magnitude unit
at the magnitude $m_K = 18.5$. The probability of finding a suitable reference

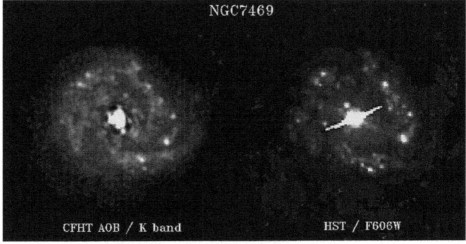

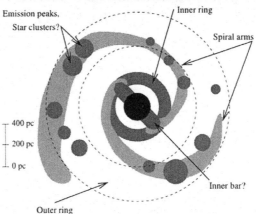

Fig. 15.22. *A circumnuclear starburst region*. The Seyfert I nucleus of NGC 7469 is observed. *Upper left*: with the CFH 3.6-m telescope and the PUEO AO system at $\lambda = 2.2$ μm in June 1996 and (*upper right*) with the HST at $\lambda = 0.5$ μm. Each image is $7'' \times 7''$. North is up, east is left. The AO image uses the nucleus ($m_V = 13.5$) for reference and is deconvolved with a PSF directly derived from the WFS measurement while observing the nucleus (see Chapter 14). The HST image shows a horizontal strike due to saturation. As outlined in the sketch (*lower*), the striking features are: (i) a previously identified ring now resolved in tightly wound spiral arms; (ii) an inner ring connected to the nucleus by an inner bar oriented NE–SW; most of the hot spots seem to be aligned along this axis. A comparison between HST and PUEO images reveals that some of these are extremely blue sources (Lai 1996)

star in the isoplanatic field at $\lambda = 2.2$ μm is greater than 10%. A first attempt of a systematic search for deep galaxies, which are missed by seeing-limited surveys because of their faintness, was undertaken in 1993, using stars near the North Galactic pole as references (Sams, private communication).

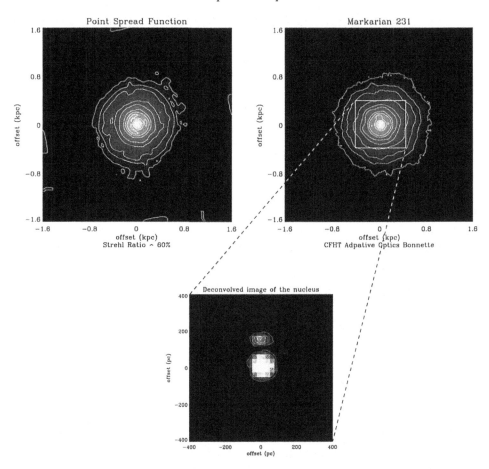

Fig. 15.23. *The distant galaxy Markarian 231.* This galaxy is observed with the CFH 3.6-m telescope and the PUEO AO system at $\lambda = 2.2\,\mu$m. North is up, east is left. Image size (*upper*) is $4'' \times 4''$ and pixel size is $0.035''$. Total integration time is 1 s! Contour levels are as square-root of intensity, with the lowest at 10^{-4} of maximum intensity and Strehl ratio reaches 0.6. *Upper right*: Image of the galaxy. *Upper left*: The PSF deduced from wave-front and servo-loop data obtained while observing the nucleus of magnitude $m_V = 13.8$ (see below). *Lower*: Blow-up of the central $0.5'' \times 0.5''$ area, deconvolved with a maximum likelihood algorithm. Comparison of the unresolved nucleus profile with the PSF reveals a source at $0.12''$ north of the nucleus, located on the first diffraction ring at a level of 0.006 of the nucleus intensity. Deconvolution enhances this feature which may be interpreted as a dense 'super-starburst' close to a quasar in formation. The faint source west of the double nucleus may be an artefact. This image of Markarian 231 shows the limits of what can be achieved in terms of morphological structures on very distant objects: this ultra-luminous IRAS galaxy ($L_{IR} \simeq 3 \times 10^{12} L_\odot$) is 160 Mpc away, and at this distance, the diffraction-limited resolution at $2.2\,\mu$m of the CFH 3.6-m telescope corresponds to 100 pc (Lai *et al.* 1998a).

Asteroids, already mentioned (Ribak and Rigaut 1994), may also provide temporary references with a large sky coverage.

AO imaging may prove extremely valuable for imaging gravitationally lensed galaxies, as the typical thickness of arclets is below 0.5″.

15.3 Acknowledgements

The authors warmly thank students and colleagues, especially D. Rouan, F. Rigaut, and F. Roddier, who all provided data prior to publication, granted permission for reproductions or exchanged ideas on the practical use of adaptive optics at the telescope. None of this could have been written without a long-lasting collaboration with our ONERA and industrial partners. It is based on the accumulated expertise, since 1987, with the European Southern Observatory COME-ON/ADONIS and the Canada–France–Hawai PUEO adaptive optics systems: teams at ESO and CFHT played a decisive role in these observations and are thanked for their cooperation.

References

Brandl, B., Sams, B. J., Bertoldi, F., Eckart, A., Genzel, R., Drapatz, S., *et al.* (1996) Adaptive optics near-infrared imaging of R136 in 30 Doradus: the stellar population of a nearby starburst. (1996). *Astrophys. J.* **466**, 254–73.

Close, L. M., Roddier, F., Hora, J. L., Graves, J. E., Northcott, M., Roddier, C., *et al.* (1997) Adaptive optics infrared imaging-polarimetry and optical HST imaging of Hubble's variable nebula (R Mon/NGC 2261): a close look at a very young active Herbig Ae/Be star. *Astrophys. J.* **489**, 210–21.

Combes, M., Vapillon, L., Gendron, E., Coustenis, A., Lai, O., Wittemberg, R., *et al.* (1997) Spatially resolved images of Titan by means of adaptive optics. *Icarus* **129**, 482–97.

Currie, D., Kissel, K., Shaya, E., Avizonis, P., Dowling, D. and Bonnacini, D. (1996) Star formation in NGC 6611 with Adonis. *The Messenger* No. 86, pp. 31–6.

Davidge, T., Simons, D. A., Rigaut, F., Doyon, R. and Crampton, D. (1997) The stellar content near the Galactic center. *Astron. J.*, **114**, 2586.

Dumas, C. and Hainaut, O. (1996) First ground based mapping of the asteroid Vesta. *The Messenger* No. 84, pp. 13–16.

Dumas, C. (1997) Nouveaux résultats d'optique adaptative: imagerie de l'astéroide Vesta. *Journal des Astronomes français* **52**, 20–5.

Heydari-Malayeri, M. and Beuzit, J. L. (1994) Further decomposition of a 'very massive' star. *Astron. Astrophys.*, **287**, L17–L120.

Lai, O. (1996) The Canada–France–Hawaii adaptive optics system and its use in the study of the inner-most regions of starburst galaxies. Thèse de doctorat. Université Paris VII.

Lai, O., Rouan, D., Rigaut, F., Arsenault, R. and Gendron, E. (1998a) Adaptive optics observations of ultra-luminous infrared galaxies: I. J, H, K images of Mkn231 *Astron. Astrophys.* **334**, 783–88.

Lai, O., Rouan, D., Rigaut, F. Doyon, R. and Lacombe, F. (1998b) Adaptive optics observations of ultra-luminous infrared galaxies: II. Arp 299. Submitted to *Astron. Astrophys.*

Marco, O., Encrenaz, T. and Gendron, E. (1997) First images of a comet with adaptive optics. *Planet. Sp. Sc.* **46**, 547–53.

Mouillet, D., Larwood, J. D., Papaloizou, J. C. B. and Lagrange, A. M. (1997) A planet on an inclined orbit, as an explanation of the warp in the β Pictoris disc. *Mon. Not. Roy. Astr. Soc.* **292**, 896–904.

Poulet, F. and Sicardy, B. (1996) The 1995 Saturn ring-plane crossings: ring thickness and small inner satellites. *Bull. Astr. Amer. Soc.* **28**, No. 3, 1124.

Ribak, E. and Rigaut, F. (1994) Asteroids as reference stars for high resolution astronomy. *Astron. Astrophys.* **289**, L47.

Rigaut, F., Doyon, R., Davidge, T., Crampton, D., Rouan, D., Nadeau, D. and Beuzit, J. L. (1998) In preparation.

Roddier, C. (1996) Adaptive optics observations of the solar system. In Proc. 'Journée scientifique: *Perspectives de l'optique astronomique*', held on Nov. 20, 1996 at the French Academy of Sciences, eds. P. Léna and V. Coudé du Foresto, *C. R. Acad. Sci. Paris*, 325, Série IIb, 109–14.

Roddier, C., Roddier, F., Northcott, M. J., Graves, J. E. and Kim, J. (1996) Adaptive optics imaging of GG Tau: optical detection of the circumbinary ring. *Astrophys. J.* **463**, 326–35.

Roddier, F., Roddier, C., Brahic, A., Dumas, C., Graves, J. E., Northcott, M. J., *et al.* (1997a) First ground-based adaptive optics observations of Neptune and Proteus. *Planet. Sp. Sc.* **45** No. 8, 1031–6.

Roddier, F., Brahic, A., Dumas, C., Graves, J. E., Han, B., Northcott, M. J., *et al.* (1997b) Adaptive optics observations of solar system objects. *Bull. Am. Astr. Soc.* **29**, No. 3, 1023.

Rouan, D., Field, D., Lemaire, J.-L., Lai, O., Pineau des Forêts, G., Falgarone, E., *et al.* (1997) The power of adaptive optics: a close look at a molecular cloud in NGC 2023. *Mon. Not. R. Astron. Soc.* **284**, 395–400.

Simon, M., Close, L. M. and Beck, T. L. (1998) Adaptive optics imaging of the Orion Trapezium cluster. *Astron. J.* (submitted).

Smith, P. H., Lemmon, M. T., Lorenz, R. D., Sromovsky, L. A., Caldwell, J. J. and Allison, M. D. (1996) Titan's surface revealed by HST imaging. *Icarus* **119**, 336–49.

16

Future expectations

FRANÇOIS RODDIER

Institute for Astronomy, University of Hawaii, USA

Covering all the aspects of adaptive optics (AO) in astronomy is a challenging task. Inevitably there have been omissions as well as redundancies. Moreover, the field is still rapidly evolving. Techniques which have been described in detail may become obsolete, whereas others barely mentioned in this book, may gain importance. Nevertheless, we hope this book will be found useful by both engineers who need to build AO systems for astronomy, and astronomers who want to observe with them.

A highly debated topic is the use of laser guide sources (LGS) instead of natural guide sources (NGS). We use here the word 'sources' rather than the more widely used word 'stars', because not only LGS, but also many NGS are not stars. In view of recent developments, it seems fair to say that the use of NGS has given better results than many people anticipated. Most of the astronomical results published to date have been obtained with NGS systems, and as seen in Fig. 15.2, the number of publications obtained with them is growing very rapidly. Significant image improvement can now be obtained in the infrared with guide sources as faint as V = 15 or 16. Also, – at least on good astronomical sites – the isoplanatic patch size was found to be larger than originally anticipated. At the CFH telescope, it is not uncommon to observe only a 10% loss of Strehl ratio in the H band, 20″ away from the guide star. This means that more objects than previously thought can be observed by using a nearby star as a guide source. It also means more stringent constraints may have to be put on the off-axis optical quality of AO systems to fully benefit from this advantage.

Still, the ultimate performance of NGS systems has not yet been achieved. At the time of this writing (Autumn 1997), a new adaptive optics system has been constructed at the University of Hawaii, and operates with actively quenched avalanche photodiodes now available from EG&G. Compared with the previous system, a twofold gain (almost a magnitude) is observed in photon detection efficiency. Moreover, recent progress in infrared detector arrays now allows infrared wave-front sensors (IR WFSs) to be built with a sensitivity approaching that of visible sensors. As emphasized in Chapter 14, IR WFSs will further extend the use of AO to the dark absorbing regions of the Galactic plane. To date, the astronomical use of AO has been limited to 3.6-m or smaller telescopes. However, 8–10-m class telescopes are now becoming equipped with AO. With such large telescopes, images up to 5 or 10 μm are no longer diffraction limited, and hence require AO compensation. In the thermal infrared, the wave-front coherence areas (Fried's r_0 parameter) become so large, that NGS compensation becomes possible over nearly the whole sky. It will undoubtedly open a new area for NGS compensation. It will also require new developments such the use of deformable secondaries, or cooled AO systems to minimize thermal emission from optical components, and perhaps a chopping capability.

On the other hand, as one aims at observing at shorter wavelengths down to the visible, the drop in limiting magnitude and isoplanatic patch size severely impedes our ability to find a suitable guide source. The possibility of creating artificial sources anywhere on the sky by means of laser beacons looks therefore extremely attractive. However, the use of laser beacons has proved to be more difficult than initially anticipated. The reasons for this are discussed in detail in Chapter 12. The main problem comes from errors in the low order wave-front term, especially the inability to measure tip/tilt with an LGS. Unfortunately, most astronomical observations require accurate tracking of the object over long exposures. As the angular resolution reaches the diffraction limit accurate tracking becomes even more difficult. Because an LGS does not provide the information, one must still require the simultaneous use of an NGS. The higher the angular resolution, the brighter the NGS has to be. Detailed analysis (Chapter 12) has shown that because one still needs an NGS, full sky coverage cannot be achieved in the visible as originally hoped for. Although LGS + NGS compensation has been experimentally demonstrated, it has not yet performed significantly better than pure NGS compensation. Whether it can do so remains to be seen.

Given the success of NGS AO, one should expect most of the large telescopes to become equipped with it, with or without provisions for future LGS systems. Initially applied to imaging, AO will increasingly be also used for more sophisticated applications such as polaro-imaging, 3-D spectral-imaging, or even ordinary spectroscopy. It will considerably boost the performance of stellar coronographs. It is also essential to the progress of ground-based optical interferometry, an area which is expected to make significant advances in this coming century. A possibility which has not been discussed in this book is the use of a space-borne laser as a guide source. The idea, originally promoted by Alan Greenaway (1992), consists of pointing a space-borne laser towards an observatory, as the laser's apparent position drifts near to an astronomical source of particular interest. As space to ground optical communication develops, a number of space-borne lasers are likely to be launched. They probably could be made available for astronomical use. They would also be ideal sources to help co-phase telescope arrays.

An important question is how adaptive optics can compete with telescopes in space. To date results have been quite complementary, the Hubble Space Telescope (HST) providing high resolution images in the visible, and AO providing a similar resolution in the near IR. Although HST is now equipped with a near-IR camera, AO has still the advantage of using bigger telescope apertures. Moreover, the near-IR HST capability is likely to end in 1998, beyond which AO will be the only way to produce high resolution images in the infrared, at least until the Next Generation Space Telescope (NGST) is launched in a decade. Even then, AO observations with large ground-based telescopes will remain cheaper and telescope time will be easier to obtain. However, the NGST will usefully complement AO by providing images with a lower background as well as access to wavelengths that are absorbed by the Earth's atmosphere.

16.1 Acknowledgement

I wish to thank here all the authors who contributed to this book, as well as my co-workers who helped me proofread the manuscripts. All of them made considerable effort to help me in the editing process. It has been a great pleasure for me to work with them all.

References

Greenaway, A. H. (1992) Satellite borne guide star. In: Proc. *Laser Guide Star Adaptive Optics Workshop*, ed. R. Q. Fugate, pp. 663–8. Starfire Optical range, Phillips Lab., Kirtland AFB, Albuquerque, NM.

Glossary of acronyms

ADC	Atmospheric Dispersion Compensator. An optical device used to compensate the chromatic dispersion produced by the atmosphere.
ADONIS	ADaptive Optics Near Infrared System. An improved version of the ESO COMEON + system.
AGB	Asymptotic Giant Branch. A class of evolved stars.
AIS	Artificial Intelligence Software.
ALFA	Adaptive optics with a Laser For Astronomy. An AO system built by Adaptive Optics Associates for MPIA and MPIE.
AMOS	Airforce Maui Optical Station. Pioneered AO for surveillance applications.
AO	Adaptive Optics.
AOS	Adaptive Optics System.
APDs	Avalanche Photo-Diodes. Are used in high quantum efficiency photon counting detectors.
AR	Anti-Reflection (coating).
ARC	Astrophysical Research Consortium. Operates a 3.5-m telescope at the Apache Point Observatory (Sunspot, New Mexico).
ASSI	Active Stabilization in Stellar Interferometry. An active beam recombiner for the two-telescope interferometer at CERGA.
BIM	BIMorph mirror (see Chapter 4).
CC	Command Computer.
CCD	Charged Coupled Device. A type of detector array used to record images (mostly in the visible).
CERGA	Centre d'Etudes et de Recherches en Géodynamique et Astrométrie (French). Observatory located in the south-east of France.
CFH	Canada-France-Hawaii. An international corporation located in Hawaii. Operates the CFHT.
CFHT	Canada-France-Hawaii Telescope. A 3.6-m telescope located on Mauna Kea (Hawaii).
CGE	Compagnie Générale d'Electricité (French). A French company.
CILAS	Compagnie Industrielle des LASers (French). A French company.
CLEAN	An image deconvolution algorithm developed by radioastronomers.
COAT	Coherent Optical Adaptive Techniques. An early adaptive optics technique used to focus laser beams through the atmosphere.
COME ON	Cilas/Observatoire de Meudon/ESO/ONera. Name of the consortium who built the first astronomical AO system. Also name of the system.

CP	Computing Power.
CPU	Central Processor Unit.
CS	Curvature Sensor (see Chapter 5).
CVF	Continuously Variable Filter. A tunable optical filter used in cameras.
CW	Continuous Wave.
DAC	Digital to Analog Converter.
d.c.	Direct Current.
DM	Deformable Mirror (see Chapter 4).
DOD	Department of Defense (US).
DSP	Digital Signal Processor.
EBCCD	Electron Bombarded CCD.
ESO	European Southern Observatory.
FA	Focus Anisoplanatism (see Chapter 12).
FAB	Full Aperture Broadcast.
FAO	Full Aperture Observation.
FASTTRAC	A tip/tilt compensation system built for the 2.3-m telescope of the Steward Observatory (Kitt Peak, Arizona).
FASTRAC II	A version of FASTRAC built for the MMT.
FFT	Fast Fourier Transform.
FOV	Field-of-View.
fwhm	Full Width at Half Maximum. The width of an intensity peak at half its maximum intensity.
HRCaM	High Resolution Camera. A tip/tilt compensated imager built for the CFHT.
HST	Hubble Space Telescope.
HVA	High Voltage Amplifier.
ICCD	Intensified CCD.
IR	InfraRed.
IRCCD	Infrared CCD.
ISO	Infrared Space Observatory.
ITEK	A US company who pioneered adaptive optics.
JOSA	Journal of the Optical Society of America.
KL	Karhunen-Loève.
LC	Liquid Crystal.
LCD	Liquid Crystal Display.
LEP	Laboratoire d'Electronique Philips (French). Philips Electronics Lab located in France.
LEST	Large European Solar Telescope (project).
LGS	Laser Guide Star.
LIDAR	LIght Detection And Ranging. An optical equivalent of the RADAR.
LLNL	Lawrence Livermore National Laboratory.
LSI	Large Scale Integration.
MAMA	Multi-Anode Micro-channel Array. A photon counting detector array.
MCP	Micro-Channel Plate. An image intensifier.
MHV	Modified Hufnagel-Valley (model of atmospheric turbulence).
MIT	Massachusetts Institute of Technology.
MMK	Modified Maua Kea (model of atmospheric turbulence).
MMT	Multi-Mirror Telescope on Mt Hopkins (Arizona).
MPIA	Max Planck Institut fur Astronomie (Germany).
MPIE	Max Planck Institut fur Extraterrestrische Physik (Germany).
MPM	Monolithic Piezoelectric Mirror (see Chapter 4).

NGS	Natural Guide Star.
NICMOS	Near Infrared Camera and Multi-Object Spectrometer. An instrument developed for the HST. Also name of the IR detector array developed for this instrument.
NOAO	National Optical Astronomy Observatory (US).
NSF	National Science Foundation (US).
NSO	National Solar Observatory (US).
OHP	Observatoire de Haute Provence (French). Astronomical observatory located in France.
ONERA	Office National d'Etudes et de Recherches en Aeronautique (French). A French public laboratory for aerospace research.
OPD	Optical Path Difference.
OPM	Observatoire de Paris-Meudon (France). Astronomical observatory located in France.
OSA	Optical Society of America.
OTF	Optical Transfer Function.
PID	Proportional-Integrator-Derivative.
PMN	Lead Magnesium Niobate. An electrostictive material.
PMT	Photomultiplier Tube.
PSD	Power Spectral Density.
PSF	Point Spread Function.
PUEO	Probing the Universe with Enhanced Optics. A CFHT user AO system named after the Hawaiian owl Pueo, a night bird with a sharp vision.
PZT	Lead Zirconate-Titanate. A piezo-electric material.
QSO	Quasi-Stellar Object or Quasar.
rms	Root Mean Square.
RTC	Real Time Computer.
SAM	Stacked Actuator Mirror.
SAT	Société Anonyme de Télécommunication (French). A French company.
SH	Shack-Hartmann. A type of wave-front sensor (see Chapter 5).
SHARP	System for High Angular Resolution. Infrared camera built by the MPIE for Adonis.
SI	Shearing Interferometer. A type of wave-front sensor (see Chapter 5).
SLC	Submarine Laser Communication. The acronym here designate a model for the vertical distribution of turbulence in the atmosphere.
SNR	Signal to Noise Ratio.
SOR	Starfire Optical Range (US). An Air Force laboratory which pioneered laser guide star adaptive optics.
SPIE	Society of Photo-Optical Instrumentation Engineers.
SW	Spot-Wander.
TA	Tilt Anisoplanatism (see Chapter 12).
THEMIS	Télescope Héliographique pour l'Etude du Magnétisme et des Instabilités Solaires (French). A solar telescope located in the Canarie islands.
UH	University of Hawaii.
UnISIS	University of Illinois Seeing Improvement System.
UT	Universal Time.
UV	Ultra Violet.
VLT	Very Large Telescope. Array of four 8-m telescopes built by ESO on Paranal (Chile).
VLTI	Very Large Telescope Interferometer. Interferometric mode of the VLT.
VME	VersaModule Eurocard. A computer standard for addressing information.

VTT	Vacuum Tower Telescope (solar telescope).
WF	Wave Front.
WFC	Wave-Front Corrector.
WFS	Wave-Front Sensor.
YAG	Yttrium Aluminium Garnet (normaly Nd:YAG, or Neodymium:Yttrium-Aluminum-Garnet). A type of laser based on this material.

Index

List of astronomical objects used in the illustrations:

Printed in the United States
By Bookmasters